Readings in
Modern Biology

Jay M. Templin
Widener University

West Publishing Company

Minneapolis/St. Paul　　　New York　　　Los Angeles　　　San Francisco

WEST'S COMMITMENT TO THE ENVIRONMENT

In 1906, West Publishing Company began recycling materials left over from the production of books. This began a tradition of efficient and responsible use of resources. Today, up to 95% of our legal books and 70% of our college texts and school texts are printed on recycled, acid-free stock. West also recycles nearly 22 million pounds of scrap paper annually—the equivalent of 181,717 trees. Since the 1960s, West has devised ways to capture and recycle waste inks, solvents, oils, and vapors created in the printing process. We also recycle plastics of all kinds, wood, glass, corrugated cardboard, and batteries, and have eliminated the use of Styrofoam book packaging. We at West are proud of the longevity and the scope of our commitment to the environment.

Production, Prepress, Printing and Binding by West Publishing Company.

ISBN 0–314–02378–X

*PRINTED ON 10% POST
CONSUMER RECYCLED PAPER*

Contents

Part One: Principles of Biology - Evolution, Homeostasis, and the Chemistry of Life

Part Two: Understanding the Structure and Function of Cells

Part Three: Chromosomes, Cell Division, and Heredity

Part Four: Homeostasis and Principles of Structure and Function

Part Five: The Continuation of Life - Reproduction and Development

Part Six: The Evolutionary Legacy - Unity and Diversity of Living Things

Part Seven: Interactions - Behavior, Ecology, and Environment.

Part One

Principles of Biology - Evolution, Homeostasis, and the Chemistry of Life

On initial inspection the water molecule seems deceptively simple. Consisting of only three atoms, it exists in three, familiar physical states: solid, liquid, and gas. However, this molecule has many strange, unexplained properties. It flows faster under high pressure than low pressure when in liquid form. Most liquids behave just the opposite. As it flows, its apparent gentle character can forge crevices as deep as the Grand Canyon. Its molecules form weak bonds with each other. Although it is a common solvent for many solutes, it avoids many others rather than dissolve them. How can all of these properties be explained? Accurate computer images of the water molecule hold the key to explain its characteristics in depth, unlocking the mysteries of a substance that covers three-quarters of the earth and composes roughly two-thirds of the body weight of most organisms.

Wet, Wild, and Weird

by Carl Zimmer

You never hear someone say, "Bartender, another round of C_2H_5OH for my friends here," or "Do you want coffee with $C_8H_{10}N_4O_2$ or decaf?" The only substance commonly known by its chemical formula, is good old H_2O. Water is not only ubiquitous, its simple: two hydrogens perch on a lone oxygen like ears on an atomic mouse. When schoolteachers search for the perfect molecule to introduce elementary concepts of chemistry, they naturally turn to water. But first impressions are deceiving. Underneath that surface simplicity, water's mysteries run deep. The Physics of Simple Liquids, a standard, 500-plus-page guide that sits on many researchers' bookshelves, mentions water only once—in the preface, explaining that water is too complicated to be discussed in the book. "It's the hardest liquid," says physicist Eugene Stanley of Boston University.

Water's subtle complexity manifests itself in many ways. Unlike other liquids, it flows faster under high pressure than under low pressure. As a solvent, it's gentle

enough to serve as the cradle of life yet brutal enough to carve the Grand Canyon. This contradictory character results from an atomic arrangement that makes water almost promiscuously friendly. While other liquids are generally aloof, water's oxygen and hydrogen atoms like to form bonds with practically anything that crosses their path, even other water molecules.

During the past decade computers have been helping researchers visualize the invisible choreography of H_2O molecules that produces water in all its familiar and exotic forms. As in all chemical reactions, the macroscopic effects we see are determined by the arrangements of molecules and the changing configurations of the electric fields they create. Water's simple structure makes it possible to simulate these features with exacting detail. Research put their imitation water through its paces, creating analogues of complex situations, observing how the molecules behave; then they compare the computer results with lab experiments using the real thing. These

computer-simulated close-ups of water as ice, liquid, and vapor are drawn from some of the most recent research. Perhaps they'll inspire you to gaze with somewhat more respect at the mystery dripping from your kitchen faucet.

FLOWING WATER

Judging by its chemical structure, water ought to be stiff and syrupy, more a gel than a liquid. The chemical bonds that hold water molecules together are extremely powerful; and a molecule in liquid water can make four bonds with other molecules—its two hydrogen atoms can latch onto other oxygen atoms, while its own oxygen atom can link with two hydrogen atoms. Just about every molecule in a glass of water is bonded at a given instant to four other molecules.

So why is water so, well, watery? Eugene Stanley and his colleagues have found an explanation in their computer simulations. They filled an imaginary cube with 216 water molecules and had the computer track the changes in energy that each one experienced as it moved around. They then made a movie showing how the molecules rearranged themselves over a millionth of a millionth of a second. It turns out that into each five-molecule cluster a sixth molecule insinuates itself. The interloper does so by forcing one of the molecules in the cluster to share a hydrogen bond with it. A shared bond becomes weakened and unstable, so that the intruding molecule can eventually push out another. Clusters are therefore continuously rearranging themselves in a microscopic game of musical chairs that keeps liquid water flowing.

The image on the preceding pages is a snapshot from the movie created by Stanley's team. The hydrogen bond between two molecules is represented by a stick connecting the centers of each. Together the sticks create a sort of three-dimensional net. Two clusters of actual molecules are superimposed on the mesh. The one in the lower left-hand corner is made up of five molecules. The one in the upper right-hand corner has been joined by a sixth.

WATER CAGES

Chemists have discovered that in vapor form, water molecules can arrange themselves into elegant cages. These images, generated by A. Welford Castleman and his colleagues at Penn State, depict clusters of molecules that the chemists created in their lab.

Researchers had long known that water vapor often contains clusters of molecules, as many as 21 in a single clump. But they didn't know what shape those clumps took. To determine that, they had to know how many unbonded hydrogens there were in such a grouping. To find out, the Penn State team spiked a beam of vapor with a chemical compound that glued itself to any available hydrogen atom.

The more unbonded hydrogen atoms on a cluster's surface, the chemists reasoned, the heavier it would become as more molecules hung on. In fact, the mass of a tagged cluster turned out to be an exact measure of the number of hydrogen atoms facing outward. Using a computer, they discovered that the only possible arrangement ten unbonded hydrogens could take was the soccer-ball shape shown in these pictures (red spheres are unbonded hydrogen atoms).

Such clusters seem to carry an extra proton that snags passing molecules to build the cage. (This stray hydrogen nucleus is probably a remnant of an H_3O molecule that broke up as two of its hydrogens and an oxygen—H_2O—hooked onto the cage, leaving the lone H behind.) On a 20-molecule cluster (opposite page, left), the molecules pass this proton to one another like a hot potato, so that it races around the surface, binding briefly with the oxygen atoms before moving on. When a twenty-first water molecule joins the cage, however, the structure becomes unstable. Then both the free proton and the new molecule get sucked into the center, where they remain imprisoned (right, page 104). Using beams of other compounds, Castleman and his colleagues have discovered how to trap atoms of metal inside water cages as well. On the preceding page, a water cage traps a cesium atom.

In the real world, such water cages are common in the very cold, sparse regions of the atmosphere 50 miles up. There they probably act as seeds for noctilucent clouds, the eerie, glowing formations that you can sometimes see far to the north.

ICE FROM SPACE

Before there was Earth, there was ice. It floated in the cloud of primordial matter that formed the sun and the planets. Made sluggish by the extreme cold of space, some of these super-chilled H_2O grains bumped into each other, stuck, and eventually grew into the cosmic snowballs we know as comets. Many researchers believe that icy comets colliding with Earth deposited much of the water contained in the early oceans. At -420 degrees, such space ice is not the everyday stuff you find in your freezer. On Earth, when ice forms, water molecules try to organize themselves into the most stable arrangement possible: each hydrogen atom binds to an oxygen atom to form an orderly, hexagon-shaped lattice. Of course, molecules randomly bumping into each other tend to be highly disorganized, and it takes energy for them to twist and turn to fit into the lattice. However, though an ice cube in a freezer is cold, it is still warm enough for the water molecules to jostle themselves into a preferred position—like a sleeper wriggling to get comfortable under the covers.

But an ice grain numbed by the extreme cold of space is more like a sleeper completely knocked out by sleeping pills. It can't muster enough energy to even get properly settled. When stray water molecules hit one of these grains, they freeze in that position for good. As a result, the grains grow long tentacles that reach out and grab more passing molecules. The process is so complex that researchers need to simulate it on a computer to get an idea of what's happening. Chemist Victoria Buch of the University of Illinois at Chicago generated these images by letting the grains aggregate, molecule by molecule, under conditions of deep space.

In the images above, the craggy bit (1) was formed on the computer at a typical -420 degrees. To illustrate how temperature determines the shape of ice, Buch then had the computer warm the ice, giving it the energy to start rearranging itself into a much more compact form (2). A compact grain has less surface area and fewer unbonded atoms to lure passing molecules; therefore, it doesn't form the tentacles typical of an ice grain in space.

Much of an ice grain's surface is open territory for stray matter. The blue oxygen atoms and yellow hydrogen atoms (3) are not bonded to water molecules. As a result, other molecules can latch onto the ice (4). The non water molecules on this dog-shaped grain are shown as the blue spheres. Once attached to the ice, the chemicals can react with each other to create new molecules. Their union releases excess energy that jettisons the molecules off the grain. These serendipitous recombinations are responsible for all sorts of unexpected compounds that are found in interstellar gas clouds, such as cyanide and ammonia—chemicals that also show up in comets.

The actual docking of an atom or a molecule on an ice grain is no simple matter. Here a hydrogen atom (5, blue sphere) approaches the ice. It dances over the surface of the ice grain, too excited with kinetic energy to land. Gradually the ice absorbs the excess energy. Exhausted, the atom comes to rest.

DROPS AND FILMS

Spill water on paper and it spreads out into a thin film, but spill it on rubber and it beads up into little drops. The reason becomes clear if we look at the molecular level. Water molecules in a drop hold on tightly, pulling each other into the most compact shape possible—a sphere. But if other, surrounding molecules exert an even stronger force than neighboring water molecules, these strangers can tear the drop apart. A surface like paper is hydrophilic ("water loving"), while rubber is hydrophobic ("water fearing"). The paper pulls water molecules away from their fellows, dispersing the drop; the rubber leaves them intact. Hydrophilia and hydrophobia are more than annoyances in mixing oil-and-vinegar dressing; these forces are central to life. Cell membranes are layered with hydrophilic and hydrophobic surfaces that precisely control the flow of chemicals in and out.

A typical drop of water contains trillions of molecules. But what happens if you spill just a few? Until recently, researchers couldn't say; they expected the familiar patterns to break down. Now a computer simulation by Michael Klein and Joseph Hautman

at the University of Pennsylvania suggests that a minuscule number of molecules behave just like the more familiar multitudes: films as tenuous as one molecule thick form on a hydrophilic surface (right, top); on a hydrophobic surface, the water still beads into drops, even though as many as half the molecules may reside on the surface (right, bottom).

For their hydrophobic surface Klein and Hautman used hydrocarbon molecules fringed with methyl groups, providing a neutrally charged frosting that couldn't attract water. Their hydrophilic surface consisted of a layer of OH groups (one oxygen atom bonded to one hydrogen atom), which exert a strong pull on water. The researches spread 90 molecules of water on each surface. When water was spread on the OH-studded hydrophilic surface, the molecules slowly made bonds with the OH groups. On the methane-covered hydrophobic surface, however, the water bunched itself up from a thin film to the droplet shown here in less than a billionth of a second. This microdrop reflects the form of its macroscopic counterpart down to some of the smallest details. The angle of the curve where the drop meets the hydrophobic surface, for instance, is the same for both macro- and microdrop. ☐

Questions: *Answers on page 197*

1.   What does it mean to describe substances as water-hating? How do they develop this property?

2.   What does it mean to describe substances as water-loving? How do they develop this property?

3.   Why is the water molecule similar to a weak magnet?

Scientists have long doubted that molecular biologists and paleontologists could locate proteins in fossils that are more than a few million years old. Such organic molecules decay, leaving no trace of their existence. However, a preserved protein from dinosaurs has been discovered. The dinosaur protein is apparently a molecule that binds to osteocalcin, a protein found in the bones of vertebrates. The protein has not been isolated from the bones, making an exact determination of its amino acid sequence impossible. This will be necessary, however, if scientists are to determine its degree of similarity to the protein in other reptiles and birds. Vertebrates with a closer protein match probably evolved from a more recent ancestor. Vertebrates with greater differences probably descended from a more remote ancestor in time. Advances in laboratory procedures may soon make this task of protein isolation and matching feasible.

Protein Identified in Dinosaur Fossils

by R. Monastersky

A team of molecular biologists and paleontologists has identified a protein preserved in dinosaur bones, opening up the possibility of using ancient molecules to help sort out the controversial relationships among dinosaurs and other vertebrates.

Scientists have long considered it highly unlikely that they would find proteins in material more than a few million years old, because such organic molecules usually decay far sooner. Yet several research teams in the past few years have reported detecting proteins in very old fossils, including dinosaur bones (SN:5/4/91,p.277). In the dinosaur case, however, the researchers did not know which proteins they had detected, and many scientists wondered whether the proteins had come from bacteria or other sources of contamination.

Now, Gerard Muyzer of Leiden University in the Netherlands and his colleagues report using immunological tests to identify a specific bone protein called osteocalcin in several dinosaur fossils that date back 75 million and 150 million years. They discuss their work in the October GEOLOGY.

"If it is indigenous, then it is the oldest protein," says Lisa Robbins, a micropaleontologist at the University of South Florida in Tampa.

Muyzer's group identified the dinosaur protein through an antibody that binds to osteocalcin, a small molecule present in the bones of vertebrate animals. The antibody test found osteocalcin in the bones of hadrosaurs, a ceratopsian, and a sauropod dinosaur. It also detected the protein in several mammal fossils and an ancient turtle bone.

The researchers believe the osteocalcin is indigenous to these fossils because invertebrates and bacteria do not produce this protein. Their tests did not show any osteocalcin present in fossilized sea shells. Another procedure showed that the dinosaur fossils contained relatively high concentrations of gamma-carboxy glutamic acid (Gla), an amino acid absent in invertebrates and microbes, say the researchers.

Other researchers, however, remain skeptical about the possibility of finding proteins from so far back. Jeffrey L. Bada from the Scripps Institution of Oceanography in La Jolla, California, says a study he did on Gla shows that it doesn't last more than 100,000 years. "I worry greatly about the stability of Gla. Why would it remain unaltered over tens of millions of years?" he wonders.

Muyzer and his colleagues had hoped to isolate the osteocalcin and then determine its amino acid sequence. By comparing that with osteocalcin sequences from birds and crocodiles, the researchers could address the long-standing question of how closely birds and dinosaurs are related. At present, paleontologists can only use dinosaur bones to make comparisons.

Muyzer's group did not succeed in isolating the protein. But advances in laboratory procedures may soon make the job easier. "The techniques are improving daily. It's just a matter of the techniques catching up with what we want to do," Robbins says. □

Questions: *Answers on page 197*

1. Why were some scientists skeptical about the authenticity of the protein in dinosaur bones?

2. Do you think that the identification of a protein that is only two million years old can shed much light on the evolutionary picture? Why or why not?

3. Why does a comparison of protein structure reflect genetic similarities and differences?

The role of actin in muscle contraction is well-documented. A newly-recognized function is that this protein is necessary for the attachment of bacterial cells to cells of a host organism. Escherichia coli, the well-studied, gram-negative bacterium, requires actin as it attaches to various intestinal cells. This gram-negative microbe is a normal flora inhabitant of the large intestine of humans. However, actin may be necessary for the attachment and attack by a pathogenic form of this microorganism. Other proteins, such as myosin, may also be necessary for the success of the invading microbe.

Actin in Cell Attachment

by Stuart Knutton, Tom Baldwin, Peter Williams, Joseph and Jean Sanger

SIR — In their News and Views article[1] discussing the potential of polycationic beads to model bacterial infection, Joseph and Jean Sanger ask whether the filaments found in intestinal cells infected with enteropathogenic Escherichia coli could be actin. The answer is definitely "yes". Since the first description of this effect[2], a considerable amount of work has clarified both the composition of this lesion and the mechanism by which it forms.

That the dense fibrous pad formed beneath enteropathogenic E. coli (EPEC) when they attach to various cell types is indeed predominantly actin was shown by staining with an actin-specific phallotoxin conjugated with fluorescein (see figure). We have proposed this fluorescence actin staining test[3,4] for rapid detection of EPEC, which were previously identified simply on the basis of serotyping. With regard to mechanism, we have also demonstrated a localized elevation of intracellular free calcium in EPEC-infected cells of various types[5], and proposed that this activates the actin-severing function of cytoskeletal villin, leading to effacement of the brush border microvilli[6]. Subsequent dissipation of local increases in calcium concentration by natural sequestration into intracellular storage would then allow dissociation of villin, generating nucleation sites for an explosive burst of actin polymerization. This would deform the plasma membrane (now denuded of its microvilli) into the characteristic pedestal structure seen in natural disease.

As Sanger and Sanger point out, the possible involvement of nucleating proteins in this process is, as yet, unclear, but obviously we need to identify a bacterial potentiator of actin polymerization and the signal pathways in target cells. But our model does answer another question posed by Sanger and Sanger, this time in the negative. If, as they speculate, EPEC were able to cruise the brush border, destroying microvilli by mopping up the actin supply, we would expect to see regions of effacement devoid of bacteria. In fact, effacement is only ever seen at sites of bacterial attachment. A more likely explanation for gross effacement of microvilli is that EPEC infection is a dynamic process in which bacteria grow and attached microcolonies of bacteria continually enlarge, so that eventually microvilli are lost from the entire apical surface of an infected cell.

The EPEC pedestal also contains other cytoskeletal proteins in addition to actin[7,8]. Of particular interest in myosin, in which the light-chain component has undergone phosphorylation by the calcium-dependent enzyme myosin light-chain kinase[8], possibly as a result of a kinase cascade. Indeed, we believe that the success of EPEC as a pathogen resides in its ability to hijack the natural processes of cytoskeletal protein recruitment that are central to the healthy functioning of a cell.

Stuart Knutton, Institute of Child Health, University of Birmingham, Birmingham B16 8ET, UK

Tom Baldwin, Peter Williams, Department of Genetics, University of Leicester, Leicester LE1 7RH, UK

SANGER AND SANGER REPLY

The evidence of Knutton, Baldwin and Williams that filamentous actin is a major component of the dense fibrous pad localized in cells where EPEC attach is quite convincing. The electron micrographs of the fibrous material[4] also resemble images of cells in the process of phagocytosing bacteria or particles. It was the short columns of fibrous material extending into the cytoplasm beneath attached EPEC in other cells[2,9] that attracted our attention and reminded us of the columns of actin beneath the beads on Aplysia[10]. Assuming that these short columns are also made of actin, perhaps they represent an earlier stage in infection than that observed by Knutton et al.

As microvilli can reform rapidly under experimental conditions, it would be more interesting to observe EPEC on the surface of living cells. For example, Goligorsky et al.[11] have demonstrated that a one-minute exposure of cultured kidney cells to parathyroid hormone initially reduces the number of microvilli, but in five minutes the microvilli reform completely. Thus, bacteria might move on the surface of the host cell and microvilli could reform in their wake. On the other hand, Knutton et al. may be correct in proposing that attachment itself leads to effacement. The ability of bacteria[12] and beads[10] to "hijack cytoskeletal protein recruitment" will enable systems to be devised that give valuable insight into both normal and pathogenic processes in cells.

J. Sanger, J. Sanger
Department of Anatomy
University of Pennsylvania,
School of Medicine
Philadelphia 19104-6058, USA ☐

1. Sanger, J. & Sanger, J. Nature 357, 442 (1992).
2. Staley, J.D., Jones, E.W. & Gorley, L.D. Am J. Path 56, 371-392 (1969).
3. Knutton, S., Baldwin, T., Williams, P.H. & McNeish, A. S. Lancet I, 1336 (1988).
4. Knutton, S., Baldwin, T., Williams, P.H. & McNeish, A. S. Infect Immun. 57, 1290-1298 (1989).
5. Baldwin, T.J., Ward, W., Aitken, A., Knutton, S. & Williams, P.H. Infect. Immun. 59, 1599-1604 (1991).
6. Matsudaira, P. & Janmey, P. Cell 54, 139-140 (1988).
7. Finlay, B.B., Rosenshine, I., Donnenberg, M.S. & Kaper, J.B. Infect. Immun. 60, 2541-2543 (1992).
8. Manjarrez-Hernandez, H.A., Baldwin, T.J., Aitken, A., Knutton, S. & Williams, P.H. Lancet 339, 521, (1992).
9. Helie, P., Morin, M., Jacques, M. & Fairbrother, J.M. Infect. Immun. 59, 814-821 (1991).
10. Forscher, P., Lin, C.H. & Thompson, C. Nature 357, 515-518 (1992).
11. Goligorsky, M.S., Menton, D.N. & Hruska, K.A. J. Membr. Biol. 92, 151-162 (1986).
12. Sanger, J.M., Sanger, J.W. & Southwick, F.S. Infect. Immun. 60, 3609-3619 (1992).

Questions: *Answers on page 197*

1.   How do actin and myosin function in muscle contraction?

2.   Can you see a parallel between the role of actin in muscle contraction and cell attachment?

3.   What do the authors mean when a pathogenic microbe "hijacks" a cytoskeletal protein?

4

A French immunologist has offered a new theory for the molecular basis of memory. Most other scientists, however, view his assertion as radical. The assertion holds that water molecules store and release information in a way not previously understood. A subtle electromagnetic language allows one water molecule to record and store the signal of a second. How the structure of water is related to such signaling is unclear. This assertion has produced one of the most bizarre disagreements in the history of science.

The Case of Ghost Molecules
by Bill Lawren

The apartment, on a quiet cul-de-sac overlooking paris' Montparnasse cemetery, is a curious mix of sheer affluence and cheerful jumble—a bathrobe thrown carelessly over an expensive Italian leather couch, a chaotic pile of papers and photographs crowning an antique mahogany desk. The apartment's owner is a study in contradictions, too.

A respected and much-cited immunologist, Dr. Jacques Benveniste directs the Immunopharmacology of Allergy and Inflammation unit at INSERM (the French National Institute of Health and Medical Research). But he has also taken a walk on the wild side. Traveling down what many of his colleagues call the path of destruction, he has conducted a series of controversial new experiments on a weird and disputed phenomenon known as molecular memory. This memory works, he says, when water molecules store and release information in a previously undetected way: through a subtle electromagnetic language that enables one molecule to record the "essence" of a second, much like a tape recorder records a sound.

If confirmed, Benveniste's work would vindicate the discredited field of homeopathy, which holds that disease can often be treated and cured using infinitesimal amounts of medicine diluted in water. More important, says Benveniste, his research could lead to "the medicine of the future," in which doctors tap into the electromagnetic molecular communication system to perform surgery without knives, prevent diseases without using vaccines, and effect cures without drugs.

"Take a simple example," Benveniste explains. "You have a toothache,. then you take a gram of aspirin, and it invades your entire body. It upsets your stomach, it blocks your blood coagulating mechanisms, so if you cut yourself an hour later you're in trouble. But if we understand the body's electromagnetic language, we could send the tooth the signal for aspirin instead of taking a pill."

These are radical assertions, and they leave most mainstream scientists cold. "A delusion," says John Maddox, editor of the respected scientific journal *Nature*, which published some of the first reports of Benveniste's work.

"Bizarre," says Henry Metzger, Director of Intramural Research at the National Institute of Arthritis and Musculoskeletal and Skin Diseases in Bethesda, Maryland. "If Benveniste were right, the French wine industry would be distraught, because all you'd have to do is put a drop of wine in a glass of water and you'd have burgundy."

But a few scientists remain agnostic. "It's a little hard to believe," says James A. Scott, a specialist in nuclear medicine at Massachusetts General Hospital in Boston, "but that doesn't mean it's not true." And at lease one researcher is supportive. "There are interactions at the molecular level that are not chemical, and we don't know very much about them," says Robert Becker, a researcher at Upstate medical Center at the State University of New York in Syracuse. "I'm sure that Benveniste is on the right track."

Right or not, Benveniste's work has already generated one of the noisiest and most bizarre controversies in the history of modern science. That controversy began in 1988 when *Nature*, the august British scientific journal,

Reprinted by permission of OMNI, June 1992, copyright 1992 by OMNI PUBLICATIONS INTERNATIONAL, LTD.

published the results of Benveniste's experiments on "water memory." The controversy culminated—but did not end—when *Nature* sent a team of "ghostbusters," including the famous iconoclastmagician James Randi, to Benveniste's laboratory to investigate his claim. During the ensuing mud fight, the *Nature* team accused Benveniste and colleagues of everything from shoddy science to delusional thinking. Benveniste responded with loud screams of "witch hunt." In the end, the shouting damaged the reputations of both parties and left Benveniste fighting not only for his job, but also for his scientific life.

It is an unlikely controversy, and Benveniste himself seems somewhat unlikely in the role of crusading scientist. Now 57, he is an elegant Frenchman with a full range of Galic expressions, at least one of which makes him look very much like Marcello Mastroianni, indeed, there is an unmistakable (and scientific?) dash to his presence—the balance of his wardrobe tips heavily toward suede and leather, and when he arrives to pick up a reporter at a paris hotel, it is in a pristine vintage Jaguar.

Yet the experiments began quietly enough when, in the early Eighties, Benveniste was approached by a young doctor at INSERM. Recalls Benveniste. "This young man said to me, 'You know, I am a homeopath.' I said, 'Is that some kind of sexual disease?'" Although Benveniste thought homeopathy was "a bunch of baloney," he gave the doctor permission to conduct a few experiments "as long as it didn't interfere with his real work."

The experiments that followed were based on a few simple scientific facts: During human allergic reactions, allergens—dust, for example, or pollens—bind to an antibody called Immunoglobulin E,

or IgE for short. The allergen-IgE combination then triggers white blood cells called basophils, causing them to release some of their contents, a process known as degranulation.

Exploiting this basic biology, the researchers created their own allergen in a serum composed of an antibody to IgE (or anti-IgE) in a solution with water. They took this solution through a series of five progressive dilutions so that little or no anti-IgE serum was left. They then exposed white blood cells to this highly dilute solution. Though the amount of anti-IgE in serum was negligible, they reported, measurable degranulation had occurred.

"As soon as I saw this," Benveniste says, "I understood that the degranulation could not be due to molecular activity because it normally takes millions of molecules of anti-IgE to produce a reaction. It had to be something else."

Curious, Benveniste decided to go further: He diluted the original solution ten times. "At ten times dilution," he explains, "there is no mathematical chance that even one molecule of anti-IgE will be left in the solution. At ten times dilution, it's just water." Yet even at ten times dilution, Benveniste says, 40 to 60 percent of the basophils degranulated, reacting to what Benveniste says was not "plain water" as if it still contained the anti-IgE serum.

In 1984 Benveniste and immunologist Elisabeth Davenas, then a student in his lab, repeated the experiment and got similar results. Benveniste wrote up his findings in a scientific report, which he submitted to *Nature* in 1986. When the journal asked for confirmation of the results, he farmed the experiments out to research teams in Italy, Canada and Israel. According to Benveniste, these teams

reported findings similar to his, and he included them in his own report to the journal.

Nature says it never saw all the results of these trials but nonetheless sent Benveniste's report to a team of expert referees. "The referees," Maddox says now, "couldn't see anything wrong with the experiments, but they didn't believe the results." This left *Nature* somewhat reluctant to publish Benveniste's report, but with no good reason not to. At the same time, Maddox says, reports of Benveniste's unusual findings were about to appear in the French press.

On the horns of a dilemma, *Nature* decided to publish the report, but with two unusual conditions: First, the published report would be accompanied by an editorial reservation noting the "incredulity" of the referees in the face of the experimental results for which there was "no physical basis." Second, after the report was published, Benveniste would allow a team of investigators recruited by *Nature* to visit his lab and witness the experiments themselves.

Benveniste agreed, and on July 4, 1988 — almost a week after the report was published—the *Nature* team arrived at Benveniste's INSERM 200 laboratory in Clamart, a suburb 15 minutes south of Paris. The three-man team, later dubbed "ghostbusters" by the press, was as unusual as the report it came to investigate. There were no immunologists on the team and only one practicing scientist. One of the investigators was Maddox himself. Another was Walter W. Stewart of the National Institutes of Health in Bethesda, Maryland, who had been one of the paper's referees and who made a reputation as a sort of scientific sheriff, pursuing a number of research fraud cases. The third member of the team was James Randi, a celebrated magician, MacArthur Foundation Fellow, and self-appointed skeptic who is perhaps best known for his ongoing and highly public battle to expose and discredit psychic Uri Geller.

This strange amalgamation of ghostbusters spent a week in Benveniste's lab. The process started quietly enough, with the *Nature* team watching four repeats of Benveniste's experiments. Benveniste says each of these yielded positive results. But the *Nature* team did not agree. In fact, Stewart declared all the results "valueless." His reason: As far as he was concerned, adequate scientific controls had not been in place.

Next came a series of three more trials with the *Nature* team taking extraordinary measures to ensure results could not be manipulated. At one point, Randi wrapped the code for the experiments—designed so that no one could know which test-tubes were yielding which results—in aluminum foil. He folded the foil into a specially sealed envelope and then taped the envelope to the laboratory ceiling.

The results of these experimental runs were negative. To Benveniste, this was not especially surprising—there were many instances, he said, in which basophils had not reacted to the anti-IgE solution at high dilutions.

But as far as the *Nature* team was concerned, the investigation had put an end to Benveniste's assertions. As Maddox put it, Benveniste's results had been "delusions" to be accounted for by sloppy experimental procedures and bad counting.

On July 28, *Nature* published the ghostbusters' findings. In the same issue, Benveniste wrote an emotional reply, labeling the investigation a "Salem witch hunt" and a "McCarthy-like prosecution" and issuing a ringing appeal to other scientists: "Never, but never, let anything like this happen [to you]," he said. "Never let these people get

in your lab."

These amounted to the opening shots in what soon became a worldwide and highly public gunfight. For the next six months the pages of *Nature*, *Le Monde*, and even *Time* bristled with opinions, suggestions, accusations, and counteraccusations. While few scientists believed that the results of the high-dilution experiments were valid, many found *Nature*'s tactics distasteful and even dangerous. "Demeaning to the scientific process," wrote Mark Johnson of MIT. "A three-ring circus," said Mass General's James Scott. "Confirmation of what I always suspected," wrote biochemist Keith Snell of the University of Surrey— "Papers for publication in *Nature* are refereed by the editor, a magician, and his rabbit."

The *Nature* team stands firm. "Had we known how poor the evidence was," says Steward, "*Nature* would never have published the paper, and we would not have gone to France."

"We would do it the same way again," says Maddox. "It was the only way to flush it out."

All the publicity hurt. Up for evaluation by his bosses at INSERM, Benveniste's job seemed to hang in the balance. In the end, two evaluating committees suggested that Benveniste stop investigating molecular memory. But INSERM director-general Philippe Lazar decided Benveniste could proceed, and in 1989, the high-dilution experiments were quietly resumed. Repeating his original trials with anti-IgE, Benveniste says he's gotten similar results.

At the same time, he launched a series of new experiments to see if highly dilute solutions could provide reactions not just in cells, but in whole organs. In these experiments, he inoculated guinea pigs with egg albumin from hens.

He then removed the guinea pigs' hearts, suspended them in a glass cylinder, and kept them "alive" and beating. Finally, he used tubes to drip highly dilute solutions of egg albumin into the disembodied hearts. If the disembodied hearts recognized the egg albumin, Benveniste knew, they would have the typical immune reaction: The coronary arteries would dilate, and blood flow would increase.

Even though the solution of egg albumin was so dilute that not a single molecule of the albumin remained, the blood vessels of the heart appeared to respond: They dilated slightly and blood flow through the heart registered a detectable increase. In other words, the heart seemed to be reacting to what was not plain water as if the egg albumin were still there.

Benveniste quietly presented these results at the April 1991 meeting of the Federation of American Societies for Experimental Biology in Atlanta. With the exception of reports in the *Journal of the French Academy of Sciences* and *New Scientist* magazine, the press did not take note.

The storm over his work temporarily at bay, Benveniste is free to ponder the basic questions posed by his curious experimental results. How could a living system—a human cell or an animal organ—react to something that isn't there? How could water "remember" a substance that's gone?

After much reflection, Benveniste could have the answer. He believes that molecules communicate via electromagnetic radiation instead of by exchanging chemicals. Like the signals transmitted by a radio station to a receiver, these electromagnetic signals have different and specific frequencies, each one prompting a different and specific biochemical reaction. "What my experiments show, in a terribly clumsy way," he says "is that when you highly dilute a solution, you separate the molecule from its electromagnetic message contained in the water."

The medium for these electromagnetic messages, according to Benveniste, is water. This powerful communication, he says, has been demonstrated in his experiments again and again. Whether you're talking about anti-IgE or egg albumin, Benveniste notes, the original molecule modulates waves originating from water molecules, causing them to emit electromagnetic signals—and to continue emitting them even after the original reagents are gone. "The message remains in the water," he declares, "just as your voice remains on a tape recording even when you're no longer talking." In other words, in Benveniste's mind, the whole vast dance of chemistry, of biology—of life itself—is orchestrated by electromagnetic signals passing through water.

Though the idea is controversial, to say the least, evidence from other researchers has begun to fall into place. Scientists such as T.Y. Tsong, a University of Minnesota biochemist whom Benveniste likes to cite, for instance, have shown that many cellular functions—including enzyme activity and synthesis of DNA and RNA—are stimulated or suppressed by electromagnetic fields. Tsong thinks that future research will show that electromagnetic radiation may

indeed constitute what he calls "the language of the cell." Tsong, of course, is not entirely comfortable with Benveniste. "My work is based on principles that we already know," he says, "and his is based on things we don't yet understand."

Robert Becker adds, "Bioelectromagnetics is still beset by an enormous amount of uncertainty, but there has to be some kind of energetic reaction to explain molecular communication, and the obvious candidate is electromagnetism." Benveniste's work, he says, is a beginning, "somebody asking what if?"

And what if Benveniste is right? What if biochemical reactions can indeed be prompted by electromagnetic signals recorded in water? "Once you're able to pick up the signal," says Benveniste, "you have a whole new biology. You can digitize the message, you can make drugs from it. You won't need the physical substance, be it ordinary aspirin or AZT, but simply the signal that constitutes its code."

Benveniste also thinks the electromagnetic language could be used to do non-invasive tests. "You stick your finger in a machine that uses an electromagnetic field to analyze your blood," he explains, and you could even use the body's own code to perform non-invasive surgery, in which a defective heart, for example, "is repaired by sending the appropriate electromagnetic signals to heal its damaged cells."

Killer diseases like cancer or AIDS could be prevented or cured by jamming the electromagnetic signals that turn normal cells cancerous or that enable the AIDS virus to find the immune system cells that it targets and eventually destroys.

Some critics think that this sort of speculation—indeed, all of Benveniste's high-dilution work—shows that the once-respected immunologist has taken leave of his scientific senses. Even Nobel laureates, notes Eugene Garfield, "have gone off the deep end in pursuit of private passions."

On the other hand, his few supporters think this kind of research may presage the dawn of nothing less than what Becker calls "a new scientific revolution" in which "present dogmatic theories of how the universe works will be replaced." In the meantime, Benveniste continues to work, refusing to let go until someone proves him wrong to his satisfaction. "My enemies say a true scientist doesn't pay attention to erratic data like this," he says. "Even my friends tell me to drop this because I am going to kill myself. But if it's true that we can have molecular activity without molecules, then we have discovered a fundamental process of life.

"I have no private passions," he concludes. "All I have are these data. What am I supposed to do with them—put them back in the drawer?" □

Questions: *Answers on page 197*

1.   Are you skeptical that the water molecule holds the molecular basis of memory? Why?

2.   What do you think is a possible biochemical basis for memory? On what do you base your answer?

3.   How do the subjects of physics and chemistry relate to the biologically-based question of this study?

A synthetic substance resembling vitamin A may stabilize a precancerous condition that can develop in the oral cavity of smokers. It may also have potential as a preventative measure in other cancers of the body. The substance is called 13-cis-retinoic acid. Along with beta carotene, it belongs to a group of compounds called the retinoids. Compounds of this family many slow the growth rate of premalignant cells. The research conducted on this family of compounds may lead to the development someday of an anti-cancer pill.

Vitamin A-like Drug May Ward Off Cancers
by K.A. Fackelmann

A synthetic vitamin A-like substance may help stabilize a precancerous condition of the mouth that often afflicts smokers. In fact, researchers believe the new drug treatment may also play a role in preventing full-blown oral cancer and other types of malignancies.

In 1986, a research group at the University of Texas M.D. Anderson Cancer Center in Houston first reported using the synthetic substance, called 13-cis-retinoic acid, to treat leukoplakia — whitish patches of skin in the mouth that can turn malignant.

However, the study also found that in many patients, the precancerous skin patches returned soon after the patients stopped taking the drug. In addition, many study participants reported troubling side effects such as redness of the eyes and dryness of the mouth.

The same team has tackled leukoplakia again. This time, Waun Ki Hong, Scott M. Lippman and their colleagues wanted to see if they could prolong the preventative benefits of the drug while minimizing its side effects.

The team designed a clinical trial in which 53 people with leukoplakia received a very large dose of 13-cis-retinoic acid for three months. then the researchers randomly assigned the study participants to nine months of maintenance therapy with either a low dose of 13-cis-retinoic acid or beta carotene, a natural vitamin A precursor found in many orange fruits and vegetables.

The Texas group tested beta carotene because it produces virtually no side effects and is converted to vitamin A in the body. Beta carotene and 13-cis-retinoic acid both belong to a family of compounds known as the retinoids. Scientists believe retinoids may help stave off cancer by slowing the growth of premalignant cells. The new study represents the first long-term comparison of 13-cis-retinoic acid and beta carotene. The researchers found that premalignant changes progressed in just two of 24 patients (8 percent) assigned to 13-cis-retinoic acid. However, the beta carotene strategy didn't pay off: 16 of 29 persons taking beta carotene (55 percent) experienced a steady advance of their premalignant condition.

Lippman and Hong reported their data last week in San Diego at the joint meeting of the American Society of Clinical Oncology and the American Association for Cancer Research.

About 8 percent of Americans will develop leukoplakia, which often strikes smokers and heavy drinkers. The standard treatment for the condition is surgery, but this approach doesn't work for premalignant patches that have spread over large areas of the mouth.

The new study is part of a body of work that suggest retinoids may help prevent cancer, comments Michael Sporn of the National Cancer Institute in Bethesda, Md. Sporn's own work suggests that retinoids may help spur immature cells to mature and thus stop the out-of-control growth that characterizes cancer.

This preventive strategy may work with other types of cancer as well. Two years ago, Hong's team reported in the September 20, 1990 NEW ENGLAND JOURNAL OF MEDICINE that 13-cis-retinoic acid helped prevent a second bout with cancer in people who had been treated for head and neck tumors. Such people, who are often smokers, run a high risk of developing another cancer after undergoing surgery to remove their primary tumor, Hong says.

Despite the promise that retinoids display, much work remains before doctors can recommend a cancer-preventing pill for their patients. In the meantime, Sporn warns against popping massive quantities of vitamin A in an attempt to ward off cancer. Large doses of this nutrient can cause liver damage and other severe side effects, he notes. ☐

Questions: *Answers on page 198*

1. Do you think that large doses of vitamin A in the diet can prevent cancer?

2. What are the drawbacks when taking large doses of vitamin A?

3. Are the retinoids a possible cure for cancer?

According to Ockham's Razor, the simplest hypothesis used to explain a collection of facts is also the best hypothesis. In other words, among competing hypotheses, favor the simplest one. Any attempt to explain data should not be any more complicated than needed. This attractive alternative now appears to have firm statistical support. The reasoning for favoring the simplest explanation is that it has a highest probability for being correct. Scientists have used this mathematical proof to their advantage when choosing the most likely hypothesis in their experiments. Often the biggest challenge is recognizing the identity of the simplest hypothesis among several possibilities. Among biologists the principle of selecting the simplest hypothesis has worked well in studies of evolutionary biology. Similarities in the DNA base sequence of two organisms can be interpreted as descent from a common ancestor. The greater the divergence in base sequence, the less likely that a common ancestry exists. This hypothesis has been tested through other lines of evidence indicating common ancestry, such as through the study of the fossil record. It has added validity to the DNA hypothesis.

Ockham's Razor and Bayesian Analysis

by William H. Jefferys and James O. Berger

The principle known as Ockham's razor has high standing in the world of science, buttressed by its strong appeal to common sense. William of Ockham, the 14th-century English philosopher, stated the principle thus: Pluralitas non est ponenda sine necessitate, which can be translated as: "Plurality must not be posited without necessity." It is not entirely certain what Ockham meant by this rather opaque saying, but later versions of the principle, which have been traced to various authors other than Ockham, have a clear enough interpretation. The idea has been expressed as "Entities should not be multiplied without necessity" and "It is vain to do with more what can be done with less"; a modern rendering might be "An explanation of the facts should be no more complicated than necessary," or "Among competing hypotheses, favor the simplest one." Over the years Ockham's razor has proved to be an effective device for trimming away unprofitable lines of inquiry, and scientists use it every day, even when they do not cite it explicitly. See Thorburn (1918) for a history of the principle.

Ockham's razor is usually thought of as a heuristic principle — a rule of thumb that experience has shown to be a useful tool, but one without a firm theoretical or logical foundation. Under some circumstances, however, Ockham's razor can be regarded as a consequence of deeper principles. Specifically, it has close connections to the Bayesian method of statistical analysis, which interprets a probability as the degree of confidence or plausibility one is willing to invest in a proposition.

Ockham's razor enjoins us to favor the simplest hypothesis that is consistent with the data, but determining which hypothesis is simplest is often no simple matter. Bayesian analysis can offer concrete help in judging the degree to which a simpler model is to be preferred. Ironically, whereas Bayesian methods have been criticized for introducing subjectivity into statistical analysis, the Bayesian approach can turn Ockham's razor into a less subjective and even "automatic" rule of inference.

Galileo's Problem

The connection between Bayesian statistics and Ockham's razor is implicit in the work of Harold Jeffreys of the University of Cambridge, whose book Theory of Probability, published in 1939, was an important landmark in the modern revival of Bayesian methods. The connection has since been made explicit by a number of others: see Good (9168, 1977), Jaynes (1979), Smith and Spiegelhalter (1980), Gull (1988), Loredo (1989) and MacKay (1991).

An example that Jeffreys discussed in 1939 provides an illuminating introduction to the problems that can arise when Ockham's razor is put to the test as an implement of scientific methodology. Suppose you are collecting some data on the motion of falling bodies, as Galileo supposedly did in his legendary experiments at the Tower of Pisa. You drop a weight and record its position, s, at several moments, t, during the fall. The challenge then is to devise a mathematical law describing the motion.

The law proposed by Galileo, and familiar to students of physics, can be expressed as a quadratic equation: $s = a + ut + 1/2gt^2$. Here a, us and g are adjustable parameters, or in other words constants that can be

Reprinted by permission of AMERICAN SCIENTIST, Journal of Sigmi Xi, The Scientific Reasearch Society

assigned arbitrary values in order to fit the empirical data. (In this case a is interpreted as the initial position of the falling object, u is the initial velocity, and g is the acceleration due to gravity.) There are straight-forward methods for finding values of a, u and g that minimize some measure of the error between the predicted and the observed positions of the body. If Galileo's task is merely to identify those optimum parameter values, then the problem is a standard exercise in estimation theory.

But Galileo did not have to confine his attention to quadratic laws. He could instead have pro-posed a cubic equation, such as $s=a+ut+1/2gt^2+bt^3$ where the coefficient b is a fourth adjustable parameter. And of course there is not reason to stop with cubic polynominals. By adding further terms the equation could be ex-tended to fourth, fifth or sixth powers of t. Indeed, an infinite sequence of equations could be formed in this way. Why is it, then, that the quadratic law is the choice of physicists everywhere?

The answer is not that the quadratic law offers closer agree-ment with the empirical data. On the contrary, for any given data set, going to a higher degree polynomial can always reduce the total error (unless the fit is already perfect). If there are n measured data points, then an equation of degree n-1 specifies a curve that can be made to pass through all of the data points exactly, so that the measured error is zero. Thus there must be something other than accuracy in fitting data that leads people to prefer the quadratic law over any of the higher degree equations.

One possible explanation is that any coefficients beyond a, u and g are generally very small, so that higher powers of t contribute little to the structure of the physical law. Another interesting point is that

even when a high-degree equation fits a given set of data exactly, the equation may do very poorly as a predictor of new data. For example, given seven experimental measure-ments, a sixth-degree polynomial can fit the data exactly, whereas a quadratic equation will generally have some residual error; but if additional measurements are made, perhaps at larger values of t, the higher-degree law is likely to yield much larger errors than the quadratic one. Looked at another way, a single quadratic law can explain a variety of data sets reasonably well, whereas many data sets would require quite different sixth-degree polynominals.

These observations might well serve as an after-the-fact justifica-tion for rejecting a law of acceler-ated motion based on a sixth-degree equation, but they fail to account for a more fundamental fact: Neither Galileo nor a modern student of physics would even consider a sixth-degree equation in the first place. They would favor the quadratic law because it is simpler, whereas all higher-degree polynominals are unnecessarily complicated.

Probabilities Prior and Posterior

Jeffreys suggested that the reason for favoring the simpler law is that it has a higher prior probabil-ity; in other words, it is considered the likelier explanation at the outset of the experiment, before any measurements have been made. This is certainly a reasonable idea. Scientists know from experience that Ockham's razor works, and they reflect this experience by choosing their prior probabilities so that they favor the simpler hypothesis. Even though scientists do not usually explain their reasoning process in terms of prior probabilities, they tend to examine simple hypotheses before complex ones, which has the same effect as assigning prior probabilities according to some

measure of simplicity. The method reflects the tentative and step-by-step nature of science, whereby an idea is taken as a working hypoth-esis, then altered and refined as new data become available.

In an earlier work Jeffreys and Dorothy Wrinch had proposed codifying the scientist's intuitive preference for simplicity in terms of a rule that would automatically give higher prior probability to laws that have fewer parameters (Wrinch and Jeffreys 1921; Jeffreys 1939). For laws that can be expressed as differential equations, they sug-gested a straightforward algorithm for counting parameters. Having sorted all possible laws according to this criterion, one can try the simpler laws first, only moving on to more complicated laws as the simple ones prove inadequate to represent the data. Thus the ordering of hypoth-eses provides a kind of rationalized Ockham's razor.

The trouble with Jeffrey's appeal to prior probabilities is that it seems to get the question. Defining the simplest law as the one with the fewest adjustable parameters is a useful strategy, but it cannot be extended to yield a clear, universal rule for assigning prior probabilities, as Jeffreys himself points out (Jeffreys 1939, pg 49). He writes: "I do not know whether the simplicity postulate will ever be stated in a sufficiently precise form to give exact prior probabilities to all laws; I do know that it has not been so stated yet. The complete form of it would represent the initial knowl-edge of a perfect reasoner arriving in the world with no observational knowledge whatever." Needless to say, no real scientist qualifies as such an unbiased perfect reasoner.

But Jeffreys also suggested a measure of simplicity that does not depend on prior probabilities; instead it is grounded in tests of statistical significance. Basically, if a law has many adjustable parameters, then it will be significantly preferred to the simpler law only if its predictions are considerably more accurate. Indeed, if the predictions of the two models are roughly equivalent, the simpler law can have greater posterior probability (the probability an observer assigns to the law after the measurements have been made and the data collected. Jeffreys never stated in so many words that this result is a form of Ockham's razor, although it seems likely that he was aware of it. The first to point out the connection explicitly appears to have been Jaynes (1979), and independently Smith and Spiegelhalter (1980), who called an "automatic Ockham's razor." automatic in the sense that it does not depend on the prior probabilities of the hypotheses. This version of the razor is not fully automatic, however, because it does depend on probabilistic modeling of the effect of the more complex law on the data.

In Berger and Jefferys (1992) we observe that even this input can often be avoided, leading to an objective quantification of Ockham's razor. We shall describe this objective version of Ockham's razor after reviewing the basics of Bayesian analysis and considering some examples of how Bayesian methods and Ockham's razor can be applied to problems in the sciences.

Probability and Plausibility

The earliest ideas in the theory of probability arose to deal with various problems in the mathematics of gambling, where a probability can usefully be defined as the frequency of a specified outcome in a long series of identical trials. For example, if a fair die is cast many times, the face bearing four pips comes up about one-sixth. This *frequentist* formulation of probability theory works well in many contexts, but there are also questions it cannot readily answer. For

example, a geologist might ask: What is the probability of an earthquake, given certain precursory seismic signals? Obviously, it is not possible to calculate this probability by performing many trials under identical conditions.

In Bayesian analysis, *probability* is used in another sense: A probability is a measure of the plausibility of a hypothesis or proposition. this alternative definition is particularly useful in the sciences. When a paleontologist states that the dinosaurs *probably* died out as a result of climatic change, or when an astronomer says that Mars is *probably* lifeless, the probabilities cannot readily be understood as frequencies, but they have a natural interpretation as indicating the speaker's degree of belief or confidence in the statement, given the available evidence.

The foundation of Bayesian statistics is a theorem proved by the Re. Thomas Bayes, an English clergyman and amateur mathematician, in 1761, the year of his death; the proof was published posthumously (Bayes 1763). At its core, Bayes's Theorem represents a way — Bayesians would argue the most consistent way — of incorporating new data into your understanding of the world.

Suppose you have a series of hypotheses about some natural phenomenon. The hypotheses are know to be mutually exclusive and exhaustive, so that exactly one hypothesis must be true. Based on all the information available to you, you assign each hypothesis a probability. These are the prior probabilities mentioned above in connection with Galileo's experiment. Now suppose some new item of data comes to your attention, such as the result of an experiment. The question is: How should you revise the probabilities you ascribe to the various hypotheses in light of the new data? Bayes's theorem offers a

mathematical procedure for answering this question.

The notation *P(X|Y)* represents a conditional probability: the probability that hypothesis *X* is true, given the available information *Y*. With probabilities expressed in this way, Bayes's theorem can be stated as follows:

$$P(H_i \mid D\&I) = \frac{P(D \mid H_i\&I)\, P(H_i \mid I)}{P(D \mid I)}$$

This equation can be used to calculate *P(H_i | D&I)*, or the probability that H_i is true, given both the prior information *I* and the new data *D*. Three factors enter into the calculation. $P(H_i|I)$ is the prior probability ascribed to hypothesis H_i, or in other words the probability of H_i given the initial information *I*. $P(D \mid H_i\&I)$ is the probability of observing the new data *D*, given the initial information *I* and assuming that H_i is true. Finally, *P(D|I)* is the total probability of observing *D* given *I*, no matter which of the hypotheses turns out to be true. Thus the final probability of *H* given both *D* and *I* increases if the prior probability of H_i increases or if *D* is more strongly predicted by H_i and *I*. Conversely, the final probability of H_i is reduced if *D* is predicted more generally, by all possible hypotheses.

The use of Bayes's theorem in statistical and scientific reasoning has had a long and controversial history; see Edwards, Lindman and Savage (1963) or Berger (1985) for discussions of the controversies. There are two main points of contention between Bayesians and traditional (frequentist) statisticians. The first is philosophical: Some argue that since only of the hypotheses H_i can be true, it makes no sense to talk about the "probability" that H_i is true. This has a certain logic if one interprets probabilities as frequencies, but the objection is beside the point if "probability" refers to the degree of plausibility of a hypothesis. This is the way most working scientists use the term.

The second point is that there are not universally accepted ways of assigning the prior probabilities $P(H_i | I)$ that Bayes's theorem requires. Hence different scientists, faced with the same data, may come to different conclusions. Bayesians have several responses to this complaint. One school believes that there is nothing inherently wrong with subjectivism, and, indeed, that the frequentist approach is really no more objective, although it has successfully disguised this fact (Berger and Berry 1988). Subjectivist Bayesians point out that it is common for scientists to disagree about the plausibility of hypotheses, and contend that this is a natural, and indeed inescapable, state of affairs.

Another school (Laplace 1812, Jeffreys 1939) has developed methods of choosing and utilizing "objective" prior probability distributions for a wide class of problems. With problems for which such methods are available. Bayesian analysis can claim to be as objective as any other statistical method. Still, there remain problems for which these objective methods do not work. Some of the examples discussed below fall into this troublesome class.

To Catch a Cheat

The key idea linking Bayesian analysis to Ockham's razor is the notion of simplicity in a hypothesis. In quantifying this notion, it is useful to observe that a simpler hypothesis divides the set of observable outcomes into a small set that has a high probability of being observed and a large set that has a small probability of being observed; the more complex hypothesis tends to spread the probability more evenly among all the outcomes. Thus the simpler hypothesis makes

sharper predictions about what data will be observed, and it is more readily falsified by arbitrary data. In the case of Galileo's problem, the more complex hypotheses have more parameters, which can be adjusted to accommodate a larger range of data. In other cases, the number of adjustable parameters is not at issue, but, nonetheless, one hypothesis restricts the possible outcomes of an experiment more than another does.

Suppose a friend who has a reputation as a prankster offers to flip a coin to decide who will perform a little chore: heads he wins, tails he loses. Knowing your friend's reputation, you might well be concerned that he would use trickery (perhaps a two-headed coin) to win the toss. The hypothesis H_{HH} that the coin has two heads is, under this understanding, a simpler one than the hypothesis H_{HT} that the coin is fair. In a series of many coin tosses, H_{HH} will be falsified if tails comes up even once, whereas any sequence of heads and tails could arise under H_{HT}.

Before the coin is flipped, you might believe that the hypotheses H_{HH} and H_{HT} are equally likely. Then the coin is tossed, and it indeed comes up heads. Your degree of belief in the two hypotheses will change as a result of this information, and (by Bayes's theorem) the posterior probability that you assign H_{HH} should not be twice what you assign to H_{HT}. Still, the evidence that your friend is trying to fool you is not very strong at this point, perhaps not strong enough to challenge him for a close look at the coin. On the other hand, if the coin comes up heads on five occasions in a row, you will be rather inclined to think that your friend is playing a joke on you. Even though both hypotheses remain consistent with the data, the simpler one is now considerably more credible.

In the days before electronic

computers, when publishing mathematical tables was still a viable business, the compiler of a table had to contend with possible copyright infringement. If someone published a table identical to your own work, how could you demonstrate to the satisfaction of a court that the new table was copies from yours rather than calculated *de novo*? To guard against plagiarism, compilers frequently took advantage of the fact that numbers ending in the digit 5 can be rounded either up or down without significantly altering the result of a calculation. By rounding such numbers randomly, the compiler could embed a secret code in the table that identified the table as his work, while not significantly affecting the accuracy of the results obtained when using the table.

For example, suppose you published a table of sines with 1,000 entries. you calculated each value to five decimal places, the rounded to four places. About 100 of the entries would have ended in the digit 5 and would have been rounded either up or down at random. Another compiler of a table would be very unlikely to happen on the same pattern of rounding, since there are 2^{100}, or approximately 10^{30}, ways to round the 5s in the table.

If you learn that a newly published table has the same rounding pattern as your own, Bayesian analysis can quantify your suspicions of plagiarism. Let H_p be the hypothesis that the second table was plagiarized from yours, and H_i be the hypothesis that the second table was generated independently and just happens to have the same pattern of roundings. On the data D that the rounding patterns are identical, we can calculate that $P(D|H_p)=1$ and $P(D|H_i)=2^{-100}$. Assuming equal prior probabilities for the two hypotheses, Bayes's theorem shows that the posterior probability of plagiarism differs only negligibly from 1.

The reason for this clear outcome is that H_p makes a precise prediction about what will be seen, and is inconsistent with almost all possible data. whereas H_l is consistent with any observation. H_l "hedges its bets" by trying to accommodate all possible data; in contrast, H_p risks everything on a single possibility. As a results, when that single possibility turns out to be turn, H_l is rewarded for the greater risk it takes by being given a very high posterior probability compared to H_p, even though H_l is also consistent with the data.

It is now routine for authors of directories, maps, mailing lists and similar compilations to deliberately introduce innocuous errors into the material. When plagiarism or other unauthorized use of the material takes place, the presence of these errors in the copies materials serves as very strong evidence of copyright violation.

David Harpp and James Hogan of McGill University have used a similar idea to detect cheating on multiple-choice tests. They wrote a computer program to compare the answers given by each pair of students in the class and look for a near-match between correct and incorrect answers. Of course as teachers they hope and expect students to know the subject material so that conclusions about cheating cannot be drawn from a student's correct answers. But if two students make the same errors, the evidence of cheating can be compelling. the analysis of the data in this problem is more complicated than it is in the case of a plagiarized mathematical table because different questions are answered incorrectly with differing frequencies and because the various incorrect responses for each question can be expected draw different numbers of responses. But there are practical solutions for these complications.

Another application of this principle comes from evolutionary biology. When the DNA of two organisms is compared, similarities in sequence can be taken as evidence of descent from a common ancestor. For DNA within a functioning gene, however, the strength of such evidence is compromised, because the nucleotide sequence could not diverge too far without impairing the function of the gene product. This constraint is removed in the case of a pseudogene, which is a region of DNA that has most of the characteristics of a gene but because of some defect does not give rise to a functioning protein or other product (Max 1986, Watson et al. 1988). A pseudogene can be passed on to an organism's progeny, even though it has lost its function. If two species have identical or nearly identical pseudogenes (as human beings and chimpanzees do, for example), this constitutes very powerful evidence in favor of the hypothesis that the species have a common ancestor. Just as with cheating on multiple-choice tests, or plagiarism of compiled materials, it is the verbatim or near-verbatim repetition of a "mistake" that gives the hypothesis of copying — in evolutionary terms, descent from a common ancestor — a high posterior probability.

A Planetary Puzzle

Our last example illustrating the predictive power that simplicity confers on a hypothesis merits somewhat more detailed analysis. It concerns a celebrated controversy in astronomy and celestial mechanics.

Beginning with the work of the French astronomer Urbain Leverrier in the 1840s, astronomers were aware of a serious problem in explaining the motion of the planet Mercury. Newtonian theory, which had been extraordinarily successful in accounting for most of the motions in the solar system, had run up against a small discrepancy in the motion of Mercury that it could not explain easily. After all the perturbing effects of the other planets had been taken into account, there remained an unexplained residual motion of Mercury's perihelion (the point in its orbit where the planet is closest to the sun) in the amount of approximately 43 seconds of arc per century.

It seemed something had been overlooked. One appealing possibility was the proposal that another planet might exist, closer to the sun than Mercury. Leverrier himself, along with the English astronomer John Couch Adams, had recently met with brilliant success by predicting that a previously unknown planet was responsible for discrepancies in the motion of Uranus. When Johann Gottlieb Galle, a young astronomer at the Berlin Observatory, looked where Leverrier suggested, the planet Neptune was discovered. It seemed possible that a similar phenomenon might explain the anomaly in Mercury's motion.

A number of astronomers duly set out to find the new planet, dubbed Vulcan in anticipation of its discovery, and some sightings were announced. The sightings could not be confirmed, however, and over time interest in Vulcan hypothesis waned.

Other mechanisms were also proposed. It was suggested that rings of material around the sun could produce the observed effect; or the sun might be slightly oblate, due to its rotation on its axis; or, finally, the Newtonian law of gravitation might not be exactly right. For example, the American astronomer Simon Newcomb (1895) proposed that the exponent in Newton's law of gravitation might not be exactly 2, but instead by be $2 + \varepsilon$.

All of these hypotheses had one characteristic in common: They had parameters that could be adjusted to agree with whatever data on the motion of Mercury existed. In modern parlance, we would call the presence of such parameters a "fudge factory." The Vulcan hypothesis had the mass and orbit of the putative planet; the ring hypothesis had the mass and location of the ring of material; the solar-oblateness hypothesis had the unknown amount of the oblateness; and all the hypotheses that modified Newton's law of gravitation had an adjustable parameter (such as Newcomb's ε) that could be chosen more or less at will.

Not all of the hypotheses were equally probable, however (Roseveare 1982). As noted above, sightings of Vulcan were never confirmed. As time went on, the hypothesis of matter rings of sufficient density also became less and less likely (Jeffreys 1921), although some still believed in them (Poor 1921). A solar oblateness of sufficient size probably would have been detectable with 19th century techniques. The hypothesis that Newton's law of gravitation needed an arbitrary adjustment to fit the data was the one explanation that could not be ruled out by existing evidence.

What happened historically is well known. In 1915 Einstein announced his general theory of relativity, which predicted an excess advance in the perihelion motion of the planets. After some confusion (Roseveare 1982, pages 154-159) it became clear that the amount of the predicted advance for Mercury was very close to the unexplained discrepancy in Mercury's motion. The amazing thing was that the predicted value, which is 42.98 seconds of arc per century using modern values (Nobili and Wills 1986), was not fudge factor that could be adjusted to suit the data but instead was an inevitable consequence of Einstein's theory.

The general theory of relativity

made two other testable predictions (the gravitational bending of light and the slowing of clocks in a gravitational field). There has been a lively debate over the years as to how important each of these phenomena has been in convincing the scientists that general relativity is the correct theory of gravity (Brush 1989). Here we shall sidestep this argument and try to put ourselves inside the mind of a Bayesian observer in the early 1920s who is trying to weigh the evidence for various explanations of Mercury's anomalous motion.

Poor v. Jeffreys

An interesting pair of papers was published in 1921 (Poor 1921, Jeffreys 1921). Charles Lane Poor was an astronomer at Columbia University who was not convinced by the evidence for general relativity and who still clung to the matter-ring theory. Unfortunately, he also made some serious errors in his assessment of how a matter ring would affect the other inner planets. Jeffreys, in response, argued persuasively that the ring theory was not viable because sufficient matter did not exist. Jeffery's paper was published before he made his major contributions to probability theory, and he does not, ironically, make the Bayesian argument that we have outlined above. And so we will make for Jeffreys the argument that he might have made had he returned to this question some years later.

Poor gives a value of $a=41.6 +- 1.4$ arc seconds per century for the observed anomalous motion of Mercury. The task for Bayesian analysis is to assign a probability based on this observation, to each of two candidate explanations of the planetary motion. Einstein's general theory of relativity and a "fudged Newtonian" theory, in which some parameter is adjusted to account for the discrepancy in the observations.

The place to begin is with the

measurement's reported uncertainty of $+-1.4$ arc-seconds per century. Although Poor's paper does not discuss the nature of this uncertainty, it is surly what statisticians designate a probable error, which is equal to 0.6745 times the standard deviation; thus the standard deviation itself is 2.0 arc-seconds per century. It is reasonable to assume that this error has a normal distribution; in other words it is described by a symmetrical, bell-shaped curve, with the total area under the curve equal to 1, and with about two-thirds of the area lying within one standard deviation of the center.

Poor reports the prediction of Einstein's theory as $\alpha_E = 42.9$ arc-seconds per century, which is quite close to the modern value. On the assumption that Einstein's prediction is in fact correct, what is the value of $P(a|E)$, the probability of observing a value of $a=41.6$ arc-seconds? The answer can be determined by evaluating the appropriate normal curve (namely the curve centered at Einstein's prediction of 42.9 and having a standard error of 2.0) at the observed data value of $a=41.6$. The resulting value, called the probability density of $a=41.6$, is about 0.16, which is reasonably high in this context. If the observed value of a were 42.9, exactly equal to the predicted value, the probability density would rise only to 0.20.

Performing the equivalent calculation for the fudged Newtonian theory is not as straightforward. For the very reason that the theory has a fudge factor, it is not easy to say exactly what it predicts. To give the theory an explicit probabilistic form, it is necessary to make some assumptions, although it will become apparent later that the outcome is quite insensitive to these assumptions.

One useful point of departure is the conservative assumption that since the Newtonian theory is well established, large deviations from it are less believable than small ones. (If gravity has an inverse-cube law instead of an inverse-square law, the difference would have been noticed long ago.) Accordingly, it is natural to choose a probability distribution for the unknown anomalous perihelion motion _ that makes _=0 the likeliest value, with the probability density diminishing smoothly as _ departs from zero. Likewise, it makes sense to give the probability density a symmetrical distribution, at least for those theories in which _ could equally well be either positive or negative, so that the perihelion motion of Mercury could be either advanced or retarded. It is important to think a priori here; we are discussing predictions of the alternative theory prior to seeing the data.

These considerations would be satisfied by a normal probability distribution with a mean of $\alpha = 0$. But the most difficult question remains: What is the standard deviation of this distribution, which determines the width of the bell-shaped curve? Again the only available guidance is the knowledge that very large values of the anomalous perihelion motion are ruled out by existing observation. For example, if some gravitational effect perturbed the perihelion motion of Mercury by as much as 100 arc-seconds per century, it would also alter the orbits of Venus, the earth and Mars to an extent that could have been detected in the 1920s. For the purposes of rough calculations, a reasonable standard deviation is about 50 arc-seconds per century, which does not contradict any observational data on the inner planets.

We now have in hand the two elements needed to calculate the probability density of the observed data $a=41.6$, assuming the validity of a fudged Newtonian theory.

Assuming that the unknown anomalous perihelion shift α has a normal distribution with mean zero and standard deviation 50, and that the observed a is equal to α plus a random error having standard deviation 2.0, standard methods of probability theory can be used to compute $P(a \mid F)$, the overall probability density of a under the fudged Newtonian theory. In this example, $P(a \mid F)$ itself turns out to have a normal distribution with mean 0 and standard deviation 50.04. This distribution is much flatter than $P(a \mid E)$, so that the probability is distributed over a much wider range. For this reason, the probability density of any one value is greatly reduced. Specifically, the probability density of the actual value $a=41.6$ is only about 0.0056, compared with the probability density of 0.16 for Einstein's theory.

What is of interest, however, is not the probability density of the data $a=41.6$ given the various theories, but rather the probabilities of the various theories being true given $a=41.6$. These latter probabilities could be calculated from Bayes's theorem if one were willing to assign prior (that is, premeasurement) probabilities to the theories. Luckily, the need to choose prior probabilities can be avoided (if desired) by use of the ratio of the probability densities of $a=41.6$ under the Einsteinian and the fudged Newtonian theories, namely

$$ B = \frac{P(a \mid E)}{P(a \mid F)} $$

It can be shown from Bayes's theorem that this ratio, called the Bayes factor, gives the odds favoring E over F arising from the data. When B is greater than 1, the data favor E, and when B is less than 1, they favor F. The overall odds of E over F are found by multiplying B by the prior odds, which is the ratio

of the prior probabilities of E and F. The point here is that it may suffice to consider only B; the Bayes factor may well answer the question without the need to formally involve the prior odds.

Plugging in the numbers yields a value of $B = 28.6$, which is moderately strong evidence in favor of the Einsteinian hypothesis. Ironically, the data that Poor himself provides in his paper against general relativity favor the Einsteinian hypothesis over the fudged Newtonian hypothesis.

The calculations leading to this conclusion involve several factory. There is, first, the matter of how well the data fit each hypothesis. Obviously, if the observed data differ sharply from the predictions of a hypothesis, one would expect that hypothesis to be assigned a low probability. In many cases such goodness-of-fit considerations are decisive in choosing among hypotheses. In this instance, however, the predictions of both theories are consistent with the data. Nevertheless, Bayes's theorem offers a clear choice between the theories.

The factor that contributes most to the outcome of the calculations is the width of the probability distribution of a for the fudged Newtonian hypothesis. Because this distribution is relatively wide, the fudged Newtonian hypothesis has to waste a considerable amount of probability on hypothetical values of a that are far from the actual $a=41.6$.

The fudged Newtonian hypothesis has an additional degree of freedom that allows it to accommodate a much larger range of hypothetical data than does the Einsteinian hypothesis. As a result the fudged Newtonian hypothesis must spread its risk over a larger parameter space in order not to miss the region supported by the data. In this sense, it is a less simple theory than the Einsteinian hypothesis. Einstein's hypothesis makes a sharp prediction about Mercury's perihelion motion, which depends only on the known values of the constant of gravity and the speed of light. Any measurement of the perihelion motion that is not close to the predicted value contradicts Einstein's theory. In contrast, a broad range of data — every value of the anomalous motion not ruled out for other reasons — is consistent with the fudged Newtonian hypothesis.

An Objective Ockham's Razor

One step in our analysis of Poor's argument may seem rather doubtful: the choice of a specific prior probability distribution for the fudged Newtonian hypothesis. And this aspect of the analysis turned out to be particularly important, since it is the great width of that distribution that makes the difference between the two hypotheses. We suggested first that the probability distribution should be symmetric about $\alpha = 0$ and decreasing as the absolute value of α increases; these are reasonable constraints on the shape of the distribution. But the final choice of a normal distribution with a specific standard deviation of 50 arc-seconds per century seems rather arbitrary. The method by which we arrived at the figure of 50 would be difficult to generalize to other problems.

The need to specify a specific standard deviation for $P(\alpha \mid F)$ is easy to overcome. One can simply consider an arbitrary standard deviation — call it τ — and then graph the Bayes factor B as a function of τ. Of considerable interest is the finding that B has a minimum value; it is always greater than 27.76. Thus there is strong evidence in favor of the Einsteinian theory no matter what value of τ is chosen.

It is less obvious how to overcome the rather arbitrary choice of a normal distribution for $P(\alpha \mid F)$. The solution is given in Berger and Jefferys (1992), where it is shown that the Bayes factor has a lower limit even if the distribution for $P(\alpha \mid F)$ is not a normal one, provided only that the distribution obeys certain rather mild conditions. Specifically, for any $P(\alpha \mid F)$ that is symmetric about $\alpha = 0$ and decreasing in the absolute value of α, B is always less than or equal to the following expression:

$$\sqrt{\frac{2}{\pi}} \left(|D_F| + \sqrt{2 \ln \left(|D_F| + 1.2 \right)} \right) \exp\left(\frac{-D_E^2}{2} \right)$$

Here D_E is the number of standard deviations that α deviates from the Einsteinian prediction; for the data under consideration $D_E = -0.65$. D_E is the number of standard deviations that α deviates from the base Newtonian prediction of $\alpha = 0$; in this case $D_E = 20.8$. Adopting this "worst-case" value gives every benefit of the doubt to the hypothesis F; if F is not favored under these conditions, then it is not favored at all. For the present case, the lower bound on B is 15.04, which remains fairly strong evidence in favor of the Einsteinian theory.

Conclusions

Ockham's razor, far from being merely an ad hoc principle, can in many practical situations in science be justified as a consequence of Bayesian inference. Bayesian analysis can shed new light on what the notion of the "simplest" hypothesis consistent with the data actually means. We have discussed two ways in which Ockham's razor can be interpreted in Bayesian terms. By choosing the prior probabilities of hypotheses, one can quantify the scientific judgement that simpler hypotheses are more likely to be correct. Bayesian analysis also shows that a hypothesis with fewer adjustable parameters automatically has an enhanced posterior probability, because the predictions it makes are sharp. Both of these ideas are in agreement with the intuitive notion of what makes a scientific theory powerful and believable. □

Acknowledgements

We wish to thank our colleagues who have commented on drafts of this paper: Raynor L Duncombe, David M. Hillis, Myra Samuels and Mark Jefferys. This work was supported by the National Aeronautics and Space Administration, under NASA Contract NAS5-29285, and by the National Science Foundation, Grant DMS-8923071.

References

Bayes, Thomas. 1763. An essay toward solving a problem in the doctrine of chances. Philosophical Transactions of the Royal Society 53:370-418.

Berger, James O. 1985. Statistical Decision Theory and Bayesian Analysis. New York:Springer-Verlag.

Berger, James O. and D. Berry. 1988. Statistical analysis and the illusion of objectivity. American Scientist 76:159-163.

Berger, James O. and William H. Jefferys. 1992. The application of robust Bayesian analysis to hypothesis testing and Occam's razor. To appear in Journal of the Italian Statistical Society.

Brush, Stephen. 1989. Prediction and theory evaluation: The case of light bending. Science 246:1124-1129.See also the response to this article in Science 248:422-423.

Drake, Stillman, and James MacLachlan. 1875. Galileo's discovery of the parabolic trajectory. Scientific American 232:102-110.

Edwards, Wl. H. Lindman and L.J. Savage. 1963. Bayesian statistical inference for psychological research. Psychological Review 70:193-242.

Good, I.J. 1968. Corroboration, explanation, evolving probability, simplicity, and a sharpened razor. British Journal of the Philosophy of Science 19:123-143.

Good, I.J. 1977. Explicativity: a mathematical theory of explanation with statistical application. Proceedings of the Royal Society A 354:303-330.

Gull, S. 1988. Bayesian inductive inference and maximum entropy. In G.J. Erickson and C. R. Smith (eds.) Maximum Entropy and Bayesian Methods in Science and Engineering (Vol. 1), 53-74. Dordrecht: Kluewer Academic Publishers.

Harpp, David. 1991. Quoted in "Big Prof is Watching You." Discover 12(April):12-13.

Jaynes, E.T. 1979. Inference, method, and decision: Towards a Bayesian philosophy of science. Journal of the American Statistical Association 74:740-41.

Jeffreys, Harold. 1921. Secular pertubations of the inner planets. Science 54:248.

Jeffreys, Harold. 1939. Theory of Probability. (Third Edition 1983.) Oxford:Clarendon Press.

Laplace, Pierre de Simon. 1812. Theorie Analytique des Probabilities. Paris:Courcier.

Loredo, T.J. 1990. From Laplace to Supernova 1987A:Bayesian inference in astrophysics. In P. Fougere (ed.) Maximum Entropy and Bayesian Methods, 81-142. Dordrecht:Kluwer Academic Publishers.

MacKay, David J.C. 1991. Bayesian interpolation. Submitted to Neural Computation.

Max, Edward E. 1986. Plagiarized errors and molecular genetics: Another argument in the evolution-creation controversy. Creation/Evolution XIX:34-46.

Newcomb, Simon. 1895. The elements of the Four Inner Planets and the Fundamental Constants of Astronomy. Washington:Government Printing Office.

pages 109-122.

Nobili, A.M., and C.M. Will. 1986. The real value of Mercury's perihelion advance. nature 320:39-41.

Poor, C.L. 1921. The motions of the planets and the relativity theory. Science 54:30-34.

Smith, A.F.M., and D.J. Spiegelhalter. 1980. Bayes factors and choice criteria for linear models. Journal of the Royal Statistical Society B 42:213-220.

Roseveare, N.T. 1982. Mercury's Perihelion from Le Verrier to Einstein. Oxford: Clarendon Press.

Thorburn, W.M. 1918. The myth of Occam's razor. Mind 27:345-353.

Watson, James D., Nancy H. Hopkins, Jeffrey W. Roberts, Joan A. Steitz and Alan M. Weiner. 1988. Molecular Biology of the Gene, 4th Edition. Menlo Park:Benjamin/Cummings Publishing Company. pages 649-663.

Wrinch, D., and H. Jeffreys. 1921. On certain fundamental principles of scientific inquiry. Philosophical Magazine 42:369-390.

Questions: *Answers on page 198*

1. Can you think of another area of biology besides evolutionary biology where Ockham's Razor could be used successfully?

2. Can you thing of an area of biology where Ockham's Razor might not be as applicable?

3. What is the value of statistics toward posing a hypothesis as part of the scientific method?

Part Two

Understanding the Structure and Function of Cells

7

One of the major tenants of modern biology is the cell theory. By this theory all organisms consist of cells and all cells are produced from pre-existing cells. The exception to this theory occurred when the first cells originated through evolution. Formed over 3.5 billion years ago from organic molecules present in the early environment of the Earth, all succeeding cells were derived from these original cells. These original cells, resembling the bacteria that inhabit the planet today, lived mainly in the oceans and represented the dominant forms of life for over two billion years. Jack Szostak, a microbiologist, is trying to duplicate the conditions of the primitive Earth in the laboratory, producing simple cells from their molecular building blocks. In an attempt to create life at the cellular level, scientists and all students of biology can gain a better understanding of the structure and function of the fundamental unit of life, the cell.

The Handmade Cell
by David H. Freedman

If you want to really understand the nuts and bolts of life, argues Jack Szostak, why not build a cell from scratch?

Getting hold of cells for research is not difficult. You can grow colonies of them in a petri dish, clone them in a test tube, or lift some right out of your own skin, for that matter. But microbiologist Jack Szostak has opted for a less conventional route: he is building his own cell from scratch. "I'm trying," he says casually, "to create life."

Without question, a cell passes any test for being alive. A single-celled organism like a bacterium can eat, grow, adjust to the environment, reproduce, evolve. So if Szostak succeeds, he will indeed have created a living organism of sorts, albeit one not quite like any other known. His creation will have a bubble-like membrane akin to the outside of a natural cell, a noteworthy development in itself. But Szostak's real triumph will be the tiny package within the membrane, a package consisting of several strands of custom-build RNA. In natural cells RNA often serves as a messenger molecule. It carries

blueprints from the cell's library of instructions — its DNA — to factories where proteins are made according to those instructions. But Szostak's RNA will do something that seemed impossible until just a few years ago: it will not only carry instructions but also execute them. And those instructions will tell each strand of RNA to build copies of itself—a function widely considered to be the very essence of life.

To make his scheme work. Szostak has to simplify the cellular machinery that exists in nature. Instead of a bagful of specialized molecules—DNA for storing blueprints, RNA for transferring instructions, specialized proteins for a variety of cellular construction jobs—he needs one RNA molecule that can do everything. It has to carry its own blueprint, transfer the blueprint to copies of itself, and use parts of itself as the copying tools. Puttering away inside its membrane, reproducing, and possibly even evolving, Szostak's RNA it seems safe to say, will be the most life-like thing ever to emerge from a test tube.

Yet oddly enough, Szostak may

not be the first to create this strange entity. Nature beat him to it by several billion years, according to a theory now widely accepted among biologists. A self-replicating strand of RNA, many believe, was a precursor to all the more complex, DNA-based forms of life that exist on Earth today. "It's generally believed that there was a time roughly four billion years ago when RNA was running the show," says Gerald Joyce, a leading RNA researcher at the Scripps Research Institute in La Jolla, California. "We just don't know how an RNA life-form came about." However it may have happened, that original RNA has long since disappeared. The all-purpose molecule gave way to specialized offspring that were far more efficient, when acting in concert, at maintaining and reproducing themselves.

Just how nature managed to come up with the portentous first strand of self-replicating RNA has been a deep and long-standing mystery. One might think that viruses could provide a clue, because many viruses carry all the self-reproduction information they

Reprinted with permission from DISCOVER, August 1992, vol. 13, no. 8, pp. 46-52 copyright 1992 by The Walt Disney Company

need on single strands of RNA. But those strands can't act on their own: they insert their instructions into the complex machinery of host cells they've invaded.

Today every known type of replicating RNA needs the help of enzymes. These flexible molecules enfold the single strand of RNA and, using this original strand as a template, string together building-block molecules called nucleotides into a complementary copy of it. Where the original has a blue, the complement will have an orange; where it has a purple, the complement will have a yellow. The complementary strand becomes like half a zipper, interlocking with the original RNA all along its length. Eventually the two strands come apart and another enzyme repeats the process on the complementary strand, stringing together nucleotides that are now a precise copy of the original RNA. The complement's orange is now matched by a nucleotide that reproduces the original's blue; its yellow by one that reproduces the original's purple.

The enzymes that do all this work are a type of protein, which means that they must be assembled according to instructions embedded in RNA. In other works, RNA must direct the assembly of the enzymes that put together more RNA. It's a nice system—but how could it ever get started? Without enzymes, something like today's RNA couldn't have any enzymes. It is biology's ultimate chicken-and-egg question.

Nearly 30 years ago researchers posited an elegant way out: a primitive RNA molecule that could also act as an enzyme for its own replication. Two of these identical molecules acting in concert—one as an enzyme, the other as template—might have been able to churn out a third, complementary molecule unaided. The complementary

molecule could then have turned out a replica of the original RNA, which would then crank out another complementary molecule, and so on. Eventually, after many mistakes that didn't survive and a few luck improvements that did, the RNA could have evolved into more complex versions capable of synthesizing separate enzymes to carry out the job more efficiently. And the rest, as Darwin might have said, is natural history.

The only problem was that no one had ever found a form of RNA that could do double duty as an enzyme, and so this particular origin-of-life scenario came to be regarded as no more than an interesting speculation. But in 1981 biochemist Tom Cech found a strand of RNA capable of performing some simple enzymatic functions, a strand he had isolated from a paramecium-like pond creature called Tetrahymena. That discovery earned Cech a Nobel Prize three years ago and seemed to give the self-replicating-RNA theory the boost it needed.

But the sense of triumph was soon muted. "When RNA enzymes were first discovered," says Joyce, "there was a flurry of papers and commentary saying, 'Well, that solves it, we can see there was an RNA life-form.' But the difference between what these molecules could do and self-replication is actually quite significant." Specifically, it had quickly become clear that Cech's RNA, along with 80 or so related enzymatic RNAs that were discovered in the next few years among a variety of microorganisms, plants, and fungi, couldn't manage much more than hacking apart its own strands in a few specific places and perhaps sticking a few nucleotides onto one end—an enzyme's handiwork, to be sure, but a far cry from replication. And no one saw a way to prove that some as yet undiscovered version of RNA could do better. That's when Jack Szostak

decided to get involved.

Szostak doesn't fit the mold of his specialty. Many origin-of-life biochemists are flamboyant, outspoken characters. The soft-spoken 39-year-old Szostak, in contrast, exhibits little affinity for the limelight and politics of high-powered research. He works in the maze-like interior of the new building in Boston that serves as Massachusetts General Hospital's research digs, and it would be easy to mistake him, in his casual attire, for one of the many thousands of graduate students slaving away in the bowels of Greater Boston's bevy of universities.

Seven years ago Szostak began the hunt for an RNA enzyme that could be coaxed into self-replication. He hadn't picked the world's easiest chore; there was a biochemical chasm between a molecule that could cut itself apart and one that could deftly weave nucleotides into a precise copy of itself. "The general feeling was that I was trying to do the impossible," say Szostak. "But I was optimistic. I really did believe that the origin of life happened this way and that if it did, I should be able to do it in the lab."

Szostak's overall strategy was dazzlingly unexciting: Think small. "If you looked at this goal in terms of a big jump, it really did seem impossible," he says. "But if you looked at it as the sum of a lot of little, manageable steps, it didn't seem as hard. So we decided to simplify the problem by breaking it down into several smaller ones. Of course, we knew that any one of these steps could have been a roadblock that would have stopped us cold. We just tried not to think about that." Joined by graduate student Jennifer Doudna in 1986, Szostak got to work.

Step one was to take Cech's Terahymena RNA enzyme, with its ability to snip itself apart, and get it to perform its chopping trick on a separate molecule, since an RNA strand serving as a replication enzyme would have to act not on itself but on another strand that would be acting as a template. To that end, Szostak and Doudna divided the Terahymena RNA into two pieces: a large chunk of about 300 nucleotides to perform the splitting action, and the remaining chunk of some 40 nucleotides that was the target site for the splitting.

After synthesizing the two strings separately in a test tube, they confirmed that the larger string on its own could indeed slice apart the smaller, unattached target string. Even better, researchers were able to make the reaction run in the other direction—that is, they showed that the larger string of RNA, like other splicing enzymes, could not only cut but join, or "ligate," the errant halves of the target back together. Thus the team had come up with an RNA enzyme that could convert two small strings, called oligos, of about 20 nucleotides each into a single longer strand. "It was a very important, very exciting first step," say Szostak.

Next Szostak and Doudna had to find a way to get their enzyme to do more than just put back together the two severed halves of the target RNA. What they were looking for, after all, was replication, and for this the enzyme would have to join free-floating oligos—oligos that would be synthesized by the researchers—into a sequence that would be complementary to a target strand. (That is, when the assembled oligos were laid alongside the target, everywhere the target had a se-quence of nucleotides that ran, say blue-yellow-orange, the complemen-tary oligo would have a string of nucleotides running orange-purple-blue.) The enzyme couldn't join together just any two oligos; it had to use the target as a template. Otherwise the oligos would be strung together randomly instead of

in a precise sequence—hardly a mechanism for accurate copying.

If the enzyme was to follow a template faithfully, it would have to weld together only oligos that were complementary to the template at every point along their length, and to refrain from joining those sequences that didn't quite match. There are four types of nucleotide in RNA (corresponding to our four color-coded versions), and they have preferred parings: adenine (A) prefers uracil (U) as its complement, while guanine (G) prefers cytosine (C). These two happy couples, A-U and G-C, zip right up; they are called Watson-Crick pairs, while other, less felicitous combinations are called wobble pairs. When an enzyme comes by, all it has to do is make sure that each nucleotide on the oligos is lined up next to its Watson-Crick pal on the template, so that there are no wobble pairs, and then weld together the ends of the oligos. When the complementary strand in turn becomes a template, it will reproduce the original strand simply by following its natural preferences.

Unfortunately, the Terahymena RNA enzyme seemed determined to join oligos that incorporated wobble pairs, a habit that would introduce mistakes later on in replication. This was a serious problem: it would prevent the enzyme from accurately copying long chains of RNA.

To remedy the situation, the two researchers looked for a way to modify the enzyme, the template, or the oligos just enough to get matches that were entirely Watson-Crickish. But given the complexity of enzymatic reactions, this sort of fine-tuning is always a hit-or-miss proposition; Szostak and Doudna therefore decided simply to throw chemicals at the enzyme and hope one of them did the trick. One of the first substances they tried was spermidine, a small, electrically charged molecule that had been

shown in other labs' experiments to twist the molecular structure of RNA.

It was a shot in the dark, but luck was on their side. When the researchers added spermidine to the RNA mixture, the enzyme readily ligated oligos perfectly matched to the templates, without wobble pairs. The researchers are still not sure what happened. Whether the spermidine altered the shape of the enzyme, the template, or the oligos, or all three for that matter, they couldn't say. "The bottom line is that it worked," Szostak says with a shrug.

Pressing their good fortune, Szostak and Doudna then tried to get the enzyme to ligate not just oligos complementary to the small target chunk of Terahymena RNA but also a variety of oligos complementary to longer, synthetic RNA strands cooked up in their laboratory. Again the enzyme worked perfectly, joining as many as five different oligos into a single strand in a sequence that made the whole string complementary to the synthetic RNA. The chances of those oligos bumping into each other in that order by themselves, without a template, and at the same time joining end to end were almost zero. Clearly, the reaction was working just the way Szostak wanted it to: oligos were bumping into the synthetic RNA and sticking to a complementary part of it; the enzyme was then jumping in and joining those oligos end to end.

If the enzyme could perform this trick on synthetic templates, why shouldn't it be able to make copies of itself? By 1988 Szostak and Doudna were sure they were hot on the trail of self-copying RNA. "This was the first thing that really looked like replication," says Szostak. In fact, what they had come up with fell short of self-replication: their enzyme could string together only a few relatively long oligos, not individual nucleotides, according to a template. That's a critical difference because it's not likely that enough of these oligos would have been floating around in the primordial ooze. "The system was really only giving us back a longer version of what we put in," says Szostak. "We wanted to get it down to handling two- or three-nucleotide oligos because then we could put in all the random combinations and let it pick the right ones."

But the enzyme didn't have anywhere near the precision or speed necessary to match and ligate the roughly 115 three-nucleotide oligos it would have needed to construct a copy of itself. Speed is important because oligos don't bond very strongly to the template; the enzyme has to do its work before the oligos can get away. Szostak and Doudna had two options: they could reengineer the enzyme to handle a large number of oligos quickly, or they could find a shorter RNA enzyme that could be constructed from a smaller number of oligos. Szostak decided to take the second route.

"We had started with the Terahymena RNA because it was so well studied," he explains. "But we had known from the beginning it might not work out, and we were prepared to switch to a different system." Looking at 50 or so related RNA enzymes that had been identified since they had started their research, Szostak and Doudna settled on a strand of RNA called sun Y, found in a bacterial virus known as T4. Sun Y had one overriding grace: at about 200 nucleotides, it was the shortest in this class of RNA enzymes.

But though it was a bit over half the length of the Terahymena RNA and even better at ligating oligos, sun Y was still too long. It would never be able to string together reliably the nearly 70 three-nucleotide oligos it would need to

replicate itself; that would be like presenting a child with a jigsaw puzzle of too many pieces and too little time to assemble them. To improve the situation, Szostak and Doudna relentlessly hacked away nucleotides from the enzyme to see how far they could shrink it—in effect, they were simplifying the jigsaw puzzle. But it was a delicate game of trade-offs: while a shorter sun Y would make an easier-to-replicate template, the same molecule when acting as an enzyme still needed all the loops and folds critical for embracing matched-up oligos and performing its ligating stunt. If the researchers hacked off too much, sun Y would be come a simpler template but would no longer by an effective enzyme.

The two researchers eventually got down to a 160-nucleotide version of sun Y. At first it proved too weak as an enzyme, but they soon discovered they could restore its enzymatic strength by shuffling around a few of the nucleotides. Still, after all this inspired tweaking, the resulting enzyme remained much too large for its alternate role as a template. And this time the researchers were stumped.

"Every time we had come to a roadblock, Jennifer and I had sat and talked about it and come up with some strategy to get around it," say Szostak. "It seemed amazing that we had gotten this far. But this was a severe problem, and we were very discouraged. We were forced to sit back and think about whether there was some completely different way to approach it."

They needed a breakthrough—and it wasn't long in coming. "I was sitting in my office staring at sun Y molecular-structure diagrams," recalls Szostak, "when all of a sudden I had this idea. What if we could split the enzyme into pieces that self-assembled?"

Szostak had little trouble finding places where he could break

the molecule into three pieces of similar length. This novel strategy seemed to meet the team's conflicting needs. In a solution, at any given time, some of the three sun Y subunits would be linked together, and some would be floating unattached. The RNA that was completely assembled would act as an enzyme; the unattached pieces would be short enough to be copied. In theory, the fully assembled molecules would join oligos into complementary strands alongside the shorter chunks acting as templates. Then the complementary strands would act as templates themselves, churning out replicas of the original templates with the help of the original enzyme. Eventually, a lot of brand-new sun Y subunits would be floating around, and some would self-assemble to become new enzymes.

"I ran out of the office and told Jennifer," say Szostak, "and she immediately got excited about the idea and went to work on it." Just two weeks later Szostak and Doudna were watching their new strategy work on a lab bench. "It was one of the most exciting moments of the project," he says.

In a very rough sense, their three-piece RNA enzyme is indeed self-replicating. However, the researchers still need to feed it prefab oligos of about eight nucleotides each. Until the system makes accurate copies out of randomly clumped two- and three-nucleotide chunks, it can't be considered a plausible model for what nature did 4 billion years ago. Right now the enzyme is too slow and sloppy to stitch enough tiny oligos together: because it deviates from the template about one out of every three times it tacks on an oligo, the chances of its being able to get 20 right are minute. "We really need to focus on this question right now," says Szostak. "We have to go from seventy percent accuracy to ninety-nine percent accuracy."

Ideally, says Szostak, he'd like nature to do the necessary design work for him. "If the process was a little more efficient and accurate," he explains, "we could let the enzyme replicate itself, errors and all. Some of the errors would lead to mutants that would replicate even better, and these would eventually dominate and lead to other mutants that did even better, and pretty soon evolution would take over."

For now, though, he has turned to a hand-operated version of evolution. When Szostak synthesizes fresh batches of his sun Y RNA in a test tube, he purposely shuffles some of its nucleotides in the hope that a handful of the possible trillions of new combinations will turn out to be more efficient enzymes. To isolate these talented few, he has the entire batch ligate various molecules and then strains the mixture to pull out the longest strings. He picks out the mutants that produced those strings, synthesizes a fresh batch of the winners, and repeats the entire process. "Getting to real evolution is still a distant goal for us," he says, "but I'm pretty optimistic."
In addition to his work on RNA, Szostak is experimenting with phospholipids, soapy substances that ball up in water to form microscopic globules. When he adds a phospholipid to a test tube full of his RNA molecules in solution, the emerging

globules trap tine drops of liquid containing RNA molecules. Although that doesn't even begin to approximate the complex way in which modern DNA directs the construction of its own cell membrane, Szostak thinks his globules can perform many of the functions of a real membrane. He has already found ways of getting them to "grow" by combining with other globules, as well as divide by squeezing through porous materials—both of which could have happened 4 billion years ago. "A lot of people neglect the idea of compartmentalizing the RNA, but separating it from the rest of the world is crucial for evolution," he says. "You want the RNA replicating its own interesting mistakes instead of wandering off in the solution."

Some scientists aren't convinced that Szostak has found the recipe for creating life. "He'll stand right up there and tell you he's trying to build a cell, and I applaud his chutzpah," says Norman Pace, a biochemist at Indiana University. "But unless he has some new gimmick I'm not familiar with, I don't think he's anywhere close to getting a self-replicating molecule. And even if he were, there's no reason to believe it has anything to do with the origin of life. RNA is much too fragile to have survived the early conditions on the planet."

But Gerald Joyce of Scripps is

one of the many who weigh in on Szostak's side. "Jack has continued to amaze over the last several years," he says. "He's not trying to recapitulate in detail something that happened on the primitive Earth; he's trying to capture the fundamental principles of how an RNA-based system could evolve. I tend to be hard-nosed about this sort of thing, but I'm optimistic about Jack's efforts." Tom Cech, who kicked off the whole field of RNA enzymes, agrees: "That three-component system convinced me this whole thing is going to work. If they keep going at this rate, they'll have RNA self-replication within two or three years."
Szostak continues to plug away. Like many scientists these days, he's finding that his biggest challenge may be to keep his funding: Hoechst, the German chemical and pharmaceutical company that has given tens of millions of dollars to Mass General's biochemical research program, has cut back its support. "I'm going the traditional route now, which means writing lots of grant proposals," he shrugs. If he's concerned, he doesn't show it. Maybe he's just thinking that if it came down to it, he could always scrounge up the money in small chunks, one modest grant at a time. After all, it's an approach that's worked pretty well for him so far. □

Questions: *Answers on page 198*

1. Why do you think that a new line of cells will not originate naturally in modern times on the Earth?

2. How will an understanding of the structure and function of the cell promote an understanding of higher levels of biological organization?

3. What makes RNA an excellent choice for an original molecule to produce a cell?

8

Medicines and therapeutic drugs have long been administered either by injection or through oral dosages. However, a new method of administration proposes delivering them in spoonfuls of small, membrane-bound spheres called protenoids. To many patients, injections are an unpleasant experience. Oral administration, in the form of pills, is often not an effective means of delivery. Most of the newer pharmaceuticals cannot withstand the harsh pH environment of the stomach. Orally - administered pills, however, must pass through this chamber before entering the small intestine, where they are absorbed into the bloodstream. Most of the older-marketed drugs, such as penicillin and morphine, can clear this barrier unscathed. Packaging substances in small, protein spheres protects the drugs from the digestive action and destruction in the stomach. Use of these biological "smartbombs" offers an efficient, desirable means to deliver numerous therapeutic agents. For example, the protein insulin would usually be digested in the stomach before reaching the small intestine. Protected and packaged inside a spoonful of protenoids, this protein can be taken as an efficient, oral treatment for diabetes mellitus.

End of The Needle

by Rick Weiss

In the beginning Robert Rosen created a kind of primordial soup. Mimicking origin-of-life experiments performed in the 1950s, he concocted a stew of amino acids, the building blocks of proteins, and boiled them under an acrid nitrogen atmosphere. He cooked and evaporated and extracted and synthesized. Then he looked down at what he had created—shimmering microscopic spheres made of protein-like material—and he said to his boss, "it is good."

The year was 1986, and the idea was to make fish oil more palatable by packing the vile-tasting vitamin supplement inside tiny, flavorless spheres that could slip uneventfully past the taste buds. Rosen's boss at that time, Solomon Steiner, the head of a small biological testing lab in New York, thought such a product would make his company rich. But Rosen, a biophysicist working as a consultant to the lab, had other ideas. "I never took fish oil seriously," he says. "You'd have to deliver so much of it—at least a tablespoon—in order for it to be effective. So I started to think about things that might be more appropriate."

And what Rosen thought about was insulin. The hormone that regulates blood sugar normally has no hope of surviving the harsh conditions in the stomach. Enzymes in the digestive system break it down. That's why diabetics, who don't have enough insulin, must take it by injection rather than as a pill. Rosen, a diabetic himself, has definite opinions about needles: "I'd love to take a pill instead of a shot." So he filled his bubbles with insulin stead of fish oil. And Rosen looked again at what he had created, and he thought about what it might be worth (billions of dollars, according to some estimates), and he told Steiner to forget about fish oil because there are a lot more diabetics looking for ways to avoid their daily injections than there are people in need of a tastier fish-oil pill. Steiner took the insulin-filled spheres and squirted them down the throats of some laboratory rats. The rodents' blood-sugar levels plummeted. Steiner's dreams of profits whet sky-high.

If they work as well in humans as they do in animals, these microscopic spheres may signal the genesis of a remarkable new era of drug delivery—one in which medicines now administered by injection will become available as simple little pills. The key to the microspheres' potential is their ability not only to withstand the hellish environment of the human gut but also to deliver their cargo exactly where it's suppose to go. After protecting their precious contents against the onslaught of digestive enzymes, the microspheres move into the relative safety of the small intestine. Not until then do these biological smart bombs release their therapeutic payloads, which are absorbed intact into the bloodstream.

Pharmaceutical manufacturers have been desperately searching for just such a system, because most of the new drugs they are now developing cannot withstand the acid test of oral delivery. The problem stems from a key difference between the newer drugs and those developed during the past couple of decades. Most of the older pharmaceuticals, such as morphine and penicillin, are made of organic compounds that can survive in the stomach. "They aren't food," say Rosen. "We haven't evolved the ability to digest them." In contrast, the newest medicines coming down the drug-development pipeline are potent proteins derived from genetically engineered cells reared in the laboratory. And proteins are held together by shared groups of molecules known as peptide bonds, which are very susceptible to digestive enzymes. "To the stomach," says Rosen, "They're just meat."

They're much more than that, of course. Recombinant human growth factor, which helps children with hormonal imbalances grown to their full potential; immune-enhancing colony-stimulating factors and cytokines, which muster tumor-chomping white blood cells in record numbers; blood-clot-preventing compounds that rescue cardiac muscles following a heart attack—all are proteins and are therefore unlikely to survive in the human stomach. That leaves syringes—universally unpopular and entirely unaffordable in many parts of the world—as the only means of getting these drugs safely into the blood.

When Rosen, who now works at Dalhousie University in Halifax, Nova Scotia, created his insulin-filled spheres, he was consulting for Emisphere Technologies in Hawthorne, New York. The lab was then known as Clinical Technologies Associates, a fledgling, privately held company that performed animal and human testing of experimental drugs for pharmaceutical developers. In the mid-1980s the company's then president and CEO, Steiner, decided the lab ought to have a product of its own. He settled on the quirky goal of developing a good-tasting fish-oil supplement. That meant something had to surround the fish oil. Steiner started pushing Rosen for suggestions. Rosen was familiar with the experiments performed by biochemist Sidney Fox, who in the 1960s was one of several laboratory scientists attempting to create life

from scratch by throwing together amino acids in an environment that mimicked Earth's pre-biotic conditions. The amino acids didn't come together in combinations we recognize as proteins, but they did string together. Fox called these creations proteinoids.

Fox also found that proteinoids, when dumped into water, came together in tiny spheres. And when they did, these spheres spontaneously encapsulated organic material floating around in the aquatic environment at the moment of their creation.

It seemed a bit like magic, and to this day no one is sure how this happens. Rosen, along with other biophysicists, speculates that it has a lot to do with the various electrical charges carried by the proteinoids. It's a rule of thumb in nature that opposite charges attract and identical charges repel. Proteinoids have small molecules, such as hydrogen, branching off them at various points along their length, and these molecules carry charges. Some areas of the proteinoid end up with an abundance of negative charges, while others have a lot of positive charges. These areas pull on oppositely charged areas in other proteinoids, and before you know it a bunch of these man-made chains glom onto one another.

But what makes them form a sphere? "The capacity to form these spheres is almost unknown," Rosen says. "Biological proteins don't do this." Rosen suspects proteinoids behave this way because they're not all created equal. Some, he suggests, are attracted to water, others are repelled by it, and some are sort of in between. That provides them with a basis for organization. "They get driven into a geometry that everything is 'happiest,'" Rosen says. "That's what holds them together. It's not some sort of lock-and-key interaction." Rosen says of proteinoids attracted to water may

arrange themselves with their heads pointing in, toward the water in the center of the spheres; those repelled by water point out; and the indecisive proteinoids form the middle of what then turns into a relatively thick membrane.

Fox had shown that these spheres resist proteases—digestive enzymes. "He's also shown," says Rosen, "that if you formed the spheres in an environment that contained organic material, the spheres would pick it up." Fox had even shown that it was possible to take advantage of the electrical charges of the proteinoids and design spheres that reacted in different ways to different environments. Rosen built on the foundation that Fox had laid. He repeated Fox's work, but he chose his amino acids on the basis of the electric charges they carry. He knew, for example, that aspartic acid and glutamic acid are positively charged, and therefore a proteinoid made predominantly of those amino acids carries more positively charged areas than negative ones. Because of the complicated way that proteinoids fold up and join in a sphere, most of these positively charged areas end up on the outside.

That, Rosen thought, would be a big advantage in the stomach, which contains acids that have a lot of positively charged hydrogen ions. Because of their positive charges, these ions wouldn't attach to the spheres, thus preserving the integrity of the structures. But-and this was the key to his scheme—an environment with negatively charged hydroxyl ions would have the exact opposite effect. These ions would attach to the sphere and begin to tear it open. And the end of the small intestine, as well as the bloodstream itself, has a lot of negatively charged hydroxyl ions. A sphere with these characteristics, Rosen concluded, would resist attack by acids in the stomach and survive intact until it passed into the intestine and the bloodstream, where it would dissolve, releasing its contents.

The procedure Rosen used to make the first insulin-containing proteinoid spheres in 1986 is essentially the same simple technique used at Emisphere today. The centerpiece is a cylindrical glass vessel 18 inches high and 9 inches in diameter, sitting unobtrusively on a countertop. With the loose panache of a short-order cook, an emisphere technician pours a dry powder of amino acids through a valve into the vessel, which is filled with argon gas (argon, it turns out, works better than the nitrogen that Rosen used at first). The technician heats the whole thing to about 400 degrees for several hours and ends up with a dark, viscous liquid that looks a lot like honey. "That's all. It's like making chili," quips the company's executive vice-president, Sam Milstein.

This liquid is then poured into another vessel, mixed with a solvent, and finally evaporated with a vacuum; left behind are clumps of amber-colored crystals resembling brown sugar. An additional processing step pulls out any remaining oils, creating the final product: a fine, tan powder that's actually a collection of proteinoids.

Technicians dissolve this powder in water; while this is going on, they mix into citric or acetic acid whatever drug they wish to encapsulate. When the proteinoid-saturated water and the acid solutions are brought together, the proteinoids spontaneously fold into seamless, microscopic spheres, enveloping some of the drug-containing fluid. All that remains is to run the solution through a filter that traps the spheres, which are then freeze dried. Millions of these microspheres will fit in a single teaspoon-or in a pill, which is how the company packages them.

Emisphere has been quick to capitalize on the discovery. The company is conducting tests not only on insulin but also on heparin, the widely used clot-dissolving compound. It already has formal agreements with three major pharmaceutical manufacturers to develop oral delivery vehicles for various injectable drugs, and it's negotiating with at least ten others.

It remains to be seen whether proteinoids represent an ideal solution. With the exception of a few technicians who have swallowed mouthfuls of empty ones (nothing happened), they remain untested on humans. Moreover, they will have to compete with other alternative means of drug delivery now under development, including nasal sprays, battery-powered transdermal patches, and implantable pumps. But because proteinoid microspheres are relatively easy and inexpensive to make and, most important, can be taken orally, many researchers rank them among the more promising products currently under investigation by drug delivery specialists.

Much as it would like to, however, Emisphere may not develop the first oral delivery system for traditionally injectable drugs to be approved by the Food and Drug Administration. In Cambridge, Massachusetts, researchers at a biotechnology company called Enzytech are experimenting with a chemical component of corn gluten-a naturally occurring protein—as an acid-resistant coating for standard doses of otherwise injectable drugs. These corn-gluten containers, called nanospheres, are less than a micron in diameter.

Unlike the smart-bomb spheres made by Emisphere, Enzytech's corn-gluten derivative doesn't specifically sense the acidity or alkalinity of its environment; it simply protects it contents long enough to allow a substantial portion of the dose to get absorbed into the bloodstream. Nanospheres have potential advantages over proteinoids. For one thing, the natural grain protein is already widely used in foods and as a coating on some pills, making the FDA approval process likely to proceed more smoothly. Preliminary studies of nanosphere-encapsulated insulin and erythropoietin (a red blood cell growth factor that helps patients with kidney diseases) in monkeys have been "very impressive," says MIT drug delivery specialist Robert Langer, who co-founded Enzytech. Human trials could begin as early as this fall.

But Emisphere's proteinoids, which can be custom designed for different types of drugs with different release points, may ultimately prove more adaptable than nanospheres. The company has already tested nearly 400 varieties. Also, Rosen points out, proteinoids have potential applications beyond oral drug delivery. When filled with immune-stimulating fragments of bacteria or viruses, for example, the microspheres may prove useful as oral vaccines. "With traditional vaccines you need somebody with a needle," Rosen says. "this could completely change the way people look at vaccination."

Only time and a lot of testing will tell whether proteinoid technology represents the dawn of a new era in drug delivery, says Robert Silverman, chief of the Diabetes and Digestive and Kidney Diseases. But if proteinoid spheres prove safe and effective in clinical trials, he says, then one thing's for certain: "They'll sell like hotcakes." ☐

9

A liposome is a hollow structure defined by a phospholipid membrane, the same type of biochemical layer surrounding the animal cell. This similarity to the structure of biological membranes allows liposomes to mingle with the cells and tissues of animals. When these spheres are filled with various substances, they serve as vehicles that can deliver those substances to the cells and tissues. The substance, for example, could be a drug that has therapeutic value. Scientists are now investigating various techniques that can prepare liposomes as a means for treating humans and for other therapeutic uses. Other possible uses involve the transfer of genes in genetic engineering and the delivery of cosmetics to the deep layers of the skin.

Liposomes
by Danilo Lasic

In the lotions aisle of the local pharmacy you may see a number of skincare products that advertise the presence of liposomes — "microscopic spheres that help reduce the signs of aging," as one manufacturer promises. Liposomes also have a variety of less frivolous applications, ranging from basic research on the properties of biological membranes to the delivery of therapeutic drugs. Whether the underlying motive is vanity or scientific curiosity, however, one cannot help being impressed by the utility of liposomes. What are these little spheres, and why do they command so much attention?

The Greek roots of the word liposome mean fat body, but a more precise definition might describe them as hollow structures made of phospholipids — the same molecules that form the membranes of all animal cells. It is this similarity to the structure of biological membranes that allows liposomes to mingle with the cells and tissues of our own bodies. But liposomes have another quality: Like little lipid bonbons, liposomes can be filled

with a variety of interesting centers. The similarity of liposomes to cell membranes and their ability to carry substances forms the basis of their scientific and industrial applications.

Liposomes were discovered in the early 1960s by the British scientist Alec Bangham. In the course of his research on the effect of phospholipids on the clotting of blood, Bangham would create reagents by adding water to a phospholipid film. He soon recognized that the phospholipid films would form closed spherical structures that encapsulated part of the liquid medium in their interior. These were liposomes.

Almost immediately after the discovery of liposomes, biophysicists and biochemists began to investigate their possible applications. These scientists had long sought a model system for the study of biological membranes and membrane proteins. The simple structure of liposomes now offered the possibility of exploring the behavior of membranes of known composition.

In the late 1970s and early

1980s the potential applications of liposomes in the pharmaceutical and medical industries produced an increasing interest that bordered on euphoria. Overly enthusiastic research programs attempted to achieve goals that are now known to be unrealistic or even impossible. Too frequently the scientific credibility of the research was suspect because the results were buoyed by enthusiasm or commercial interests rather than rigorous experimental or theoretical work.

More recently the objectives of liposome technology have been revised and refined. There are currently about a dozen liposome-based therapies that are being investigated in clinical trials. This is considerably less than was anticipated 15 years ago. Other applications of liposomes have shown much more stable growth. Here I shall review some of these developments and consider the future outlook.

Making Liposomes
The primary constituent of a liposome is a lipid: a lollipop-like

Reprinted by permission from AMERICAN SCIENTIST, Journal of Sigmi Xi, The Scientific Research Society.

molecule, consisting of a polar, hydrophilic "head" attached to a long, nonpolar, hydrophobic "tail." The hydrophilic head typically consists of a phosphate group (hence phospholipid), whereas the hydrophobic tail is made of two long hydrocarbon chains. Because the lipid molecules have one part that is water-soluble and another part that is not, they tend to aggregate in ordered structures that sequester the hydrophobic tails from water molecules.

One of the simplest structures that can be formed from lipids is a bilayer, in which the heads form the surfaces of a sandwich protecting the tails from interacting with the water. A flat sheet of lipid bilayer is not stable, however, because the edges are exposed to the water. As a result, the lipid bilayer tends to wrap itself into closed spherical structures, or liposomes.

At one time or another, all of us have probably made liposomes inadvertently in the kitchen, Beating eggs to make an omelet, for example, produces the right set of circumstances to make a batch of liposomes. Egg yolk contains the lipid molecule phosphatidyl choline (Lecithin), which can aggregate to form liposomes. However, making liposomes in the course of a Sunday brunch is not especially useful; the yield is low, the liposomes are heterogeneous and the results would be difficult to replicate. In the laboratory the challenge is to make liposomes that are consistent in size, structure and content.

Three attributes interest a scientist preparing a batch of liposomes: the chemical composition of the lipid bilayer, the size distribution of the liposomes and the number of layers in each liposome. There are many ways to prepare liposomes while maintaining control over these properties. The classical methods, still used today, involve adding water to thin films of dry lipids. Agitating the wetted lipids produces a suspension of large, many-layered liposomes of various sizes. These liposomes known as multilamellar vesicles or MLVs, can be further treated to produce liposomes of a desired size that are made of a single lipid bilayer. Such unilamellar vesicles may be small (SUVs) with a diameter of a few hundredths of a micrometer, or large (LUVs), as much as 100 micrometers in diameter.

For many years the formation of liposomes was enigmatic: no biophysical mechanism appeared to be common to the different methods of preparation. Recently, however, we have made some progress (Lasic 1988). It is now clear that all liposomes form from existing bilayers in one of two ways: A small section of a flat bilayer may break off and close upon itself, or part of the bilayer may bud off from the large aggregate if it is forced to curve. It is possible to force a curve into the bilayer by introducing agents that increase the area of the outer lipid monolayer relative to the area of the inner layer. One way to accomplish this is to increase the size of the polar heads of the outer lipids. For example, ionizing the outer monolayer attracts water molecules that bind to the polar heads, effectively increasing their size. The area of the outer monolayer can also be increased by intercalating certain amphophilic molecules, such as detergents, among the existing lipids. Both of these processes must be performed relatively quickly for a liposome to form. Otherwise the inner monolayer will ultimately begin to incorporate both water molecules and lipids over time, and the planar symmetry of the bilayer will return (Lasic 1991).

Liposomes are usually made of a mixture of phosphatidyl choline, cholesterol and some electrically charged lipids, such as phosphatidyl serine and phosphatidyl glycerol. The inclusion of charged lipids improves the physical stability of the liposomes because electrostatic forces prevent the vesicles from making contact and agglutinating.

The tails of the phospholipids may also be tailored to suit a particular need. The hydrocarbon chain is usually from 14 to 18 carbon atoms long, and rarely is there more than one double bond between carbon atoms. In general, chains that are greater than 14 carbon atoms long and have no double bonds give rigid gel-like properties to the lipid membrane at body temperature; other molecules yield more nearly liquid membranes. The gel-like liposomes are usually lest prone to leak their cargo.

Drug Delivery

One of the basic aims of chemical therapeutics is to deliver the medicinal substance efficiently and specifically to the site of the disease or disorder. In some instances this can be achieved by administering the drug in a pure or "free" form. In many cases, however, the effectiveness of a drug can be improved by encapsulating it in some form of carrier. Ideally the carrier would deliver the drug to the site where it is needed and nowhere else. This method of directing medicine to specific sites is a modern version of the "magic bullet" idea first proposed by the German bacteriologist Paul Ehrlich at the turn of the century.

The notion of using liposomes as magic bullets that carry drugs to their target sites was envisioned not long after their discovery. One of the first such proposals exploited a defense and sanitation network in the body called the mononuclear phagocyte system. Cells in the network, called macrophages, patrol the circulatory system in search of unwanted particles. In the course of their travels, macrophages ingest wick and dying cells, invading pathogens and anything else that appears to be out of place — including liposomes.

The observation that macrophages ingest particles suggested that diseases involving the mononuclear phagocyte system might be treated by encapsulating drugs within liposomes (Alving et al. 1978; Black, Watson and Ward 1977; New et al. 1978). In particular macrophages are specifically infected by protozoa of the genus Leishmania — parasites that can spend part of their life cycle within a human host. (An estimated 100 million people are infected throughout the world; Ostro 1987). Leishmaniasis can be treated with intravenous injections of compounds that include the element antimony; unfortunately these substances are highly toxic in their free form. Antimonial drugs enclosed within liposomes act like Ehrlich's magic bullets because they interact almost solely with macrophages. Indeed, when animals infected with a disease that models leishmaniasis are treated with liposome-encapsulated antimonial drugs, toxicity is substantially reduced, while therapeutic efficacy increases 700 times (Alving et al. 1978, New et al. 1978). Nevertheless, liposome-based antimonial drugs have yet to make a significant impact on the treatment of leishmaniasis — possibly owing to a lack of commercial interest in developing a drug for a disease that is prevalent mostly in poor, third-world populations.

Despite such setbacks, other liposome-encapsulated drugs show considerable promise as medical and commercial ventures. Among these drugs is the antibiotic amphotericin B which is used for the treatment of systemic fungal infections. Patients whose immune system has been compromised — as a result of chemotherapy for cancer or from an immunodeficiency disorder such as AIDS — frequently suffer from widespread infections of fungi such as Candida albicans. Such infections can be deadly; for leukemia patients who become infected, the death rate is about 20 percent (N_ssader et al. 1990). The standard treatment of free amphotericin B — which preferentially binds to fungal membranes — must be limited because the therapeutic doses are toxic to cells in the kidneys, the central nervous system and the blood-forming tissues of the body. Moreover, the use of free amphotericin B is hindered because it is not soluble in water.

The problems associated with free amphotericin B can be obviated by encapsulating the drug in liposomes. Liposomes help to dissolve the drug in the blood and reduce its toxicity because they do not enter the kidneys — the organs most severely affected by the free drug. In addition, the tendency of macrophages to ingest liposome-encapsulated drugs confers an advantage because organs of the mononuclear phagocyte system — the liver, the spleen and the lungs — are frequently the targets of systemic fungal infections. As a result, the liposome-encapsulated drug tends to accumulate in the diseased organs. These advantages have already proved their worth. In several cases liposome-encapsulated amphotericin B has saved the life of a patient when all other treatments had failed (Lopez-Berenstein et al. 1985).

Encapsulating drugs within liposomes in order to avoid sensitive sites in the body has also been attempted in the treatment of certain types of cancer. Most anticancer drugs are extremely toxic, and their dosage must be limited to reduce undesired side effects (loss of hair, damage to the heart and suppression of bone-marrow function). The most commonly prescribed drugs for cancer patients are agents that prevent the growth of tumors, such as the compound doxorubicin. Unfortunately, when free doxorubicin is administered to cancer patients, it can severely damage the heart (Gabizon et al. 1982, Storm et al. 1987). However, since liposomes injected into the circulatory system do not accumulate in the heart, it was suggested that doxorubicin could be encapsulated in liposomes to ameliorate the drug's toxic effects (Juliano and Stamp 1978, Forssen and Tokes 1981). Studies in laboratory animals have, in fact, demonstrated that doxorubicin-induced cardiotoxicity is reduced (van Hoesel et al. 1984). Moreover, clinical trials with liposomal doxorubicin show that patients experience less nausea and hair loss compared to patients treated with the free drug (Gabizone et al. 1989; Tread, Greenspan and Rahman 1989).

These early results for the treatment of certain diseases show considerable promise for liposome-based drugs. In some respects, however, these are relatively easy cases for liposomes. Liposomal drugs, for example, may not be equally effective against all forms of cancer. Moreover, although the uptake of liposomes by macrophages may be useful for the treatment of leishmaniasis and systemic fungal infections, it is an obstacle to the delivery of drugs to other cells of the body. As a result, some investigators are attempting to make liposomes that can evade macrophages and other parts of the immune system.

Stealth Liposomes

One might think that avoiding a macrophage in the circulatory system would be an easy task for a liposome. Surely there must be somewhere to hide among the many red blood cells and white blood cells, the plasma proteins, amino acids, vitamins, hormones and lipids circulating in the blood. Alas, macrophages are very efficient hunters. Success in hiding liposomes has been achieved only recently.

Early endeavors attempted to disguise liposomes as miniature red blood cells by mimicking the lipid composition of the red cell's membrane. Although these experiments did not produce evasive liposomes, some encouraging results suggested an avenue for further exploration. In particular, the incorporation of lipid molecules that had certain sugar molecules attached to their polar heads seemed to be most effective (Papahadjopoulos et al. 1991). This result is consistent with the manner in which the cells of our own bodies avoid being engulfed by macrophages or attacked by other components of the immune system.

As with most human cells, red blood cells avoid activating the immune response by coating themselves with molecules conjugated with oligomers of a sugar called sialic acid. Any cell that is coated with sialic acid can avoid activating the immune system. In fact, certain pathogenic bacteria deceive the body's immunological defenses by coating themselves with sialic acid. In principle, this form of mimicry should also provide liposomes with a means to evade macrophages.

The development of evasive liposomes has been based on this principle, but a new generation of evasive liposomes has gone beyond the use of sugar oligomers. Polymers of ethylene glycol (the main ingredient of automotive antifreeze) seem to be even more effective. Liposomes coated with chains of polyethylene glycol 50 to 100 monomers long are able to stay in the blood hundreds of times longer than regular liposomes, with half-lives of over a day.

My colleagues and I have recently proposed a hypothesis to explain the ability of these evasive cargo-carriers, called Stealth

liposomes, to avoid the immune system's defenses (Lasic et al. 1991). In the normal course of events, a foreign particle — such as a regular liposome — is identified by a macrophage because immuno-logical substances called opsonins adhere to its surface. Opsonins act as flags that identify the liposome as a foreign particle, and thus facilitate its ingestion by the macrophage. Conventional liposomes are also destroyed by molecules such as plasma lipoproteins that remove the lipids from the liposomes. We proposed that the polyethylene glycol or the Stealth liposomes prevents the lipoproteins and opsonins from adhering to the lipid surface of the liposome. As a result, Stealth liposomes are less likely to be identified as foreign particles, and they remain in the circulation longer. We have recently confirmed this hypothesis by showing that components of the blood plasma react less with phospholipid bilayers that include these polymers.

Our first tests of an anticancer drug loaded into Stealth liposomes have been very encouraging (Mayhew et al., in press). Tumors are often endowed with a rich blood supply — which helps the tumor grow — but the walls of the blood vessels that feed the growth are not very tightly constructed. As a result, small Stealth liposomes that evade the immune system leak through the openings of the wall, directly into the tumor. When we treated laboratory mice that had solid colon tumors with Stealth liposomes filled with the anticancer drugs doxorubicin and epirubicin, the mice were completely cured. It is notable that such tumors are usually resistant to treatment by these drugs in their free form or even when they are encapsulated in conventional liposomes (Vaage, Lasic and Mayhew, submitted).

The diagnosis and treatment of arthritic inflammation offers another

possible use for Stealth liposomes. Arthritis is usually accompanied by edema, an indication that the vasculature is permeable to fluids and minute particles (such as small liposomes). In such instances, Stealth liposomes can be used as carriers of radioactive markers or anti-inflammatory drugs for the diagnosis and treatment of the disease. In general, Stealth liposomes should be useful for applications where the cargo should circulate in the blood for a relatively long period or where the target is something other than the cells of the mononuclear phagocyte system.

Topical Applications

The application of ointments and salves to the surface of the skin has been part of the healer's routine for centuries. Modern medicine has continued the practice by incorporating drugs and other ingredients into lotions, pastes and crease. In general these preparations serve not only as carriers of the drug — allowing it to adhere to the skin — but also as agents that increase the absorption of the drug. In some instances, however, the drug may pass through the skin into the circulatory system and be disseminated throughout the body. Such a widespread distribution has the unwanted effect of "medicating" healthy tissues — possibly causing some toxic side effects.

In the past decade, liposome-encapsulated drugs have been investigated as a possible alternative to the standard topical preparation. The first experimental studies used a model drug, triamcinolone acetonide (TRMA), embedded in a liposomal preparation that was applied to depilated rabbit skin (Mezei and Gulasekharam 1979). In comparison to the conventional preparation, the liposomal form delivered four and a half times as much drug to the target layer of the skin, and one-third as much to the thalamus, a part of the brain that could be adversely affected by the drug. It was the first suggestion that the topical application of liposomal drugs could serve as a selective delivery system.

Topical applications of liposomal drugs offer not only a more selective distribution of the drug but also an increase in the amount of time that the drug is present in the skin. Patel (1985) found that the drug methotrexate (another antitumor agent) encapsulated in liposomes was retained in the skin of hairless mice two to three times as long as the free drug. The sustained release of the drug suggests the possibility of a more uniform distribution of the medication throughout the duration of its application.

Topical liposomal drugs have also been investigated in clinical trials on human subjects. The first studies were performed on women with hirsutism, the excessive growth of body hair (Rowe, Mezei and Hilchie 1984). Excessive hair growth is often a consequence of abnormally high levels of androgenic hormones such as testosterone. In theory, female hormones such as progesterone (which does not stimulate the growth of hair) can compete with androgens for receptor sites, and thus block androgen activity. The challenge is to deliver the progesterone to the hair follicles in the skin. The ability of liposomes to pass through the outer, horny layer of the skin into the layer that contains the hair follicles suggested the possibility of treating hirsutism with liposomal progesterone. After applying liposomal progesterone to patches of skin for a three-month period, the investigators found as much as five fold reduction of hair growth. The liposomes not only penetrated the surface layer but also were able to release the progesterone to the necessary sites.

Since these early studies, the topical application of liposomes has experienced a sustained and gradual increase. Topically applied liposomes can be used for the administration of antibiotics and anti-inflammatory drugs. Moreover, the use of topical preparations of liposomes has extended beyond the skin to the eyes, lungs and other tissues. Currently, topical liposomal preparations are being developed for the treatment of dry eyes, glaucoma and asthma.

The ability of liposomes to deliver their contents into the superficial and deep layers of the skin has also made them important in the cosmetics industry. Even without any active ingredients, liposomes can deliver moisture and supply lipid molecules to the horny layer. Water molecules can also be delivered to the deeper layers, either by hydrated lipids or by ingredients such as sugars or amino acids, which actively bind water. In addition to moisturizers, liposomes can carry many other substance such as sunscreens, vitamins, proteins and antibiotics. Some manufacturers are even developing liposome-encapsulated perfumes, which should increase the duration of the fragrance.

The effectiveness of these treatments, however, is still a matter of controversy; many of the commercial claims have not been substantiated by rigorous scientific experiments. For example, it is difficult to interpret some results suggesting that the active ingredients have been delivered to the desired location. Also, it is not entirely clear how the liposomes might pass through the relatively small pores of the skin — a feat comparable to stuffing a basketball into a billiards pocket. (It is possible, however, that the liposomes squeeze through the pores by flattening their shape.) In spite of these caveats, it is true that liposomes are a biodegradable and nontoxic vehicles for the delivery of substances to the skin. In this respect they offer an alternative to other carriers that use special solvents and detergents to dissolve water-insoluble drugs and cosmetics.

Gene Transfer

The ability of liposomes to transport and protect encapsulated substances may also play a role in genetic engineering. In recent years there has been great interest in inserting new or foreign genes into a cell. In some cases this allows scientists to explore some of the basic properties of the gene in question by investigating how its activity alters the recipient cell — which normally does not express the gene. In other cases, this scheme paves the way for "gene-replacement" and "protein-replacement" therapies for diseases. The rationale here is that one can compensate for a faulty gene by either inserting a functioning gene or, alternatively, its protein product into a cell. In all of these instances, however, the cell membrane presents a barrier to large molecules such as proteins and nucleic acids.

One way to traverse the barrier is to inject the genes or proteins directly into the cell — a strategy that does not meet with success very often. Few cell types are large enough to be injected with any kind of precision, and many cannot withstand such extreme intrusions.

Ultimately, the success of gene-insertion schemes relies on a deception of the cell. Packaging genes and proteins within a carrier that the cell recognizes facilities their incorporation. Once inside the recipient cell, these packages dump their molecular contents in to the cell's cytoplasm. From there, it is hoped that the molecules will be transported to the correct location within the cell and will perform their appropriate function.

Throughout the 1970s and the early 1980s, scientists expected liposomes to fit the bill as molecular carriers. It was shown that they could encapsulate very large molecules — as large as an entire chromosome. Furthermore, since the components of the liposomes could be controlled, so too could the vesicle's fluidity, permeability, charge and size. This raised the hope that liposomes for the delivery of genes and proteins could be tailor-made to deliver their contents with high specificity to particular target cells. For example, by including a molecule for which the target cells have a receptor, one can direct the liposomes quite specifically to cells bearing that receptor. In this way, liposomes that incorporate the sugar galactose will go to liver cells. On the other hand, liposomes that carry the sugar mannose will be recognized and absorbed by white blood cells.

Even when molecules have been taken up by the target cell, however, their subsequent fate is a matter of great concern. Depending on the features of the liposome, its contents can either be deposited into enzyme-filled intracellular vesicles called lysosomes or released into the cytoplasm of the cell. One of the ways a liposome can enter a cell is by endocytosis, in which the cell engulfs the carrier and its cargo whole. Once inside the cell, the liposomes are slowly digested in lysosomes and then release their contents into the cytoplasm. In contrast, liposomes containing viral fusion-promoting proteins can fuse with the outer cell membrane and release their contents directly into the cytoplasm. Consequently, inseting proteins into lysosomes is a task for conventional liposomes, whereas the delivery of a molecule such as RNA into a cell's cytoplasm is best accomplished with liposomes that can fuse with the cell's membrane.

One problem arises, however, with the delivery of DNA, whose biological activity is achieved not in lysosomes or in the cytoplasm but in the nucleus of the cell. As yet, liposomes can only deliver DNA into the cytoplasm, where much of it is degraded. (New methods of forming a complex between a positively charged liposome and the negatively charged strand of DNA appear to increase the number of successful deliveries of genetic material into the cell.) The incorporation of DNA into the nucleus is dependent on other intracellular transport processes and cannot yet be reliably controlled. Nevertheless, some DNA does end up in the nucleus. In one dramatic example a human X chromosome encapsulated in a liposome was successfully inserted into mouse cells, and a gene on that chromosome was properly expressed (Mukherjee et al. 1978).

Apart from the problems associated with the delivery of DNA to the nucleus and its degradation in the cytoplasm, the usefulness of liposomes is limited by their inability to entrap large amounts of genetic material. As a result, liposomes have been supplanted for gene transfer by other carriers. Viruses, for example, not only target specific cells but also integrate the genetic material into existing chromosomes, thereby ensuring the gene's replication and proper function. But the viruses are considered potentially harmful, and liposomes may offer a safer alternative.

The future role of liposomes for the transfer of genetic material appears to be more hopeful in the botanical world. In plants, the technology for gene transfer has lagged far behind that for animal cells for lack of appropriate transfer vehicles. Since viral vectors have not been found for most plant species, liposomes still constitute a promising avenue for gene transfer. Research has shown, for example, that genetic material can be inserted into protoplasts, which are plant cells stripped of their outer cell wall. Since it is possible to develop mature plants from a single proto-plast, it is theoretically possible to make genetically altered plants. This technique has already proved successful for introducing genes that allow plants to assimilate atmospheric nitrogen (Lurquin 1979).

Conclusion

Liposome research is a thriving field at the confluence of biophysics, cell biology and medicine. Our scientific understanding of liposomes and the state of the technology provides an excellent example of how interdisciplinary endeavors can yield valuable information and products. The trend continues as new liposomal products are marketed and liposomal drugs are tested in the clinic.

Many institutions throughout the world are investigating the physical and behavioral properties of liposome. As we begin to appreciate these properties, the achievements of the technology are bound to increase. After all, in only 30 years liposomes evolved from an academic curiosity to a practical fact of everyday life. □

Bibliography

Alving, C.R., and R.L. Richards. 1991. Liposomes containing lipid A: a potential nontoxic adjuvant for a human malaria sporozite vaccine. Immunology Letters 25:275-280.

Alving, C.R., E.A. Steck, W.L. Chapman, Jr., V.B. Waits, L.D. Hendrick, G.M. Swartz., Jr., and W.L. Hanson. 1978. Therapy of leishmaniasis: superior efficacies of liposome encapsulated drugs. Proceedings of the National Academy of Sciences 75:2959-2963.

Bangham, A.D., M.M. Standish and J.C. Watkins. 1968. Diffusion of univalent ions across the lamellae of swollen phospholipids. Journal of Molecular Biology 13:238-252.

Batzri, S., and E.D. Korn. 1973. single bilayer liposomes prepared without sonication. Biochemica et Biophysica Acta 298:1015-1019.

Charvolin, J. 1990. Crystals of fluid films. Contemporary Physics 31:1-17.

Cornelius, F. 1991. Functional reconstituting of the sodium pump. Biochemica et Biophysica Acta 1071:19-66.

Deamer, D., and A.D. Bangham. 1976. Large volume liposomes by an ether vaporization method. Biochemica et Biophysica Acta 443:629-634.

Egbaria, K., and N. Weiner. 1991. Topical applications of liposomal preparations. Cosmetics and Toiletries 106:79-93.

Fendler, I.H. 1987. Atomic and molecular clusters in membrane mimetic chemistry. Chemical Reviews 87:877-899.

Fielding, R.M. 1991. Liposomal drug delivery. Clinical Pharmacology 21:1-10.

Fischer, T.H., and D.D. Lasic. 1984. A detergent depletion technique for the preparation of small vesicles. Molecular Crystals and Liquid Crystals 102:141-153.

Forssen, E.A., and Z.A. Tokes. 1981. Use of anionic liposomes for the reduction of chronic doxorubicin-induced cardiotoxicity. Proceedings of the National Academy of Sciences 78:1873-1877.

Gabizon, A., A. Dagan, D. Goren, Y. Barenholz, Z. Fuks.1982. Liposomes as in vivo carriers of Adriamycin reduced cardiac uptake and preserved antitumor activity in mice. Cancer Research 42:4734-4739.

Gabizon, A., A. Sulkes, T.Peretz, S. Druckmann, D. Goren, S. Amselem and Y. Barenholz. 1989. Liposome-associated doxorubicin: Preclinical pharmacology and exploratory clinical phase. In Liposomes in the Therapy of Infectious Diseases and Cancer, G. Lopez-Berenstein and I.J. Fidler, eds., pp. 391-402. New York:Alan R. Liss, Inc.

Gabizon, A., and D. Papahadjopoulos. 1988. Liposome formulations with prolonged circulation time in blood and enhanced uptake by tumors. Proceedings of the National Academy of Sciences 85:6949-6953.

Gregoriadis, G. 1988. Liposomes as Carriers of Drugs. Chichester:J. Wiley and Sons. Hauser., H., H.H. Mantsch and H.L. Casal. 1990. Spontaneous formation of small unilammelar vesicles by pH jump: a pH gradient across the bilayer membrane as a driving force. Biochemistry 29:2321-2329.

Helfrich, W. 1974. Size of sonicated vesicles. Physics Letters A50:115-116.

Juliano, R.L., and D. Stamp. 1978. Pharmacokinetics of lipsome-encapsulated antitumor drugs: studies with vinblastine, actinomycin D, cytosine, arabinoside and daunomycin. Biochemical Pharmacology 27:21-27.

Lasic, D.D. 1982. A molecular model for vesicle formation. Biochimica et Biophysica Acta 692:501-502.

Lasic, D.D. 1987. A general model of vesicle formation. Journal of Theoretical Biology 124:35-41.

Lasic, D.D. 1988. The mechanism of liposome formation. A review. Biochemical Journal 256:1-11.

Lasic, D.D. 1990. On the thermodynamic (in)stability of liposomes. Journal of Colloid Interface Science 140:302-304.

Lasic, D.D. 1991. On the formation of membranes. Nature 351:163.

Lasic, D.D., F.J. Martin, a. Gabizon, S.K. Huang and D Papahadjopoulos. 1991. Sterically stabilized Stealth liposomes: a hypothesis on the molecular origin of the extended blood circulation times. Biochimica et Biophysica Acta 1070:187-192.

Lautenschlager, H., and J. Roding. 1989. Kosmetiche Formulierungen mit phospholipiden und liposomen: Umfeld und Zussamenhänge. Parfumerie und Kosmetik 12:757-764.

Leibler, S. 1988. Membranes, surfaces, and microemulsions. Physics Today January:S25.

Lipowsky, R. 1991. The conformation of membranes. Nature 349:475-481.

Lopez-Berenstein, G., and I.J. Fidler. 1989. Liposomes in the Therapy of Infectious Diseases and Cancer. New York: Alan R. Liss.

Lopez-Berenstein, G., V. Fainstein, R. Hopfer, K. Mehta, M.P. Sullivan, M. Keating, M.G. Rosenblum, R. Mehta, M. Luna, E.M. Hersh. J. Reuben, R.L. Juliano and G.P. Bodey. 1985. Liposomal amphotericin B for the treatment of systemic fungal infections in patients with cancer. Journal of Infectious Diseases 151:704-710.

Lurquin, P. 1979. Entrapment of plasmid DNA by liposomes and their interaction with plant protoplasts. Nucleic Acids Research 6:3773.

Luzzati, V., H. Mustacchi, A. Skoulios and F. Husson. 1960. La structure des colloides d'association. Les phases liquide crystallines des systemes amphiphile-eau. Acta Crystallographica 13:660-667.

Martin, F.J., and G. West. 1988. High encapsulation liposome processing method. United States Patent 4,752,425.

Mayhew, E., D.D. Lasic, S.Babbar and F.J. Martin. In press. Pharmacokinetics and antitumor activity of epirubicin encapsulated in long-circulating liposomes incorporating polyethylene glycol derivatezed phospholipid. International Journal of Cancer.

Mayhew, E., R. Lazo, W.J. Vail, J. King, and A.M. Green. 1984. Characterization of liposomes using a microemulsifier. Biochimica et Biophysica Acta 775:169-174.

Mezei, M. 1985. Liposomes as a skin drug delivery system. In Topics in Pharmaceutical Sciences, D.D. Breimer and P. Speiser. eds.. pp.345-358. Amsterdam:Elsevier Science Publishers.

Mezei, M. 1988. Liposomes in the topical applications of drugs: a review. In Liposomes as Carriers of Drugs, D. Gregoriadis, ed.. pp.663-677. Chichester:John Wiley and Sons.

Mezei, M., and Gulasekharam, V. 1979. Liposomes -- a selective new drug delivery system. Abstracts of the 39th International Congress of Pharmaceutical Sciences.

Mezei, M., J.C. Hilchie and T.C.Rowe. 1985. Formulation and evaluation of liposomal progesterone products. Cinical Investigations in Medicine 8:c3.

Mukherjee, A.B., S. Orloff, J.D. Butler, T. Triche, P. Lalley and J.D. Schulman. 1978. Entrapment of metaphase chromosomes into a phospholipid vesicles (lipochromosomes):carrier potential in gene transfer. Proceedings of the National Academy of Sciences 75:1361.

Nässander, U.K., G. Storm, P. Peeters and D. Crommelin. 1990. Liposomes. In Biodegradable Polymers as Drug Delivery Systems, M.Chasin and R. Langer, eds. pp.261-338. NewYork:Marcel Dekker.

Needham, D., T.J. McIntosh and D.D. Lasic. Submitted. Interactive and mechanical properties of lipid membranes containing surface-bound polymer. Biophysical Journal.

New, R.R.C., M.L. Chance, S.C.Thomas and W. Peters. 1978. Anti-leishmanial activity of antimonial entrapped in liposomes. Nature 272:55-56.

Nicolau, C., and A. Cudd. 1989. Liposomes as carriers of DNA. Critical Reviews in Therapeutic Drug Carrier Systems. 6:239-271.

Ostro, M.J. 1987. Liposomes. Scientific American 256:103-111.

Papahadjopoulos, D. 1978. Liposomes and their uses in biology and medicine. Annals of the New York Academy of Sciences 308:1-412.

Patel, H.M. 1985. Liposomes as a controlled-release system. Biochemical Society Transactions 13:513-516.

Poznanksy, M., and R.L. Juliano. 1984. Biological approaches to the controlled delivery of drugs: a critical review. Pharmacology Review 36:277-336.

Rowe, T.C., M. Mezei and J. Hilchie. 1984. Treatment of hirsutism with liposomal progesterone. The Prostate 5:346-347.

Senior, J.H. 1987. Fate and behavior of liposomes in vivo: A review of controlling factors. Critical Reviews in Therapeutic Drug Carrier Systems 3:123-193.

Storm, F., D.H. Roerdink, P.A. Steerenberg, W.H. de Jong and D.J.A. Crommelin. 1987. Influence of lipid composition on the antitumor activity exerted by doxorubicin liposomes in rat solid tumor model. Cancer Research 47:3366-3372.

Szoka. F.C., and D. Papahadjopoulos. 1978. Procedures for preparing liposomes with large internal space and high capture by reverse phase evaporation. Proceedings of the National Academy of Sciences 75:4194-4198.

Treat, J., A.R.Greenspan and A. Rahman. 1989. Liposome encapsulated doxorubicin preliminary results of phase I and Phase II trials. In Liposomes in the Therapy of Infectious Diseases and Cancer, G. Lopez-Berenstein and I.J. Fidler, eds., 353-365. New York:Alan R. Liss, Inc.

Van Hoesel, Q.G.C.M., P.A. Steerenberg, D.J.A. Crommelin, A. van Dijk, W. van Oort, S. Klein, J.M.C. Douze, D.J. Wildt and F.C. Hillen. 1984. Reduced cardiotoxicity and nephrotoxicity with perservation of antitumor activity of doxorubicin entrapped in stable liposomes in the Lou/M Wsl rat. Cancer Reserach 44:3698-3705.

Virchow, R. 1854. Ueber das ausgebreitete Vorkommen einer dem Nervenmark analogen Substanz in der thierischen Geweben. Virchows Archiv 6:562-573.

Voage, J., D.D. Lasic and E. Mayhew. Submitted. Therapy of primary and metastatic mouse mammary carcinoma with doxorubicin in long circulating liposomes. Cancer Reserach.

Williams, B.D., M.M. O'Sullivan, G.S. Saggu, K.E. Williams, L.A. Williams and J.R. Morgan. 1987. Synovial accumulation of technetium labelled liposomes in rheumatoid arthritis. Annals of Rheumatic Diseases 46:314-318.

Woodle, M.C., and D. Papahadjopoulos. 1989. Liposome preparation and size characterization. Methods in Enzymology 171:193-217.

Questions: *Answers on page 198*

1.   How can liposomes be used as "magic bullets" in therapy?

2.   How does a knowledge of biochemistry help understand the makeup of liposomes? Cite a few examples.

3.   Can you think of another use of liposomes in addition to those suggested in the article?

10

Apoptosis, the programmed death of cells, is a recognized phenomenon to most biologists. Events such as the development of human fingers and toes, or the metamorphosis of a caterpillar into a butterfly, could not occur without apoptosis. This process is also part of the normal turnover and replacement of worn-out tissues in adult organisms. Scientists are now finding, however, that diseases such as AIDS and autoimmune reactions may result from the disruption of this process which is usually orderly. A gene has also been identified that can trigger cell death if conditions in the outside environment are not favorable for the proliferation of the cell. If favorable, it signals the healthy cell to keep dividing. This discovery means that the cell may have a built-in, self-destruct mechanism. This "cell-death" gene may be one of several switches programming the lifespan of the cell.

A Time To Live, A Time To Die
by Carol Ezzell

"First, you murder," Michael O. Hengartner forthrightly told a horde of expectant faces.

"Next, you get rid of the body. Then, you hide the evidence," he explained, pacing back and forth in the dimly lit room.

Hengartner wasn't instructing a group of apprentice hit men. Instead, the Massachusetts Institute of Technology (MIT) biologist was addressing a gathering of cancer researchers, detailing the functions of a recently identified set of genes that controls life's only inevitable process: death.

Together, Hengartner and his MIT colleagues constitute one of scores of research teams around the world who are reviving scientific interest in the molecular mechanisms of a phenomenon called apoptosis, or programmed cell death. Among other things, this phenomenon (pronounced apa-tosis, with the second "p" silent) prevents humans from having webbed fingers and eliminates cells of the immune system that can't tell "self" from "nonself." It also underlies meta-

morphosis — the magic wand that turns caterpillars into butterflies and tadpoles into frogs. In adults, it phases out old body cells so they can be replaced by new ones.

Over the past year, biologists from a range of disciplines have uncovered evidence that this seemingly salutary process has a dark side. Several new studies suggest that apoptosis can play roles in AIDS and autoimmune diseases; other indicate that disruptions in the usual orderly progression of apoptosis lead to the uncontrolled cell growth of cancer.

Apoptosis — which means "dropping off" in Greek — was first described in 1951 as a step in animal development. The process takes its name from its appearance as it unfolds under the microscope: Within minutes, cells undergoing apoptosis shrink and shed tiny, membranous blebs that neighboring cells quickly gobble up, mirroring Hengartner's colorful description. In contrast, during necrosis — cell death arising from injury — cells swell for hours and then burst,

spraying their contents about as a chemical signal that attracts immune-system cells to fight the injurious microbe or substance.

In the late 1960s and early 1970s, researchers began gathering evidence that apoptosis occurs as part of the normal turnover and replacement of worn-out tissues in adult organisms. They discovered that apoptosis resembles suicide in some ways: Old cells actively participate in their own demise by turning on genes and making new proteins that will shortly cause their death.

Since the mid-1980s, cell biologists and geneticists have started sorting out the causes and implications of apoptosis in a wide range of animals, including humans. Last spring, they began reporting evidence of the role played in apoptosis by the cancer-causing c-myc gene — named for its initial discovery in myeocytomas, tumors consisting of tightly packed bone marrow cells.

The c-myc oncogene becomes overactive in a wide range of

mammalian tumors, including cancers of the breast, bladder, colon, lung, and cervix (SN:6/1/91,p.347). In many cases, c-myc's hyperactivity begins when a cell inexplicably creates extra copies of the gene, reproducing it over and over within the cell nucleus.

Because cells with such c-myc amplifications grow and divide nonstop — and further, because the c-myc gene encodes protein-containing regions that can bind to DNA — scientists hypothesize that c-myc regulates other genes involved in cell division. Ironically, Gerard I. Evan of the Imperial Cancer Research Fund Laboratories in London and his colleagues reported in the April 3 CELL that c-myc can also cause apoptosis under certain conditions.

Evan's group found that while laboratory-cultured cells with hyperactive c-myc genes can grow faster than cells with less active c-myc genes, they also die faster than those deprived of growth medium. Moreover, the researchers noted, the cells with overactive c-myc genes died with all the visible hallmarks of apoptosis.

Evan and his co-workers conclude that c-myc functions as a two-edged sword: While it usually acts to keep a healthy cell dividing, it can also trigger cell death if outside conditions aren't right for continued cell proliferation or if the cell has become genetically damaged. In this way, c-myc can function as a built-in cellular self-destruct mechanism.

So how does c-myc cause cancer? According to a model developed by Evan and his co-workers, damage to the c-myc gene — caused either by skips on the DNA-repair machinery or by environmental injury — usually results in cell death. But some cells sustain such genetic damage and go on to develop a mutation that activates, or turns on, a second gene.

This second gene somehow overrides c-myc's death command, allowing the cells to grow into tumors.

Two papers in the Oct. 8 NATURE provide evidence that this second gene is bcl-2, an oncogene named for its initial discovery in human immune-system cancers called B-cell lymphomas. In the first paper, a team led by Douglas R. Green of the La Jolla (Calif.) Institute for Allergy and Immunology reports that death-prone cells containing extra c-myc genes survive much longer following insertion of an activated bcl-2 gene, which produces a protein with unknown function.

"In the absence of bcl-2, c-myc induces death," summarizes Green, "but in the presence of bcl-2, there's no death." Cancer results, he asserts, "not just because the [mutated] cells grow faster, but also because they die more slowly."

In the second paper, a team led by the Imperial Cancer Research Fund's Evan reports similar results and provides evidence suggesting that the bcl-2 mutation can help cancer cells resist the deadly effects of chemotherapeutic drugs. Many such drugs kill cancer cells by causing them to undergo apoptosis.

Evan's group administered the anti-cancer drug etoposide, also known as BP16, to death-prone rat cells genetically engineered to contain the activated bcl-2 gene. The researchers found that the bcl-2 gene prevented many of the cells from undergoing apoptosis and delayed its onset in others.

Further evidence that bcl-2 increases the resistance of cancer cells to chemotherapy is published in the Oct. 1 CANCER RESEARCH. Toshiyuki Miyashita and John C. Reed of the University of Pennsylvania School of Medicine in Philadelphia inserted copies of the activated human bcl-2 gene into mouse lymphoid tumor cells. They

found that the genetically engineered cells survived a dose of the steroid drug dexamethasone roughly 100 times larger than that required to kill cells lacking the bcl-2 gene. Moreover, the cells resisted death induced by several other chemotherapeutic drugs, including the widely used cancer therapies vincristine and methotrexate.

The findings "may open the door to a whole new approach for the treatment of cancer," says Reed, who is now at the La Jolla (Calif.) Cancer Research Foundation. "If you could use drugs to reduce the expression of bcl-2 [in cancer cells], you might make the cells more sensitive to existing chemotherapeutic drugs," he suggests.

Reed and his colleagues are working with Genta, Inc., a San Diego-based biotechnology company, to develop so-called antisense drugs to block the activity of bcl-2. Antisense drugs — which consist of the same chemical building blocks that make up the genetic material DNA — turn off specific genes by binding to and inactivating messenger RNA, the intermediate compound that genes use to tell a cell to make a given protein (SN:2/16/91,p.108).

Reed says initial tests in laboratory-cultured cells show that antisense drugs that target bcl-2 make cancer cells more vulnerable to apoptosis induced by chemotherapeutic drugs. "We're hoping to get our [bcl-2] antisense drug into clinical trials soon," says Reed. "We'd love to see if we could get it to work [in cancer patients]."

"There's a possibility that in all [the processes that turn cells cancerous] there may be mechanisms that favor cell death," adds Green. "If other genetic changes override that, you get full-scale transformation [into a cancer cell]."

In the meantime, MIT's Hengartner has found that the tiny roundworm Caenorhabditis elegans has a gene that resembles human bcl-2. He reported last month that the structure of bcl-2 is similar to that of a roundworm gene called ced-9, for C. elegans death (SN:10/10/92,p.229). Moreover, like bcl-2, ced-9 protects cells from programmed cell death, Hengartner and his colleagues reported in the April 9 NATURE.

Hengartner says that ced-9 regulates the activity of two other genes, ced-3 and ced-4, that actually cause cells to undergo apoptosis. When ced-9 is "on," it shuts off ced-3 and ced-4, allowing a cell to live. But when ced-9 is inactivated by a mutation, ced-3 and ced-4 start up, prompting a cell to commit suicide.

This feedback mechanism ensures that so-called stem cells in a developing roundworm die when they are no longer needed, says Hengartner. Scientists know that the minuscule roundworm generates 1,090 cells during its embryonic development. However, 131 of these cells die, so an adult roundworm consists of exactly 959 cells.

Hengartner's team has shown that roundworms with an abnormally activated ced-9 gene develop superfluous body parts, presumably because the extra 131 cells never die. In contrast, the researchers report, the off-spring of roundworms lacking functional ced-genes die as embryos, evidently because the ced-3 and ced-4 genes functioned unchecked, killing all of the young organism's cells prematurely.

"Ced-9 is the switch between life and death" in the developing roundworm, concludes Hengartner.

Two studies published earlier this year demonstrate that the mammalian immune system may employ a similar set of cell-death genes. In the first study, a group led by Shigekazu Nagata of the Osaka Bioscience Institute in Osaka, Japan, has found that mice genetically predisposed to an affliction resembling the human antoimmune disease lupus erythematosus (SLE) have defects in a protein required for apoptosis in white blood cells.

Accordingly, Nagata and his colleagues report in the Mar. 26 NATURE, the mice fail to purge themselves during embryonic development of white blood cells that attack their own tissues. As a result, the animals develop the swollen lymph glands, lethargy, and tissue damage characteristic of lupus.

The results reported by Nagata's team "are the first hint of a cell-death link with a real disease model," comments Green. "It looks like a gene that is involved in the programmed cell death process is defective in this strain of mouse with horrendous autoimmune problems."

In the second paper, which appear in the July 10 Science, a group led by Frank Miedema of the University of Amsterdam in the Netherlands reports evidence that AIDS resembles the other side of the same coin. Miedema and his colleagues took white blood cells called T-cells from male AIDS patients. When they stimulated the cells' CD3 receptors using antibodies, up to one-fourth of the cells committed suicide by apoptosis. In contrast, the antibody treatment failed to induce significant levels of apoptosis in T-cells isolated from men not infected with the AIDS-causing HIV virus.

Miedema and his colleagues suggest that HIV infection "Hyperactivates" T-cells, giving them a hair-trigger tendency toward suicide. They say this mechanism may explain why AIDS patients show a decrease in all types of T-cells, not just those bearing the CD4 receptor that HIV used to enter and infect some T-cells.

Developments such as these signal renewed interest in the study of cell death among researchers from a variety of fields, say many biologists. "This will be a very fruitful area of research for some time to come," predicts Reed. □

Questions: *Answers on page 199*

1. Do you think cell death is more genetic or environmentally-related?

2. How is the study of cell divison - cell death related to cancer?

3. Why is cell death necessary for some developmental patterns?

Part Three

Chromosomes, Cell Division, and Heredity

11

The Y chromosome and chromosome 21 are the two smallest human chromosomes. Now two groups of researchers have mapped the genes on each of these structures. Inheritance of the Y chromosome in humans determines maleness. One research group has reconstructed 98 percent of the part of the Y chromosome that contains genes. Some of these genes are responsible for the development of male characteristics. Knowledge of the makeup of chromosome 21 is important to the development of certain human diseases. It contains the genes for amyotrophic lateral sclerosis and some forms of Alzheimer's disease. A third copy of this chromosome, as present in trisomy 21, causes the Down syndrome. The mapping of each chromosome is a major step toward mapping all of the estimated 100,000 genes found on all 46 human chromosomes.

Two Human Chromosomes Entirely Mapped
by C. Ezzell

In two of the earliest major advances in the mammoth international effort to identify and decipher all of the estimated 100,000 human genes, two groups of researchers have taken apart and put back together the smallest human chromosomes: the Y chromosome and chromosome 21.

The exercises have yielded for each chromosome a set of overlapping segments of DNA assembled in the correct order. Scientists expect both of these socalled physical maps to help them find new genes more quickly. They also predict that the map of the Y chromosome will shed new light on human evolution. Scientists at the Whitehead Institute for Biomedical Research in Cambridge, Mass., constructed the physical map of the Y chromosome. The team, led by David C. Page, began by examining the Y chromosomes of individuals who had inherited only fragments of this rod-shaped structure, which bears the genes that make a male.

By comparing the different-sized Y chromosome fragments of 96 such individuals, Page and his colleagues discovered naturally occurring breakpoints that they could use as molecular probes. This comparison also allowed the researchers to organize the probes into the order in which they would occur in an intact Y chromosome. Page's group then used the probes to isolate long pieces of the Y chromosome from a man who had three extra Y chromosomes, which provided the researchers with an abundance of material for study. By assembling the pieces in the order of the probes, Page and his colleagues reconstructed 98 percent of the part of the Y chromosome that contains genes. They describe their work in two papers in the Oct. 2 SCIENCE. Douglas Vollrath, a key member of Page's group, says the discovery should speed the Human Genome Project.

"Until recently, the most difficult part was finding the DNA that you thought contained a particular gene," says Vollrath. "Now the problem shifts...you can go to the freezer and pull out a vial that contains the DNA you want."

Simon Foote, another key group member, adds, "This map and future maps will serve as the substrate for large-scale sequencing efforts" to read every letter in the encyclopedia of DNA that makes up the human genetic complement.

Vollrath and Foote say they plan to use detailed maps of the Y chromosome to shed light on the male side of human evolution. Several teams of evolutionary biologists have already used sporadic mutations in mitochondrial DNA — genetic material located outside the cell nucleus and inherited only from the mother — in attempts to trace human origins.

A team of 36 researchers from Europe, the United States, and Japan collaborated on the physical map of chromosome 21. Led by Daniel Cohen of the Paris-based Center for the Study of Human Polymorphism, they discovered 198 equally spaced landmarks on chromosome 21. These landmarks allowed the researchers to divvy the chromosome up into manageable chunks and to assemble the chunks in the correct order, they report in the Oct. 1 NATURE. Chromosome 21 is particularly important to human disease, Cohen and his colleagues state, because it contains the genes for amyotrophic laterial sclerosis (Lou Gehrig's disease) and for some forms of Alzheimer's disease and epilepsy. An extra copy of chromosome 21 causes Down's Syndrome.

The maps of the two chromosomes "represent a massive body of work," Peter Little of Imperial College in London comments in an editorial in the October 1 NATURE.

"The important message is that [such mapping] can be done and it is now only a matter of time (and money) before all human chromosomes are completed." □

Questions: *Answers on page 199*

1. How would you define chromosome mapping?

2. What is the importance of chromosome mapping?

3. Does all DNA on the chromosome compose genes?

12

Samples of DNA, taken from the bones of human fossils from Florida, are being studied to reconstruct the family tree of a lost tribe. The DNA is estimated to be nearly 8000 years old. Remarkably, this molecule has remained preserved, thus offering the biological code for the cells of this extinct Florida tribe. It is unusual for such a hidden treasure of DNA to survive, as the acid conditions found in most soil samples destroys it quickly. However, its encasement inside the skulls and other bones of the ancient Florida inhabitants protected it from this threat. Through an amplification technique, scientists can duplicate a small fragment of DNA into a billion copies. This technique has made the DNA easier to read and analyze.

Dark Bogs, DNA and the Mummy
by Gurney Williams III

For more than seven millennia, nothing disturbed the ancient tribal burial site in the hot Florida countryside except the ground-shaking rocket launches from the space center 15 miles to the east.

Then one day in 1982, Steve Vanderjagt's back hoe carved a ten-foot-deep hole in a Brevard County bog, digging up dozens of femurs, ribs, and skulls. Vanderjagt, a contractor working on a new road for a 1,500-acre housing development called Windover Farms, had no idea what he'd uncovered. No one found out until years later that many of the skulls carried a cargo as precious as any probe we've sent to the stars.

Hidden inside the dark-brown bones were gray, clay-like brains. They in turn contained 7,000 to 8,000-year-old samples of DNA, the basic biological code for building every human cell.

In a remarkable coincidence, the Windover Farms discovery came just a few years before scientists learned to "amplify" DNA, making it easier to read. Applied to the buried brains, the new technology has given researchers the beginnings

of a family tree for Florida's "lost tribe."

The amplification technique, called the polymerase chain reaction (PCR), allows scientists to make a billion copies of a mere fragment of DNA. Like detectives looking at a computer-enhanced photograph enlarged many times, researchers using PCR can analyze the hereditary messages inside plants, animals, and human cells. When first developed, the technique helped a Wayne State University geneticist delve into the genes of leaves that dropped into a pond 17 million years ago. Scientists today call on the same technique to explore questions about the history of ancient tribes. Using PCR, scientists have also shown human DNA to be unimaginably durable, able to survive our living bodies by thousands, perhaps millions, of years. Traipsing through the centuries with this extraordinary time machine, anthropologists are unearthing secrets about the ancient human brain and immune system and scrutinizing the sweeping course of evolution itself.

These possibilities never

occurred to Vanderjagt on the day his hoe cut ten feet into the old burial ground. He didn't even know at first that he had unearthed part of a body—he thought he'd stumbled on a ball-shaped rock that had rolled out of the soggy black-and-redbrown peat mound.

"But he had enough sense to know that we don't have round rocks in Florida," says Jim Swann, the developer who hired Vanderjagt to clear out the bog to make way for the sand base of a new road. When Vanderjagt got off his back hoe and picked up the strange piece, he came face to face with eye sockets. He and his foreman soon found other bones, washed them, and telephoned Swann. The developer arrived to find several skulls stored in a bucket of water.

"I called the county attorney and said, 'What do you do when you find human bones?'" Swann recalls. The sheriff's department packed the bones in body bags and carried them in a car trunk to the county coroner, "And he proclaimed they were more than a hundred years old," Swann says. The finding closed the book on a possible murder investigation.

Swann got his bones back. He himself had become intrigued by the mystery surrounding them and paid for an analysis of their age. No one yet knew about the brain tissue enclosed in the old skulls.

By the time the original bones came back, "It had rained on the pile of peat," Swann says, "and there were bones all over the place." For several days, the foreman had carted some of them around in his truck. To reduce the threat of vandalism, Swann took as many of the bones as he could home with him and left them in buckets in his back yard in Cocoa, Florida. His wife thought he was crazy, he says.

"I had buckets everywhere," he says. "It got to be spooky. You'd look down and see these sockets looking up at you. I wanted to get rid of them." Swann donated the bones to Florida State University. And the state's legislature, at Swann's urging, appropriated $200,000 to carry on with archaeological precision what Vanderjagt's rugged back hoe had begun.

By careful digging, Florida State researchers led by anthropologists Glen Doran, Ph.D., and David Dickel, Ph.D., were able to develop a clear picture of the ancient tribe's burial procedures. And they learned some things about the way the tribe lived day to day.

Objects and cloth found in the burial site show that these ancient people used spears for hunting and bones from deer, dog, and bobcat to make awls and needles. These early Americans were also surprisingly skilled weavers, evidence that they were good enough at their workaday hunting and gathering to have time for crafts.

There were other messages, as eloquent as a churchyard stone, in the way the ancient residents arranged graves in what was then a shallow pond. Most of the clothing-shrouded bodies lay on their sides, legs tucked into a loose fetal position. It appears that mourners covered the bodies with peat and wood and built a stick frame like the skeleton of a tent over the remains. Offerings—bone and antler tools, an oak bowl, a double-ended pestle—accompanied some of the bodies, and the gifts were most generous for younger tribe members.

About ten feet of peat accumulating over the site during subsequent millennia provided fortress-like protection for the burial ground and its hidden treasure of DNA. Most peat bogs destroy genetic material even while preserving the bodies in which it resides, according to William W. Hauswirth, Ph.D., a professor of microbiology at the University of Florida College of Medicine.

That's because most bogs are acidic, "very bad on DNA," Hauswirth says. "In fact, DNA can't survive for more than a few days in the typical kind of acidity that peat bogs have." But the particular blend of plant debris at the Windover Farms site neutralized the acidity. Additionally, the lack of oxygen in the peat and minerals it contained inhibited most kinds of bacterial and fungal organisms that prey on buried bones and brains. The skin and flesh dissolved over years. But many of the bones, including half the skulls and a number of brains, came out of the natural time capsule well preserved.

So solid were the skulls that it wasn't until two years after Vanderjagt make his initial contact with the bones that scientists found the first of the brains.

From the beginning, researchers had noticed that the skulls were heavy. The scientists had always assumed that the weight resulted from clumps of peat in the crania. But by December 1984, some of the Florida State researchers were beginning to suspect that pieces of tissue might have survived.

One skull fragment was attached to material that didn't look like peat. When researchers scraped the mystery substance and gave it a crude test, "it seemed to be human tissue," Hauswirth says.

They suspected they might find more brain tissue in some of the intact skulls—one in particular. The 8,000-year-old skull was heavy, Hauswirth says, "and there was something thumping around inside." More peat, lab workers suspected. Using a saw from the lab's tool chest, Phil Laipis, Ph.D., an associate of Hauswirth's, removed the top of the skull. And a brain fell out.

"It just plunked into a student's hands," Hauswirth says. "Somebody said, 'It's a brain!'"

It had shrunk to about a quarter of its original size, the reason it had rolled around freely inside the skull. The brain lacked the usual protective membranes. Blood vessels had disappeared. The narrow slits of fissures in the brain were filled with peat. But the overall appearance was startlingly similar to a modern human brain.

After the initial shock, researchers rushed the brain into a sealed glass jar. They filled the jar with inert argon gas to prevent any further decomposition from contact with oxygen, and refrigerated the whole package. At that point, the scientists didn't know what to do with it. PCR was in its infancy. "We didn't know if we could ever rescue a gene," Hauswirth says.

Some lab work at the time had raised hopes that it might be possible. California researchers led by the late Allan Wilson, Ph.D., at the University of California at Berkeley, reported that they had been able to isolate and copy DNA from museum remains. Their specimen was an extinct horse-like animal called a quagga. Assisted by Russell G. Higuchi, Ph.D., Wilson accomplished the cloning by inserting pieces of quagga hereditary code into bacteria. While the bacteria multiplied, they multiplied the quagga DNA as well. The researchers eventually extracted that DNA—genetic clones of the original quagga genes—from the bacteria. Then they compared the cloned quagga genes with genes from modern animals. To the scientists' surprise, the evidence showed that the quagga was actually a closer relative of the zebra than the horse.

Another breakthrough came that same year at the Cetus Corporation in Emeryville, California. There, a researcher developed a new technology that would supplement bacterial cloning with a biological "copy" machine.

The polymerase chain reaction (PCR) machine was the invention of biochemist Kary B. Mullis, Ph.D., at Cetus. [See Omni Interview, April 1992.] Mullis, now a consultant, says the idea for the technique came to him one Friday night in May of 1983. By the following fall, he was ready to try his first experiment. It failed. "I was shooting for the moon," he says, with an overly ambitious test of the idea. He scaled back and ran the first successful PCR experiment in December. When it worked, "I felt terrific," Mullis says. That night, he dropped by the home of Fred Faloona, a technician who had worked with him, and over beers told Faloona that the experiment "was going to change molecular biology." He was right.

Mullis had invented a comparatively simple way to amplify any small piece of DNA. The easy-to-use device, Hauswirth says, works like an oven that alternatively heats and cools the molecules of heredity. "It's just a temperature cycler," he says. "Just a block with a few holes in it for test tubes. There are no moving parts."

The catch is making the recipe for what goes into the test tube. Chief ingredient is a sample of the DNA under study. In its natural

state, it's shaped like a long ladder twisted into a spiral. Under heat close to the boiling point of water, the hydrogen bonds that hold the two strands of the spiral together break vertically, down the middle of the "rungs." The heat-sundered parts now simmer in a soup, rich in parts needed to make copies of the original.

What's clever is that the soup contains tiny fragments of laboratory-produced DNA called oligonucleotides, far smaller than the sample DNA strands. These are comparable to short chains of letters a word processor operator types on a keyboard to begin a search for a passage in a document. After the heat goes on and breaks down the sample, the oligonucleotides find the beginning and end of a specific portion of the broken, ancient DNA. They bind with and mark off portion after portion, eventually reconstructing each of the two broken strands. Then other ingredients in the "soup" fill out two new double-stranded spirals that perfectly match part of the original sample.

Researchers don't have to know the pattern of the DNA between the oligonucleotides any more than a word-processor operator needs to know all the words between markers in a block of copy. But once the strands have been "tagged" with the oligonucleotides, Mullis says, "you can take a portion of DNA out of its context and stick it somewhere else, as in word processing. Or you can make a lot of copies of it," each exactly the same length and therefore easy to identify and study in later analyses. Researchers make the duplicates by repeatedly turning up the heat—breaking up the double-stranded spirals and forcing chemicals from the soup to pair up with each single strand, forming two new double strands. The process enables its users to reproduce the ancient gene and to scan for similarities and differences between ancient and

modern hereditary code.

None of this takes a long time. "You can amplify one gene from one brain in about four hours," Hauswirth says, although analysis of results may take another week or two.

The technique doesn't always work perfectly, Hauswirth acknowledges. "Each sample required different PCR conditions to get it to amplify . . . It's not just a routine thing to crank out a whole bunch of samples." But when it works, PCR is so sensitive, it can detect one cancer cell out of a million healthy cells. And for archaeologists, it's like a time machine: The technique enables them to reconstruct far older and hence more-time ravaged samples than ever before.

Hauswirth is now applying the laboratory "time machine" to a mother lode of such samples. Out of 177 bodies found so far in the bog, 91 yielded brains or substantial brain tissue. If we could reconstruct their memories, the collection would give us centuries of tribal history. "These people were buried over a period of at least one thousand years," Hauswirth says.

But the brains may yield up something almost as good as an oral history. Stored in a locked freezer at -80 degrees Celsius, the neuronal tissue may give researchers genetic information virtually impossible to obtain by studying modern populations. "We have a genetic sampling of fifty generations of humans," Hauswirth says, "very likely related to each other. We may have the great-great-great—forty-seven times great—grandfather of the youngest individual. We'll try to work out those relationships as well as we can.

Hauswirth says PCR probably won't reveal a full family tree spanning a millennium. But preliminary work shows that portions of chromosome 1, the largest of 23— each made up of thousands of genes—remains unchanged throughout most of the 50 generations. It'll probably take five years to trace all the chromosomes, he says. The PCR work should disclose some of the most important personal events in tribal history. "It will tell us things like how isolated this population was—how much contact these people had with other populations with different genetics, and how much mixing there was."

Another major study is designed to compare the Windover genetic material with genes from several other ancient, contemporaneous populations—probably one other Northern American tribe and another from Middle or South America. The work should help answer some old questions about our past. If the genetic profiles are different, for example, that might mean that waves of immigrants crossed the Bering land bridge at different times. Or it might mean that just one group of nomads entered the hemisphere, and then, over thousands of years, their common genetic pool dispersed and changed.

The PCR trail may also lead to reconstructions of tribal catastrophe. Prior to Columbus' arrival here in the late fifteenth century, the native population amounted to more than 20 million, according to some estimates. In the subsequent century, contact with Europeans cut the population to fewer than 5 million, perhaps from diseases like smallpox, influenza, or measles. In theory, these survivors passed through a genetic "bottleneck," leaving behind the rich hereditary diversity that had developed on this continent since the first inhabitants arrived.

The PCR studies of several Native American populations could confirm the devastation. The test is straight-forward, Hauswirth says. "We should see a higher diversity—more different versions of a given

gene—in ancient populations than in modern."

PCR might even find messages in the DNA about what killed the earliest Americans. In research published last year, David Lawler, Ph.D., a postdoctoral fellow at Stanford University, and his associates, working with Hauswirth, reported amplifying genes from one of the Windover brains named SS325 (for "special sample"). The genes were partly responsible for immune-system defenses against illness. No one can say yet whether SS325 was more or less likely than we are to catch colds or any of the New World maladies imported by European settlers. But researchers predict that within the next five to ten years, PCR will give us the tools to read prehistoric medical records: DNA evidence of disease resistance or susceptibility. Comparisons with our own immune systems may reveal the nature and extent of the changes over 7,000 years.

And scientists may be able to peer back not just 7,000 years, but literally millions. The reason for this estimate: the work of a Wayne State researcher who's been able to make copies of plant DNA millions of years old. At first, the claim of Edward Golenberg, Ph.D., an ecological geneticist, drew open skepticism from other researchers who argued that DNA couldn't survive intact for more than a few thousand years. Golenberg's repeated success has blunted the criticism, although he says plant DNA has a clear longevity edge over the DNA of animals and humans.

The plants Golenberg analyzed came from the ranch of Francis and Vickie Kienbaum of Clarkia, Idaho. The leaves never would have surfaced if Francis wasn't a snowmobile buff. In the fall of 1971, he borrowed a bulldozer from a friend to make a four-acre oval racetrack

on his 100-acre land. He was amassing dirt from a 60-foot-high knoll at the south end of the track late one night when he happened to hit a slick patch of soil about 16 feet under the surface. "I guessed it was just crap," Kienbaum says.

He couldn't see it clearly under the lights of the bulldozer, and he left the wet dirt before sunrise. When he returned to the site a week later, he was surprised to find a pile of dead leaves where he had been digging. "It smelled like an old lake bottom," Kienbaum says. He puzzled over the find for a few days and then called the geologists at the University of Idaho.

"I told a lady there that there were all kinds of black leaves on the ground," Kienbaum says. "They were sitting in clay, and they came loose and blew all over. She started quizzing me. She says, 'Well, how old are the leaves?' I say, 'About two weeks.'"

University of Idaho geologists led by C. Jack Smiley, Ph.D., soon determined that Kienbaum's estimate was wrong. The researchers using radioisotope techniques estimated that the leaves were alive during the Miocene period. Then, the land that became the Kienbaum snowmobile track—today used for dirt-bike races—was a warm forest. The leaves fell from oak, beech, tulip, and magnolia plants through humid air into a large lake.

The dead plants Kienbaum found were probably at least 17 million years old. Today, anyone can scoop up a carton of the leaves by paying Kienbaum $5. (Call ahead to Clarkia: 208-245-3608.) A couple thousand people a year pick up some, wrap them away in a cardboard case. Kienbaum says the leaves often appear red, yellow, or green when you shovel them from the ground but turn black within minutes when exposed to air.

They're not much good as

compost, he says. "It's harder than hell to get something to grow" in the sticky, clay-filled soil that racers on his track call "dinosaur dung."

But starting in 1988, PCR analysis of the leaves has found and amplified traces of what Golenberg argues is ancient DNA. He started with seven samples of frozen plant material, and within 24 hours, working in the laboratory of Michael T. Clegg, Ph.D., at the University of California at Riverside, suspected he had copied one of their genes. It took six to nine months to complete the work confirming that the DNA was recognizable—as the biological instructions for making part of a magnolia, 17 million years ago.

Golenberg's find has raised new speculation about whether the genes of dinosaurs—who lived some 48 million years earlier than the ancient magnolia—might have survived as well. "There's a chance, yeah," Golenberg says. But in general, plants have an edge on dinosaurs and us, he says. One reason is that for millennia, flowers and greenery have produced protective chemicals that are distasteful to herbivores. Some of these, like tannins, are natural preservatives. So plants often carry with them nature's version of a do-it-to-yourself mummification kit. Animals and humans are compara-

tively riddled with bacteria that often live on after their host's death and consume the genetic code, particularly in soft tissue. That's why our best hope for retrieving dino DNA, Golenberg says, is probably finding ways to pull it out of bones.

One widely recognized expert on bone DNA, though, says it's unlikely. "I think it's worthwhile trying," says Erika Hagelberg, Ph.D., a research scientist with the Institute of Molecular Medicine at the University of Oxford in England. "There's material that's very well preserved. But I don't think that the PCR techniques developed up to the moment would be really sufficient to be successful with anything as old as that. It would need quite a bit of work and refinement still."

Over the past three years, Hagelberg has been successful in amplifying DNA from seventeenth-century skeletons found in a mass grave in Abingdon, England, and from a 5,000-year-old thighbone found in a cave in the Judaean Desert. The genes she found were certifiably human.

Some critics suggested that her results might be due to contamination. PCR is so sensitive that simply handling the bones, rubbing off a few flakes of skin, could result in a

false blossoming of human DNA. "I was getting more and more worried, thinking this is a real possibility," Hagelberg says. So she decided to put PCR to a challenging test.

She ran her technique on a white, well-preserved bone taken from the Mary Rose, a ship that sank and silted quickly in the English Channel in 1545. She knew before she started that the bone was a leg of pork, one of many found among the cutlery, leather shoes, and medicinal salve dug from the silt. Fortunately for PCR, the results, published last year, were unequivocal. The DNA in the bone was pure pig.

Hagelberg was one of thousands of researchers in hundreds of laboratories today using a technique that didn't exist nine years ago. Thanks to the radical new technique, the scientists say, we now have the ability to go back and look at genes that, in essence, no longer exist on the face of the earth. In fact, the PCR breakthrough also marks a turning point for the human race. From now on, even after we die, with proper tending in the modern equivalent of the Brevard bog, we can supply the template for copies of our genes. And someone can reproduce portions of the code that made us. □

Questions: *Answers on page 199*

1. Why can DNA from bone cells alone show the genetic makeup of all cell types of the extinct Florida tribe?

2. How does the structure of DNA represent a complex code? Cite several examples.

3. How does the concept of pH relate to the study of DNA in this investigation?

13

Inheritable defects in a gene named p53 have appeared in cancer patients with no previous history of this disease in their family tree. Specifically the gene causes breast cancer and malignancies of the bones and soft tissues. The protein normally encoded by this gene informs a cell to stop dividing. Presumably the defective form of the gene makes a different protein signal, producing the rapid, uncontrolled division of cells indicative of cancer. This defective gene is not the only gene causing breast cancer, as only one percent of women with this cancer have the defective p53 gene. Overall, it is responsible for only a fraction of all cancers.

Cancer Gene May be Relatively Common
by C. Ezzell

A year and a half ago, two separate groups of researchers found defects in a particular gene among several families with a rare, inherited syndrome of breast cancer and malignancies of bone and soft tissues. "Breast Cancer Gene Found," trumpeted some of the headlines of news stories on the development, "Screening Test Available Soon." But at the time, scientists could only speculate that the same gene underlies cases of breast cancer in families without a history of the unusual disease, called Li-Fraumeni syndrome (SN: 12/1/90, p.342).

Now, two other research teams — one including some members from an earlier group — have verified that hypothesis. In back-to-back papers in the May 14 NEW ENGLAND JOURNAL OF MEDICINE, they report that heritable defects in a gene named p53 can show up in cancer patients with no previous family history of the disease.

Although the finding still does not confirm p53's role as the breast cancer gene, experts agree that it has important implications for the prevention and treatment of many types of cancer. They say it suggests that physicians could use tests for defective p53 genes to identify patients at high risk of developing cancer, particularly if those patients have a close relative with cancer. They add that it might also help doctors single out and follow more closely those cancer patients likely to get a second, different type of cancer later in life.

The normal protein encoded by an intact p53 gene is known to be involved in telling a cell when to stop dividing. Researchers have found defective p53 genes in most types of malignant tumors. They believe the p53 gene mutates in some people only after the process of cancer has begun, whereas in others — including those with Li-Fraumeni syndrome — the p53 mutations are present from birth in every cell in the body.

In the first study that contributed to the new finding, a team led by cancer researcher David W. Yandell of the Massachusetts Eye and Ear Infirmary in Boston analyzed DNA taken from the blood cells of 196 people with various types of sarcoma, or cancer of the bone or soft tissues. Fifteen of the patients had either a family history of cancer or more than one type of cancer in their lifetimes. For controls, the researchers analyzed DNA from 175 healthy volunteers and from 25 people with noncancerous bone or soft-tissue tumors.

They found that eight of the 196 sarcoma patients (4 percent) had mutations in the p53 gene, while none of the 200 controls had such mutations. Moreover, three of the eight sarcoma patients had no family history of cancer, suggesting that their mutations originated in their mother's egg or father's sperm. Two of these patients had had two different types of cancer during their lives: the third died from his first cancer. One of the surviving patients went on to have a daughter with an identical mutation who developed cancer during childhood.

Yandell concludes that p53 mutations "look much more extensive than just Li-Fraumeni syndrome." Although researchers must perform further studies to determine the proportion of cancers caused by the defective gene. "It looks like there are probably lots of patients

Reprinted with permission from SCIENCE NEWS, the weekly newsmagazine of science, May 16, 1992, vol. 141, no. 20, pp. 324, copyright 1993 by Science Service, Inc.

out there who have multiple primary cancers in their lifetime that can be traced to p53," he asserts.

In the other study, investigators led by pediatric oncologist Stephen H. Friend of Massachusetts General Hospital Cancer Center in Charlestown examined whether p53 might underlie the cancer risk of children and young adults with multiple cancers. Friend also led one of the earlier groups that studied Li-Fraumeni syndrome. The new study turned up p53 mutations in four of 59 patients (7 percent). Although none of the four had a family history of cancer, close relatives of three of these patients had the same p53 mutation and were diagnosed with cancer during the study period.

Friend says his group's study "gives us a clue that [p53] ... might be useful in screening people who had no reason to believe they were at an increased genetic risk for cancer." But he cautions that p53 is probably responsible for only a fraction of all cancers: An unpublished study by his group indicates that only 1 percent of women with breast cancer have a defective p53 gene. Friend's team is now studying all new cases of childhood sarcoma in the United States to determine whether p53 accounts for a greater fraction of these cancers.

Alfred G. Knudson of the Fox Chase Cancer Center in Philadelphia calls the new studies "exciting....I think it's going to make people scurry around to see why some families show more [cancers due to] the gene than others."

"I'm sure we'll learn that there are other, modifier genes involved," he adds. □

Questions: *Answers on page 199*

1. How does cancer relate to mitosis?

2. Do you think any form of a genetic-based cancer centers on only one gene?

3. How can a mutated gene be a source of cancer?

14

Blacks in the United States run a risk of developing hypertension that is twice as great as the risk experienced by Caucasians. The reason for this significant difference is not known. However, one hypothesis offers a basis that is hereditary. According to this hypothesis, some ancestral black slaves brought to this country in the sixteenth century had an inherited tendency to promote their survival. This tendency allowed them to conserve salt in their body, an advantage in the hot grueling conditions they faced as slaves. However, retention of salt also leads to the retention of fluid in the circulatory system. This fluid buildup can promote both a high systolic and diastolic pressure. At this time a gene promoting hypertension has not been identified. The entire hypothesis is controversial in the scientific world. Scientific investigation is needed to test the hypothesis. In addition, other factors can be responsible for hypertension.

The African Gene
by Kathy A. Fackelmann

When Byllye Avery took the microphone last month at a meeting in Washington, D.C., the statistics on hypertension suddenly coalesced into a painful reality faced by one black American family.

"Twenty-one years ago, my husband died of a massive heart attack," she said in a soft, determined voice. "He was 33 years old. He was hypertensive and we didn't know it."

Avery, founding president of the National Black Women's Health Project in Atlanta, recounted her experience at a September 19 editors' seminar on cardiovascular disease in black Americans, organized by the Dallas-based American Heart Association (AHA).

Blacks in the United States run a hypertension risk twice as great as that of Caucasians, but scientists still don't know why. One controversial hypothesis, proposed in 1988 by hypertension researcher Clarence E. Grim, traces the disparity back to the 16th century, when European slave traders began shipping captured Africans to plantations in the New World. According to this scenario, blacks living in the United States today may owe their high hypertension rate to a genetic trait that helped their ancestors survive the grueling conditions of slavery. That trait is an inherited tendency to conserve salt within the body, says Grim, director of the Drew/UCLA Hypertension Research Center at the Charles R. Drew University of Medicine and Science in Los Angeles.

Now Grim reports new research findings that strengthen the slavery hypothesis. His results, which remain controversial and unconfirmed, hint that physicians might one day be able to use genetic testing to identify blacks, and perhaps others, with a particularly high risk of hypertension, so that preventive measures can begin at an early age. Grim presented the new findings at the AHA's 45th annual scientific sessions on high blood pressure, held in Chicago the week after the Washington meeting.

U.S. blacks afflicted with hypertension run a greater risk of its life-threatening complications than whites. Elevated blood pressure puts extra strain on blood vessels and can damage the vessel walls, increasing the risk of stroke if a clot blocks the brain's blood supply. And, if left untreated, years of high blood pressure can inexorably damage the heart and kidneys, raising the very real possibility that these crucial organs will fail. Among hypertensives in the United States, blacks are 10 to 18 times more likely than whites to suffer kidney failure, and three to five times more likely to develop chronic heart failure, notes Edward S. Cooper of the University of Pennsylvania, who spoke at the Washington conference.

High blood pressure is defined as systolic (heart-pumping) pressure of at least 140 millimeters of mercury (mm Hg) and diastolic (heart-resting) pressure of at least 90 mm Hg. Several factors can boost a person's risk of developing the condition, including stress, advancing age, obesity, lack of exercise, alcohol use, salty diet and family history of hypertension. But in nine out of 10 cases, physicians cannot pinpoint the cause. And scientists can only speculate about what makes black Americans particularly vulnerable to this potentially deadly condition.

Some researchers have suggested that socio-economic differences may underlie the chronically elevated pressures that predispose U.S. blacks to hypertension (SN: 2/16/91, p.111). For one thing, they note, the high cost of medical care and lack of insurance can prevent low-income patients from getting regular examinations or treatment. In addition, the stress of poverty or racism may evoke a hormonal "fight or flight" response that boosts heart rate and blood pressure, says Curtis G. Hames of the Medical College of Georgia in Augusta. In earlier times, these metabolic changes helped people escape physical danger, Hames says, but today's artery-pounding pressures may spring instead from inescapable stresses of modern life.

There's no doubt that environmental factors such as stress enter into the black hypertension puzzle. However, new data gathered by Grim and a colleague suggest that a genetic component may also heighten the risk faced by so many black Americans.

Grim theorizes that Africans with a salt-conserving gene or genes were less likely to die of dehydration during the long, parched voyage across the Atlantic. The ability to hold onto salt — and thus water — also helped them weather the harsh conditions they encountered in the New World, Grim suggests. But the same trait that conferred a survival advantage upon certain African slaves may prove deadly to their modern-day descendants.

"The long-term goal of our research is to look for evidence for selective survival during slavery and its effect on health today," says Grim, who conducted the new study with Thomas W. Wilson, a medical historian and epidemiologist formerly at Drew University.

Although the historical hypothesis has provoked heated debate, some researchers think Grim and Wilson are headed in the right

direction. "I think Dr. Grim has his finger on the most important aspect of hypertension in blacks," says Cooper, who believes that research into the underlying cause of hypertension's racial differences will ultimately help physicians prevent the disease.

The Caribbean island of Barbados may seem an unlikely place to look for answers to a health problem suffered by U.S. blacks. Yet the history of Barbados includes a period in which landowners imported African slaves to work on the island's sugar plantations. And previous studies have shown that black Barbadians suffer from the same hypertension rates as U.S. blacks, Grim says.

Unlike the United States, the island's black population is in the majority and has ready access to health care. Grim and Wilson speculated that Barbados blacks might not experience as much stress as their U.S. counterparts. They reasoned that a Caribbean study might help factor out some of the environmental triggers of hypertension among U.S. blacks.

Thus, the researchers traveled to Barbados where they recruited 50 healthy individuals aged 20 to 60. All participants identified themselves as black, but to get a more accurate picture of their African heritage, the researchers turned to a test that looks for DNA of African origin. They collected blood samples from each volunteer and isolated the white blood cells, focusing on the tiny mitochondria within the cells. These energy-producing structures contain DNA inherited from an individual's mother.

When an egg and sperm unite, the sperm tail, which contains the paternal mitochondria, drops off, Wilson explains. Thus mitochondrial DNA analysis can show an unbroken line of African DNA on the maternal side of the family. While the test doesn't look at the

contribution from the father, it provides a stronger marker of African heritage than skin color, Grim notes. Previous hypertension research relied on skin color as an indication of African lineage.

Among study participants whose mitochondrial DNA suggested purely African maternal lineage, systolic blood pressures averaged 122 mm Hg. By contrast, those whose mitochondria showed non-African DNA had average systolic pressures of 115 mm Hg. The researchers discovered the highest pressures among blacks with the African marker, which suggests they may run a greater risk of developing high blood pressure later in life.

To control for non-genetic factors that might confound the results, the researchers factored in age, body mass and years of schooling (a measure of socioeconomic status). Even then, link between maternal African DNA and higher blood pressures remained.

These findings, although very preliminary, suggest that the DNA test one day may serve as a "strong predictor" of hypertension risk, Wilson says. Grim cautions, however, that before physicians use mitochondrial DNA to forecast risk, the researchers need to follow all participants to find out who will ultimately develop blood pressures that soar into the danger zone.

The team also plans to test for paternal African DNA, perhaps by looking at the Y chromosome, present only in males, Grim says. Together, the Y chromosome and mitochondrial data might provide a more sensitive indicator of hypertension threat, he says.

The data from Barbados tell only part of the story of hypertension, genes and slavery. To get the rest of the picture, the researchers traveled to West Africa, where the forced diaspora of Africans began. The team focused on rural Obod-

Ahiara. This agricultural center of about 6,000 people has no running water or electricity, but it lies in an area of Nigeria that was home to many ancestors of present-day black Americans, says Wilson, who is now at the Western Consortium for Public Health in Berkeley, California.

He and Grim conducted a preliminary study to determine whether hypertension rates in the village matches those of blacks in the United States and Barbados. Their findings, presented August 1 in Brazil at the Sixth Annual International Conference on Hypertension in Blacks, revealed very low rates of hypertension despite a relatively salty diet.

The researchers began by recruiting 140 men aged 20 to 60. They measured the men's blood pressures and asked them to collect their urine for 24 hours.

Analyses of the urine samples showed that the men ate relatively salty diets: The average 24-hour sodium excretion (an indicator of dietary salt) logged in at 7.3 grams of sodium per day.

While blacks in Nigeria and in the U.S. share a penchant for salt, there was a "striking" difference in blood pressures between the two groups. Despite their high-salt diet, only 12 percent of the Nigerian men had systolic pressures of 140 mm Hg or greater. By contrast, about 32 percent of U.S. adult blacks show such elevated pressures, Wilson notes.

Moreover, age did not significantly increase blood pressures among the Africans, whereas 60 percent of U.S. blacks aged 50 or older suffer from dangerously elevated blood pressure.

The men of Obodo-Ahiara don't appear to show salt sensitivity — a finding that supports the theory that U.S. blacks owe their high hypertension rates to selective survival, perhaps during the slavery period. But the researchers say they have a long way to go before they can prove that such selection occurred.

"The survival hypothesis is only a hypothesis," says Grim, who notes that no one has yet found a salt-retaining gene or genes that would predispose a person to hypertension.

Richard S. Cooper (no relation to Edward Cooper), a hypertension researcher at Loyola University in Maywood, Ill., is among the many critics of the slavery hypothesis. "I think it's a crock," he says. "To attribute that magnitude of evolutionary change to a fairly brief period is a kind of fantasy."

Wilson counters that 70 percent of African slaves died within four years of their capture. If those with salt-conserving genes survived more often, it's not unreasonable to expect the second generation to exhibit this salt-hoarding trait, he argues. "It's a real myth to say that evolution must take millions of years," Wilson says.

John M. Flack, an epidemiologist at the University of Minnesota in Minneapolis, brings up another criticism: The slavery hypothesis, he says, is reminiscent of controversial arguments — largely rejected by the scientific community — that genes render blacks intellectually inferior to whites.

The theory has also fueled a long-standing debate over the relative importance of genes versus environmental influences.

"Obviously, what we're looking at is a gene/environment interaction," Grim says. "We think there are people who inherit the ability to store salt well, and that this tendency leads to high blood pressure in a high-salt environment."

Indeed, Grim believes thinks salt-conserving genes may contribute to hypertension in some whites as well as in blacks.

Others doubt that genes play much of a role at all in the development of hypertension. "We know that there are powerful environmental factors that could by themselves explain all the racial differences," Richard S. Cooper says. "Whether there are genetic factors as well remains to be discovered."

Hypertension researchers do agree on one thing, however: They need more data to tease out the complex factors leading to elevated blood pressure in U.S. blacks. And until scientists solve the mystery underlying hypertension, most cardiologists advise all U.S. blacks to get their blood pressure checked regularly. For those with high pressures, doctors often recommend a low-salt diet and in some cases, medication aimed at bringing dangerously high pressures down. And because U.S. blacks face such a high risk of this deadly condition, Grim suggests a low-salt diet and plenty of exercise as a preventive strategy.

His recommendation remains a safe bet for Americans of all colors in the battle against this killer disease. □

Answers on page 200

Questions:

1. Would a recorded blood pressure of 150 over 96 be considered hypertensive?

2. How can high blood pressure shorten life?

3. From what you have learned about osmosis, how can salt retention lead to high blood pressure?

15

Does the tendency toward alcoholism have a hereditary basis? Researchers at the University of California believe that it does. They claim to have located a gene that modifies alcoholism rather than serving as a direct cause. They think it is the first of many genes that will be discovered that are associated with alcoholism. The gene interferes with the action of dopamine in the brain. Dopamine is an important chemical transmitter, called a neurotransmitter, that is necessary for the signaling between brain cells at synapses. These recent findings reaffirm earlier findings by scientists who identified the alcoholism gene at a precise spot on human chromosome 11. Larger genetic studies are planned to learn more about the genetic predisposition for alcoholics.

Gene in a Bottle
by Bruce Bower

Research into the genetics of alcoholism invariably stirs up spirited controversy. A report issued last year, describing the first evidence that a specific gene creates a susceptibility to at least one type of alcoholism, proved no exception. Critics immediately pointed out flaws in the study, and independent follow-up investigations suggested that the gene plays no role in fostering uncontrollable alcohol consumption.

But the gene will not go away. Its original proponents — who had identified the culprit as one of two genes that occupy a precise spot on chromosome 11 and direct the function of key dopamine receptors on brain cells — now report further evidence linking it to cases of severe alcoholism with medical consequences of alcoholism — rather than cause the disorder — by disturbing normal dopamine transmission. Dopamine, an important chemical messenger in the brain, normally helps to regulate pleasure-seeking behaviors.

"We may have found a gene that modifies, rather than causes alcoholism," says psychiatrist Ernest P. Noble of the University of California, Los Angeles, who co-directed the original study with psychopharmacologist Kenneth Blum of the University of Texas Health Science Center at San Antonio. "We just don't know yet. As research continues, I think we'll find many genes associated with alcoholism."

For now, though, the dopamine receptor gene stands alone. Noble's team first reported finding it in DNA from 24 of 35 alcoholics, compared with only seven of 35 nonalcoholics. All DNA samples came from the brain tissue of deceased individuals. The researchers used medical records and reports from family members to determine which individuals met the criteria for alcoholism. Because most of the alcoholics had failed in several rehabilitation efforts and had died of alcohol-related causes, the investigators concluded they had suffered from a severe form of the disorder (SN: 4/21/90, p.246).

Another study, reported last January, raised doubts about the proposed alcoholism gene. These researchers, led by psychiatrist Annabel M. Bolos of the National Institute on Alcohol Abuse and Alcoholism (NIAAA) in Bethesda, Md., examined DNA from 40 alcoholics and 127 nonalcoholic controls, including 62 with cystic fibrosis. In both groups, the dopamine receptor gene turned up in about one-third of the volunteers. Bolos and her colleagues contended that psychiatric interviews with the participants, all of whom were living, allowed for more accurate alcoholism diagnoses than those deduced by Noble's group (SN: 1/12/91, p.29).

But Noble and Blum, who have reexamined data from the Bolos study, say the results actually support a link between the dopamine receptor gene and severe alcoholism. According to their analysis, the gene's prevalence increased from 25 percent of the nonalcoholic controls (excluding those with cystic fibrosis, who often die before alcoholism has a chance to develop, according to Noble and Blum) to 30 percent of the 20 alcoholics with no medical complications and 45 percent of the 20 alcoholics with related medical conditions such as liver cirrhosis.

Bolos' group excluded alcoholics with the most severe, "acutely active" medical complications, thereby lowering the frequency of the dopamine receptor gene in their study. Noble and Blum maintain in the May 22/29 JOURNAL OF THE AMERICAN MEDICAL ASSOCIATION.

In a commentary accompanying Noble and Blum's argument, Bolos and her colleagues question the revision of their work, noting that a standard alcoholism screening test reveals no difference in symptom severity between alcoholic participants with and without the gene.

Noble and his colleagues continue to pursue the chromosome 11 offender. They have now extended their original genetic investigation to living volunteers: 43 nonalcoholics, 52 with severe alcoholism (dependency symptoms plus medical complications) and 44 with less severe alcoholism (dependency symptoms only). The dopamine receptor gene occurred in 21 percent of the nonalcoholics, 34 percent of those with less severe alcoholism and 63 percent of those with severe alcoholism, they report in the September ALCOHOL.

The researchers have also conducted a biochemical analysis of 66 of the 70 brain samples from their 1990 study. Samples from individuals diagnosed with alcoholism showed significantly fewer brain-cell binding sites for the dopamine receptor controlled by the chromosome 11 gene, as well as impaired binding function at those sites, the team reports in the July ARCHIVES OG GENERAL PSYCHIATRY.

This suggests — but does not firmly establish — that genetically disturbed dopamine activity confers susceptibility to severs alcoholism, they say.

Another report in the same issue indicates that the cerebral havoc wreaked by this gene may jack up the severity of alcoholism, rather than light the fuse of alcohol abuse. Geneticist Abbas Parsian and his co-workers at Washington University School of Medicine in St. Louis found that 13 of 32 alcoholics (41 percent) carried the dopamine receptor gene, compared with three of 25 nonalcoholics (12 percent). And among alcoholics with serious, related medical problems, six of 10 carried the gene.

However, when the same researchers performed genetic analysis of 80 individuals in 17 families with numerous cases of alcoholism, they uncovered no increased susceptibility to either mild or severe alcoholism among those carrying the dopamine receptor gene.

Since the critical gene clearly stands out among unrelated severe alcoholics with medical complications, but does not congregate in family members afflicted by alcoholism, the St. Louis scientists conclude that it probably plays a secondary role of fanning the flames of uncontrolled alcohol consumption. The gene may also speed the progression of alcohol-related diseases such as liver cirrhosis, they say.

This intriguing possibility calls for larger genetic studies that carefully partition alcoholics according to the severity of their medical problems, asserts P. Michael Conneally, a geneticist at Indiana University in Indianapolis. Conneally's plea may not go unheeded. He and seven other investigators, based at six research centers, now direct the largest-ever study on the genetics of alcoholism, Participants in the NIAAA-financed study include 600 alcoholics and thousands of their family members. Project investigators hope to determine whether certain genes produce a specific susceptibility to all sorts of compulsive behaviors.

Noble suspects the dopamine receptor gene will fall into the latter category, working in concert with several genes to promote the full spectrum of substance use and abuse.

"If the good Lord didn't have alcohol around, we'd still have this gene, and we'd still get a charge out of certain pleasurable behaviors that sometimes become compulsive," Noble says.

The recent emphasis on combing through chromosomes for offending genes linked to alcoholism cannot deny evidence of vigorous environmental influences on compulsive alcohol use, he adds. These include expectations about alcohol's effects, as well as conditioned emotional and situational cues that trigger a craving for alcohol (SN: 8/6/88, p.88).

"The environment is a tremendously powerful agent in producing alcoholism," Noble remarks. "But genes are easier to study." □

Questions: *Answers on page 200*

1.   Do you think that a tendency for alcoholism is only hereditary?

2.   How can the identification of an alcoholism gene be used to treat and prevent the development of alcoholism?

3.   Do you think that the findings of a possible genetic basis for alcoholism is conclusive?

16

Through advances in molecular biology, more precise control is being gained over the ripening of tomatoes and, therefore, their taste, color, and texture. Superior strains of tomatoes, soon to appear in supermarkets, are the result of gene splicing. In building the better tomato, genetically-engineered products with more sugars (fructose and glucose) were the outgrowth of this major advance in biotechnology. This biochemical change greatly enhanced their taste. Genetic control of ethylene production affected the ripening process of this fruit. In their continuing research, geneticists will search for tomato genes that can receive a patent. Tomatoes are probably the first of many crops that, through gene splicing, will present more desirable products for human consumption.

Better Than the Real Thing

by Daniel Pendick

Pity the supermarket tomato. Pale red and unyielding to the touch, it lacks the vibrant color and luscious flavor of a ripe, fresh-picked fruit. Yet, despite their dissatisfaction with mass-produced tomatoes, U.S. shoppers still buy an impressive 2.8 billion pounds of them every year.

Now, through advances in molecular biology, genetic engineers are gaining more precise control of tomato ripening and its effects on taste, texture, color, and shelf life.

The first of such gene-spliced products is well on its way to market. Genetic engineers from Calgene Fresh, Inc., have created the "Flavr Savr," an alternative to the beleaguered supermarket tomato. This engineered tomato, designed to resist softening, will have the fresh flavor consumers desire and an extended shelf life as well, Calgene contends. If and when federal officials issue a favorable opinion on Calgene's tomato, the Evanston, Ill.-based company will splice its genetically engineered fruit into the human food chain. This could

happen as early as next year.

Reaction to the imminent arrival of Flavr Savr on supermarket shelves has varied. To promoters of biotechnology, Calgene's tomato heralds a new cornucopia of genetically engineered fruits and vegetables. Critics of gene-spliced food seem to fear a real-life enactment of the cult science-fiction film "Attack of the Killer Tomatoes."

Consumers may soon have the chance to decide for themselves, however. Several companies besides Calgene have developed their own gene-spliced tomatoes and intend to put them on the market in the next few years.

Much of this tomato tinkering aims at remedying the shortcomings of current tomato farming practices. Growers of fresh-market tomatoes harvest their crop firm and green so it can be washed, sorted, packed, and shipped, commercial packers bathe the tomatoes in ethylene gas for several days to spur ripening.

The problem is, says Mark Stowers of the Monsanto Co. in St. Louis, these prematurely plucked

and artificially ripened tomatoes just don't have the flavor consumers desire.

From a grower's viewpoint, the ideal mass-production tomato could be left on the vine to build up the sugars and acids critical to fresh taste and aroma, yet remain firm enough to handle without damage. This is exactly what Calgene claims to have done.

Using a technique called antisense genetics, Calgene researchers permanently endowed their tomato with a backward (antisense) copy of the gene for polygalacturonase (PG), a fruit-softening enzyme. Both copies of the gene produce messenger RNA, which carries genetic information from the nucleus to protein-making ribosomes in cells. However, the antisense RNA molecules bind to the normal, "sense" RNA. This prevents the tomato from making the usual amount of PG. As a result, the tomato can remain longer on the vine without getting too soft for handling, says Calgene, and it can also hold up longer in the produce department.

Antisense genetics is more than just a tool for customizing tomatoes. Scientists have used the technique to discover some very important things about the ripening process, says plant molecular biologist Athanasios Theologis in a review article in the October PLANT CELL.

For example, Theologis and others at the U.S. Department of Agriculture's Plant Gene Expression Center in Albany, Calif. confirmed last year that ethylene is the master control hormone of ripening in many plants. Theologis' group used antisense genetics to cut production of ethylene to such an extent that green tomatoes remained on the vice for as long as five months, ripening only to a pail orange.

Tomatoes and many other plants produce ethylene to control the various biochemical processes that cause fruit to ripen, including the breakdown of chlorophyll, the synthesis of the red pigment lycopene, the buildup of sugars and acid and the softening of tomato tissue by PG and other enzymes.

Some genetic engineers have chosen ethylene control as a means of tailoring tomato ripening to growers' needs. Monsanto's agricultural research group, for example, has outfitted its tomato with a bacterial gene whose enzyme digests an acid called ACC, the raw material that tomato cells convert into ethylene gas. Stowers, who is business director of Monsanto's tomato project, says that growers can leave these ethylene-deprived fruits on the vine for three or four more days — just long enough to build up extra flavor.

In addition to the quest for a better fresh-market tomato, genetic engineers want to carve out a place for themselves in the tomato-processing industry — and the size of this market is impressive. Every year. U.S. processor transform 12 billion pounds of raw tomatoes into

juice, sauce, paste, ketchup, and other products. Part of this crop goes into the 315 million cans of Campbell's tomato soup consumed annually in North America.

The best tomato for making these products is high in solids, chiefly the sugars fructose and glucose. And tomato solids are no small potatoes: Indeed, the tomato-processing industry estimates that a 1 percent increase in tomato solids could save $70 million to $80 million a year in processing costs.

Using antisense genetics, Avtar K. Handa of Purdue University in West Lafayette, Inc., serendipitously created a tomato with 10 percent more solids than current varieties grown for processing. He and his collaborators report the results of their antisense experiments in the June Plant Cell.

Handa did not set out to build a better tomato. His group used the antisense technique to block the gene for an enzyme called pectin methylesterase (PME) so he could study its role in fruit softening. The high-solids trait, unexpected though welcomed, showed up during standard testing of the harvested fruit.

In tomatoes, PME works in concert with PG to break down pectin, a major cell-wall building block. During ripening, these enzymes slowly soften tomato tissues, leaving them susceptible to bruising and rot.

Currently, Handa does not know exactly why suppressing PME enriches the solids in tomatoes. He suspects, however, that the undigested pectin present in antisense-PME tomatoes becomes part of their solids content.

Handa expects to finish compiling the field-testing data on his high-solids tomato by year's end. At that point, a private company could begin shepherding the new tomato onto the market, he says.

Genetic engineers continue to

search for patentable tomato genes. Researchers at ICI Seeds in Berkshire, England, for example, have isolated 13 tomato genes that affect fruit quality and have patented five, says Simon G. Best of ICI Americas, Inc., in Wilmington, Del.

Other genetically engineered crops will follow, Best promises. "Tomatoes are just the first crop, from which we've identified a lot of genes that have other used in crops with related biochemistry," he says. ICI may eventually use its patented genes and techniques to create new kinds of peaches and melons.

DNA Plant Technology Corp., another company developing a genetically engineered tomato, also plans to deploy its gene-control techniques further afield. Robert Whitaker, managing director of research at the company's facility in Cinnaminson, N.J., envisions bananas, papaya, and cut flowers as logical targets for future efforts to ensure freshness through ethylene control.

If these products win consumers over, genetic engineering may generate as much new green stuff for supermarkets as it has on the stock exchange. But given public concern about food purity and past squeamishness about gene splicing, must the biotechnology industry gear up for a major defense of genetically engineered tomatoes?

Consultant Richard A. Herrett, a member of the board of directors of the Association of Biotechnology Companies (ABC) in Washington, D.C., thinks it unlikely that torch-bearing citizens will ever come looking for the Frankentomatoes of the future. On the contrary, Herrett believes consumers will respond positively if the industry makes clear the potential gains in quality and nutrition that genetically engineered foods may offer.

However, Jeremy Rifkin and his Foundation on Economic Trends in Washington, D.C., hope to generate an international boycott of genetically engineered foods. Says Ted Howard, leader of the group's boycott effort, "I really feel this is going to be an extremely hotly contested consumer issue."

Although the Foundation on Economic Trends continues to demand additional regulation, recent changes in government policy have moved increasingly toward deregulation of agricultural biotechnology companies in the United States. For example, USDA last month significantly loosened restrictions on field testing of genetically altered plants.

The U.S. Food and Drug Administration (FDA) is considering whether Flavr Savr should be treated the same as tomatoes produced by traditional breeding methods and whether the antisense gene added to the tomato poses any health risks, explains Eric Flamm in the FDA's Office of Biotechnology.

Andrew Kimberll, attorney for the Foundation on Economic Trends, has already decided on one response to a favorable FDA ruling. "The minute FDA rules that this food does not need to be labeled, we will sue them," Kimbrell promises.

Should the FDA decide in Calgene's favor, the Flavr Savr tomato will take its place on grocery shelves and in history as the first genetically engineered wholefood product available to the consumer. Considering the hundreds of genetically engineered crops now in field trials, however, it will not be the last.

Even so, it is not clear yet that people's yearning for a better mass-produced tomato will translate smoothly into widespread acceptance of genetically engineered foods. In the end, gene-spliced tomatoes may prove as ripe for controversy as the technology that is creating them. □

Questions: *Answers on page 200*

1. Will the biotechnology industry need to defend the genetic engineering of tomatoes?

2. What dangers could genetically-engineered crops present to the public?

3. How is the subject of plant hormones relevent to genetic engineering of plants?

Part Four

Homeostasis
and
Principles of Structure
and Function

17

To date two AIDS viruses have been identified throughout the world. HIV-1 is the commonest form of the virus, found throughout Europe and the United States. HIV-2 is a virus endemic to Africa. Now some scientists believe that a previously-identified, AIDS-like virus may be emerging in the tissues of humans. These subjects are ill or dying as they reflect the AIDS symptoms. However, they are not carrying either one of the recognized HIV viruses. Skeptics have labeled this virus the MTV virus (media transforming virus) in reference to the significant media attention given to its possible existence. Its existence, or proof of it as a causative agent of AIDS, has not yet been verified scientifically.

The Mysterious Virus Called "Isn't."
by Barbara J. Culliton

The first close scrutiny of reports of a new virus supposed to cause symptoms of AIDS in people with neither HIV-1 nor HIV-2 suggests that there is no new virus. Can one have AIDS without the AIDS virus? Or is it possible that a third AIDS virus (neither HIV-1, the commonest form of the virus in the United States and Europe, nor HIV-2, which appears to be endemic in Africa) is lurking somewhere in the tissues of immunosuppressed individuals who have precariously low concentrations of CD4 T cells — the lymphocytes whose progressive depletion is associated with the development of AIDS?

Worldwide concern over the possibility that a previously unidentified infectious AIDS-like virus may be causing severe immunodeficiency disease was ignited last month by an article in *Newsweek* that asked, "Is a New AIDS Virus Emerging? The patients are sick or dying and most have risk factors. What they don't have is HIV."

Last week, researchers from the US National Institutes of Health (NIH) and Centers for Disease Control (CDC), together with AIDS and infectious disease experts from institutions all over the United States, held a one-day conference at CDC in Atlanta to assess data from several laboratories reporting retroviral activity in the cells of patients who appear not to be infected by HIV.

They reached a consensus on several points.

* First, as Anthony S. Fauci, AIDS coordinator for NIH put it, "It is essential that we keep an open mind."

* Second, present data do not support the suggestion that there is a third virus, which is why some are calling it the "isn't" virus. (The mystery virus has also been dubbed "MTV" for media transforming virus in the light of all the press attention it has received.)

* Third, there is even less evidence that the alleged virus or virus particles, were they to be confirmed, cause disease.

* Fourth (and most important), members of two expert panels — one to evaluate epidemiological data, the other retrovirology — find nothing to suggest they are dealing with a new transmissible agent.

This point is central to the issue underlying current anxiety —

namely, the safety of the blood supply. The analogy has been made to the early days of the AIDS epidemic when scientists were uncertain what they were dealing with, but the analogy does not hold. From the outset, AIDS cases appeared in clusters, particularly in California and New York, and there was evidence that physicians were dealing with an infectious disease. But that is not the case here.

The *Newsweek* article, faxed to reporters in Amsterdam where the international AIDS meeting was in full swing during the week of 20 July, based its arresting question on the work of a couple of research teams who have reported CD4 patients. One team is that of Jeffrey Laurence of Cornell University Medical College in New York, who reported in the 1 August of *The Lancet* (**340**, 273, 1992) on five patients (four of them at risk for AIDS) who have low counts of CD4 T cells, various opportunistic infections and no apparent signs of HIV-1 or HIV-2 in the blood. In the *Lancet* article, not published until after *Newsweek* hit the faxes, Laurence suggests that the cases he describes "raise the question of the

Reprinted with permission from Nature, August 20, 1992, vol. 358, pp.619 copyright 1992 by Macmillan Magazines, Ltd.

existence of other agents associated with transmissible immune deficiencies that can evade current laboratory detection techniques".

One of Laurence's co-authors, Stephen Morse of Rockefeller University, acknowledges that they may be seeing an artifact, a contaminant or viral activity from an endogenous retrovirus. But, he says, the issue is still "open" and research will proceed. "Our working hypothesis is still to look for a virus". Data similar to those from Laurence's laboratory were presented in Atlanta by David Ho of the Aaron Diamond AIDS Center in New York. He has seen more than a dozen patients, several of whom do not have the usual risk factors associated with AIDS -- drug use, a history of blood transfusions or multiple homosexual partners. Two of them have signs of reverse transcriptase activity, but none is seropositive for HIV. Ho sent his material to an outside laboratory expert in retrovirus detection, but the answer came back negative.

The data that have probably received the lion's share of press attention were reported by Sudhir Gupta and colleagues at the University of California at Irvine. In a paper in the 15 August issue of the *Proceeding of the National Academy of Sciences* (**89**, 7831-7835; 1992), which was released early in response to the public furore in Amsterdam, Gupta reports detection of a human intracisternal retroviral particle associated with CD4 T-cell deficiency in a 66-year-old woman with P. carinii pneumonia and her

healthy, but CD4-depleted, daughter, (Gupta and the university have applied for a patent on the particle.)

Gupta's data are viewed with considerable skepticism by retrovirologists, several of whom have summed up their evaluations by saying, "there is nothing significant there".

Retroviral particles (particularly those known as A-type) have been studied on and off for years, in animal and human systems, without evidence that they cause disease. In his paper Gupta reports that the particle he found (and which he calls both a particle and a virus) is distinct from A-type particles. His particles do have a gag and pol genes as HIV does, but lack the envelope gene that would allow the particles to get out of a cell. So the probability that they are infectious is said to be close to nil.

A fourth report at the Atlanta meeting by Robert F. Garry of Tulane University School of Medicine recapitulated two-year old data on retroviral particles and Sjogren's syndrome — an autoimmune disease that causes dryness of the eyes and mouth. In a paper in *Science* (**250**, 1127-1129; 1990). Garry and his colleagues reported a human intracisternal A-type retroviral particle in Sjogren's patients that is antigenically related to HIV but is distinguishable from it by ultrastructural and enzymatic analysis. Garry concludes not that the A-type particle causes Sjogren's syndrome but that the data support "an association between retrovirus

infections and autoimmune phenomena that has long been suspected." An association, as yet unexplained etiologically or mechanistically (and probably unrelated to AIDS), may be what is emerging from the current reports of CD4-depleted patients.

So, where do things now stand? According to Fauci, there is general agreement that together Ho, Laurence, Gupta and others who have CD4-deficient patients may have identified a new and heterogeneous syndrome. "It could be decades old", he says. "It could just be coming to light because, for the past five years or so, we've been testing for CD4. We just don't know yet." CDC is calling the new syndrome ICL for idiopathic CD4 lymphopenia and there seems to be a consensus that it is epidemiologically and clinically real. The virology is more of a mystery but, Fauci says, the skepticism that greeted early reports in Amsterdam of a new virus "has really been reinforced by our analysis so far".

But good science and good public policy demand that the questions be more clearly resolved. NIH and CED have established disease registries for CD4-deficient patients whose immune deficiency cannot be explained.

NIH and CDC have also established a repository for blood and other tissues from CD4-depleted patients that can be distributed to laboratories around the world for studies that may help to resolve this intriguing mystery.   □

Questions: *Answers on page 200*

1. Do you think that a virus is living?

2. How would you explain the unknown cause of AIDS symptoms without apparent viral cause?

3. How do you think the controversy over the cause of unexplained AIDS symptoms will be settled?

The claim of an internal human body temperature of 98.6 degrees Fahrenheit has long been unquestioned. However the data from a new study suggests that a more accurate value is 98.2 degrees. The average temperature between individuals may vary as much as 4.8 degrees on the Fahrenheit scale. It will even vary slightly within an individual over 24 hours. Overall, in a subject the body temperature was lowest at about 6 a.m. and highest at 4 p.m. Quicker and more accurate recordings contributed to the accuracy of the data. The conclusion of 98.6 degrees arrived at years ago involved primitive equipment.

Body Temperature: Don't Look for 98.6 F
by J. Raloff

In 1868, Carl Wunderlich published a seminal paper on body temperature in 25,000 adults. His more than 1 million measurements indicated that while temperatures of healthy individuals varied, they averaged 98.6° F (37 C). Texts on fever still cite Wunderlich's study, one of the few to investigate normal temperatures, and accept as dogma the 98.6° F average-body temperature figure, notes Philip A. Mackowiak at the University of Maryland School of Medicine in Baltimore. However, his new data suggest the average should be 98.2° F.

Mackowiak's group measured oral body temperature digitally up to four times daily for three consecutive days in 148 volunteers. The healthy men and women ranged in age from 18 to 40.

Like Wunderlich, Mackowiak's team found that normal temperatures vary between individuals (by as much as 4.8° F) and even within individuals over the course of a day (by up to 1.09° F). However, the Baltimore group found that 98.6° F "was not the overall mean temperature, the mean temperature of any of the time periods studied, the median temperature, or the single most frequent temperature." Indeed, they report in the Sept 23/30 Journal of the American Medical Association, 98.6° F accounted for just 8 percent of their 700 readings.

Maximum normal temperatures varied from a 6 a.m. low of 98.9° F to a 4 p.m. high of 99.9° F. Though age did not appear to influence reading in this group, women tended to be about 0.3° F warmer than men, and blacks about 0.1° F warmer than whites. Mackowiak's team also observed a 2.44-beats-per-minute increase in heart rate with each 1° F rise in body temperature.

Mackowiak described Wunderlich's data collection as awesome, noting that thermometers then took 15 or 20 minutes to obtain temperatures and had to be read while still in place — the armpit. Today, however, physicians eschew such underarm readings because of their unreliability. □

Questions: *Answers on page 200*

1. Why do you think internal human body temperature fluctuates in a person over 24 hours?

2. Why do you think internal human body temperature fluctuates between people?

3. Is the Fahrenheit scale the most appropriate temperature scale to use in science?

A standard therapy to treat children with asthma has been the use of bronchodilator drugs. These drugs open the respiratory passages to allow the easy flow of air during inhalation and exhalation. Anti-inflammatory drugs now offer an alternative to this standard therapy. The new therapy has been far more successful in treating the symptoms of an asthmatic attack. Ongoing inflammation of the respiratory tract is a major cause of the chronic disease. Therefore the proper anti-inflammatory drug can potentially prevent the actual disease process. Although bronchodilator drugs do open the respiratory passages, they do not treat the source of the problem, an inflammatory reaction in the tissues of the air tract. Anti-inflammatory drugs, often steroids, can produce unwanted side effects. However, their potential advantages for treatment may far outweigh the drawbacks.

Anti-inflammatory Drugs May Quell Asthma

by K.A. Fackelmann

Asthmatic children treated with anti-inflammatory drugs fare better than their peers who receive standard therapy with bronchodilator drugs, Dutch scientists report. Their finding offers the hope that anti-inflammatory drugs will help some youngsters outgrow the debilitating respiratory illness.

Scientists know that exercise, cold air, or exposure to allergens can set off an asthma attack. Such triggers cause the muscles surrounding the airways to contract, leading to an inability to catch one's breath. But muscle contraction is just part of the asthma story. The other key component of this chronic disease is ongoing inflammation of the airways.

A team of Dutch scientists led by Elisabeth E. van Essen-Zandvliet of the Sophia Children's Hospital in Rotterdam decided to examine two pediatric asthma treatments, one designed around an anti-inflammatory drug and the other at relaxing tightened airways with a bronchodilator.

They began by recruiting 116 youngsters age 7 to 16 with moderate to severe asthma. Next, the Dutch team randomly assigned each young person to one of two treatment groups. One group received treatment with an inhaled bronchodilator and a placebo three times a day. The remaining participants used a bronchodilator and an aerosol version of a steroid drug called budesonide.

On average, the children in the bronchodilator-placebo group got worse, the researchers report in the September AMERICAN REVIEW OF RESPIRATORY DISEASE, a journal published by the American Lung Association. They found that such youngsters showed declining lung function and suffered more asthma attacks than did children assigned to the other treatment regimen.

Indeed, about half the youngsters relying on the bronchodilator drug alone experienced such a decline that they were forced to drop out of the study by the 22nd week. At that time, an independent review panel stopped the trial and recommended that all the children receive treatment with inhaled steroids.

"This study suggests that anti-inflammatories can actually affect the disease process," comments H.

William Kelly, an asthma expert at the University of New Mexico in Albuquerque who wrote an editorial to accompany the Dutch report. It's the first long-term pediatric study of this size, he told SCIENCE NEWS.

The new findings add to the growing unease about relying on bronchodilators as standard treatment. Many asthma experts now believe that an overuse of these drugs may cause asthma to progress. While bronchodilators do open constricted airways, they do not halt the inflammation that can ultimately cause scarring and a permanent narrowing of the bronchial tubes, Kelly notes.

The Dutch study underscores a recommendation made last year by a panel of experts appointed by the National Heart, Lung and Blood Institute. That group advised physicians to rely on anti-inflammatory drugs as the mainstay of asthma treatment (SN: 2/9/91, p.86).

Physicians remain reluctant to prescribe inhaled steroids for their pediatric asthma patients, however, perhaps because previous studies have linked these drugs to growth abnormalities, Kelly says. While the

Dutch study found no sign of impaired growth in the children taking inhaled steroids, further safety studies are needed, van Essen-Zandvliet concedes.

For now, Kelly suggests that physicians first try cromolyn sodium, an anti-inflammatory drug that is nearly free of side effects. For youngsters who continue to get worse, Kelly recommends an inhaled steroid. He believes the benefits of such therapy far outweigh the potential hazards. □

Questions: *Answers on page 201*

1. What does a vasoconstrictor substance do to the air passageways? How does this affect body health?

2. Are you convinced with the findings of this study?

3. What effects do you think steroids could have on the child taking them for asthma?

20

Excess iron stored in the human body may represent a greater risk factor for the development of coronary disease than the level of cholesterol in the blood. The study producing this claim needs further verification, as this preliminary finding is the first link of heart disease to this mineral. Measurements show the men with high concentrations of ferritin, more than 200 micrograms per liter, were twice as likely to develop heart disease as men with lower values. Red meat and other dietary trends increase the iron concentration in the body. The relationship of iron levels to the gender gap of heart disease rates, higher in men than women, remains unclear. Its effect on the use of aspirin and fish oil to prevent heart disease also remains unknown.

Excess Iron Linked to Heart Disease

by K.A. Fackelmann

High levels of iron stored in the body may boost the risk of heart disease, according to a new study by a team of Finnish researchers. In fact, stored iron may prove a more significant risk factor for coronary disease than total blood cholesterol levels, they say.

The new study published in the September CIRCULATION, provides the first empirical evidence for this theory. "It is the first time that iron stores have been looked at as a risk factor," comments Jerome L. Sullivan of the Medical University of South Carolina in Charleston. Sullivan first proposed the iron and heart disease hypothesis more than a decade ago.

Basil Rifkind of the National Heart, Lung and Blood Institute calls the Finnish findings "interesting." He points out, however, that this is the first time scientists have shown a link between iron stores and heart disease. Other researchers must confirm the finding before public health experts can make any recommendations to reduce iron stores, he adds.

Epidemiologist Jukka T. Salonen of the University of Kuopio and his Finnish colleagues focused on 1,931 middle-aged men who showed no sign of heart disease at the study's start in 1984. The researchers drew blood to test for stored iron and cholesterol and asked the men about other risk factors for heart disease. The team then estimated dietary iron intake by asking the men to record their food choices during a four-day period.

After adjustment for risk factors such as cholesterol, the data revealed that men with high concentrations of ferritin in their blood (more than 200 micrograms per liter) were twice as likely to suffer a heart attack as men with lower ferritin values. Ferritin is a molecule that stores iron in the blood and other parts of the body. The researchers found that every 1 percent increase in blood ferritin was associated with a more than 4 percent rise in the risk of heart attack.

Men who typically ate iron-rich foods faced a higher likelihood of heart attack than did those who had an iron-poor diet, Salonen says. Red meats, which also contain a lot of fat, are rich in iron.

There's no doubt that iron-depleted blood can cause anemia, a medical disorder that can result in fatigue. But Sullivan and Salonen propose that, although people need a trace amount of iron in their diet to remain healthy, too much iron can promote the formation of free radicals.

Free radicals may injure the cells lining artery walls and damage heart muscle, Sullivan says. Free radicals may also lead to the formation of a dangerous type of cholesterol known as oxidized low-density lipoprotein (LDL). Scientists believe that oxidized LDL cholesterol is more likely than nonoxidized LDL to stick to artery walls and thus to trigger the buildup of fatty plaque that can clog arteries and lead to heart attacks.

The iron theory may help explain the mysterious gender gap in heart disease rates. Cardiologists have long noticed that premenopausal women remain largely protected from the ravages of heart disease, whereas men start suffering heart attacks in their forties. Many scientists believe the sex hormone estrogen helps women ward off

Reprinted with permission from SCIENCE NEWS, the weekly newsmagazine of science, September 19, 1992, vol. 142, no. 12, pp. 180, copyright 1993 by Science Service, Inc.

heart disease until menopause, when the production of estrogen tapers off and heart attack rates go up. Sullivan remembers puzzling over that gender gap during his medical training. At the same time, he was studying normal iron metabolism. "When I saw those curves for iron acquisition in men and women, I really had a eureka moment," he says, noting that men build up iron stores steadily, while women don't start accumulating iron until menopause.

Sullivan thinks that young women are shielded from heart disease because they lose iron every month during menstruation. After menopause, the stored iron in a woman's body builds up rapidly — and women's advantage in terms of heart disease gradually disappears, he adds.

The iron theory might also explain why aspirin and fish oil help protect people from heart attacks, Sullivan adds. He notes that both substances may increase chronic blood loss through minor bleeding and thus loss of iron.

The findings, if confirmed, could force public health experts to rethink dietary recommendations for iron ingestion. Even normal levels of stored iron may prove damaging, Sullivan says. Over-the-counter vitamin supplements often contain iron, as do some enriched foods such as cereals, he notes.

Sullivan offers a few simple solutions for people worried about the iron-heart disease connection. "I think we can say that adults should avoid iron supplements unless they have iron-deficiency anemia," Sullivan says. "Also, I think people should consider blood donation." ☐

Questions: *Answers on page 201*

1.   Should all iron be avoided in a person's diet?

2.   What do you think causes the gender gap between men and women for heart disease rates?

3.   In light of this study, what advice would you offer a person concerning iron consumption?

21

Studies of a 70-year-old woman indicate that two separate knowledge systems may exist in the brain, one that is visually-based and one that is language-based. The woman studied had patches of damage to each temporal lobe of the cerebral cortex as well as damage elsewhere in the brain. From a series of pictures portraying various animals, along with presented sounds of the animals, she could not recognize the animals. Yet pictures and sounds of other living things, living and nonliving, led to recognition. She did reveal, however, an intact knowledge about the functional properties of animals. Researchers concluded that the areas throughout the brain most likely coordinate the separate knowledge systems when both are working properly.

Clues to the Brain's Knowledge System

by B. Bower

The peculiar inability of a 70-year-old woman to name animals has led scientists to propose that the brain harbors separate knowledge systems, one visual and the other verbal or language-based, for different categories of living and inanimate things, such as animals and household objects. Moreover, testing of the woman suggests that verbal knowledge about the physical attributes of members of a category exists apart from verbal knowledge about their other properties, according to neurologist John Hart Jr. and psychologist Barry Gordon, both of Johns Hopkins University in Baltimore.

The woman, referred to as K.R. by the researchers, suffered brain damage in both temporal lobes as well as patches of damage elsewhere.

K.R. showed no ability to name animals portrayed in pictures. Nor could she name animals based on recordings of easily recognizable sounds they make. Yet pictures and sounds associated with other living or inanimate things posed no problems for her.

K.R. also lacked the ability to identify physical attributes of animals, such as color or number of legs. For instance, when asked about the color of elephants, she claimed they are orange. However, she retained functional knowledge about animals, such as the realization that elephants are not kept as pets.

Further tests indicated that K.R. specifically lacked verbal knowledge about the physical attributes of animals. Her naming of animals did not improve with the aid of physical attributes as clues, such as a picture of an udder following a picture of a cow. But clues involving nonverbal perceptions of an animal, such as the sound "moo," significantly boosted K.R.'s naming accuracy.

She also correctly matched pictures of animal bodies to the appropriate heads and knew when an animal's picture portrayed the wrong color. But when asked, she still could not say which color belonged on, say, a lion pictured as gray.

A verbal system in K.R.'s brain must have mediated her intact knowledge of the functional properties of animals, while a separate visual system allowed her to recognize the physical attributes of animals in pictures, Hart and Gordon contend in the September 3 NATURE.

The verbal system contains "subdomains" of knowledge, they add, since K.R. could not identify visual physical attributes of an animal from its name, but could recall other information.

Areas throughout the brain probably coordinate these knowledge systems, the researchers conclude. □

Questions: *Answers on page 201*

1. How does the principle of brain mapping apply to this study?

2. Why does the brain remain the most mysterious organ of the human body?

3. Why are most regions of the cerebral cortex not mapped as specifically sensory or motor?

22

A new wave of advocates claim that a variety of smart drugs and nutrients improve human mental capacity. These substances supposedly improve memory, concentration, alertness, and problem-solving ability. Scientists do not support the claims of this new culture. Scientific proof to back up the claims is nonexistent. Nevertheless the ingestion of smart drugs and nutrients offers an attractive alternative to a growing number of Americans who feel that the traditional medical establishment is not meeting their needs. Pharmaceutical firms are reacting to their demand seriously, as they are currently investigating the development of over 100 cognitive enhancers. They hope that these substances may be used for the treatment of mental-based diseases, such as Alzheimer's, as well as for promoting general improved mental capacity.

Brain Gain: Drugs That Boost Intelligence
by Julie Erlich

For ravers, cyberpunks, hackers, phone phreeks, and technophiles treading the New Edge, reality is something to be enhanced, altered, or tailored to our needs. If you've got a computer, a modem, a data glove, who needs Fed Ex, faxes, or even sex? Just boot up and jack in to the matrix. You've got cyberspace at your fingertips. The only limitation is your brain, and tens of thousands will attest, there are ways to broaden your bandwidth, upload your memory, and upgrade your hardware. The millennium is almost here—and so are "smart drugs"(SDs).

He hands me a red plastic cup filled with something orange. It tastes like Tang. My first Think Drink. All it is, he assures me, is a mix of vitamins and fructose with a measure of choline, phenylalanine, and ephedra, and a "dash" of caffeine. "There's the left brain and the right brain, and this membrane which connects them called the corpus callosum," he says. He pops a piracetam from its tinfoil backing and offers me one. "Basically, we're taking about two computers in the brain linked by a data bus, and piracetam speeds up data transmission. These are drugs for the Information Age."

So they say: so he thinks; I don't know what to think. The drinks and the drugs are said to improve memory, concentration, alertness, problem-solving ability, as well as delay the cognitive effects of aging. Smart pharmaceuticals like piracetam, said to enhance cerebral metabolism, are prescribed in Europe, Japan, and China for stroke and memory impairment. Users here are calling them miracle drugs. The drinks, they say, "wake up your brain." They ought to. Ephedra (an herb), phenylalanine (an amino acid), and caffeine, are stimulants. Choline, a nutrient, is necessary for memory function. All are dietary supplements available at health-food stores.

In San Francisco, where the future happens first, kids dance all night to electronically synthesized house music at parties called raves, organized by groups like Toontown, Mr. Floppies Flop House, and The Gathering. Toontown's New Year's Eve celebration—7,000 people attended at $25 a pop-was advertised as a psychedelic apocalypse. The entertainment included a holographic gallery, a brain-machine room, virtual reality, interactive video screens, and the Nutrient Cafe run by Cat, who serves drinks with names like Intellex and Renew-You. (Other smart bars serve concoctions such as Psuper Psonic Psyber Tonic and Energy Elixir.) And Ecstasy, or MDMA, is the preferred chemical for opening the doors of perception. Ecstasy has a psychoactive effect that dissolves inhibitions, breaks down boundaries, and promotes feelings of interconnectedness and well-being.

Though rave culture has its roots in the sixties, the driving force is pure Nineties technoculture. The New Edge is where Haight-Ashbury meets Silicon Valley. It's a sensibility that owes as much to the possibilities of virtual reality and artificial intelligence as to Timothy Leary and *Neuromancer* author William Gibson.

What's really behind rave culture, Cat says, is "something spiritual which brings people together. It's kind of an internationalist, tribal, global consciousness. It's about information sharing within a global matrix." Cat's been using SDs since 1979, when he first moved to San Francisco. He'd been doing a lot of Ecstasy, smoking a lot of pot, and was beginning to feel burned out. A chemist friend suggested he try phenylalanine—a gram a day. It worked. He felt more energetic. Then he heard about tyrosine, lecithin, ginseng, choline, Ginkgo biloba. Now he uses them as well as Hydergine religiously. "Smart drugs help me focus. When I use Hydergine, I never feel scattered."

The popularity of smart drugs with Ecstasy-imbibing ravers and those who survived the Eighties with a history of drug use, might be explained in part by the theory that the brain's supply of neurotransmitters—particularly dopamine, serotonin, and norepinephrine, which have been depleted by Ecstasy, cocaine, and amphetamines—can be restored by their precursors, amino acids like tyrosine and phenylalanine.

The smart-drug movement grew out of the work of Life Extension gurus Durk Pearson and Sandy Shaw, who promote the use of nutritional supplements as a way to optimize mental function and neutralize immune-destroying, age-accelerating agents known as free radicals. The active ingredients of their powered drink mixes are a variety of amino acids. When amino acids combine, they form proteins that are necessary to the life of all cells and tissues of the body. Some amino acids and nutrients like phenylalanine and choline—both essential brain foods—are converted into the neurotransmitters norepinephrine and acetylcholine. (Neurotransmitters carry messages between brain cells.)

Durk and Sandy (as they're known) license their "designer mind foods" to companies including

Smart Products in San Francisco, and Texas-based Omnitrition, which sells franchises to the likes of West Coast marketer Jerry Rubin, and they're used as the basis for many of the Think Drink mixes sold at smart bars.

Like Jerry Rubin—and Timothy Leary—Durk and Sandy came of age in the Sixties and got rich in the Eighties. And while the Summer of Love may be over, Northern California provides fertile soil for its legacy. What began as an underground scene has evolved into a full-fledged movement. Smart-drug advocates run the gamut from computer nerds and cyberpunks to AIDS activists and life-extension enthusiasts. Users include students, yuppies, and 30- and 40-somethings who believe SDs give them a mental edge in the competition to absorb and process exponentially increasing bytes of information.

Now, Toontown has leased its own space, and when they host a rave, thousands attend. Smart bars are popping up from L.A. to New York, and books like *Smart Drugs and Nutrients* provide information for the uninitiated. *Mondo 2000*, a slick magazine with a cyberpunk slant, is a forum for New Edge consciousness.

Part Dada, part tech-head cybernetics, *Mondo* heralds an age when human and machine interact along a seamless continuum. Editor in chief R.U. Sirius (a.k.a. Ken Goffman), a self-described "decadent, soft-core, commercial anarchist," attributes *Mondo*'s success to the "cultural Zeitgeist—a kind of waiting for the apocalypse."

As we ooze toward the millennium, concepts like "getting back to nature," "consciousness expansion," and "free love" acquire a new level of meaning: consciousness, nature, and sex may one day soon be accessed via computer. They may have to be. The environment may be toxic; sex can be lethal. Biotechnol-

ogy and genetic engineering have spawned a new generation of possibilities for life extension and intelligence expansion. And the computer nerds of Silicon Valley are perfecting the technologies which will make possible the simulation of reality, the manipulation of biology, and increase our ability to process information.

"We can't cope with what we and technology are becoming without widening the bandwidth," Sirius says. "SDs widen the bandwidth." He believes these technologies are pulling us "toward something we want to become, something we've invented which will free us from biological control."

Mark Heley, one of Toontown's producers who came here from England a year ago, echoes Sirius. "Smart drugs and Virtual Reality are going to change the world," Heley says. They're concepts which are like time bombs. There's a cultural unraveling, and what's coming into existence is "post human—it's like we're becoming a different species." Heley believes the widespread use of pharmaceuticals will result in a major cultural change.

Enthusiasts are spreading the word. John Norgenthaler, co-author with Ward Dean of *Smart Drugs and Nutrients*, advocates the use of medical technology for enhancement purposes rather than for the treatment of disease. "Smart drugs—pharmaceuticals like vasopressin, Lucidril, Deaner—improve brain function. They improve the hardware." No, you won't have more profound thoughts, Morgenthaler admits, but you can improve brain-cell metabolism and oxygen availability. "It's not a software installation; it's better, faster, more powerful hardware." He ought to know—he's tried all of them.

A variety of nootropics (a class of cognitive enhancers, including piracetam and its analogs, which improve metabolism, memory, and attention) as well as other pharmaceuticals said to enhance mental functioning are marketed and sold abroad to combat the effects of senile dementia, Alzheimer's, and stroke. In Italy, for example, hospitals dispense piracetam from bubble gum machines, and people take it every day. Morgenthaler and Dean's book is a kind of users' guide, giving detailed information about how to order smart pharmaceuticals from other countries.

As a result of a loophole in the 1988 Food and Drug Administration's (FDA's) policy guidelines, which allowed the importation of non-FDA-approved drugs to combat AIDS, a number of "offshore buyers clubs" have been offering a wide variety of nonapproved pharmaceuticals as well as foreign versions of U.S. approved drugs to anyone with a check and a stamp.

And growing numbers of Americans, including people with AIDS and Alzheimer's disease, are turning their backs on the medical establishment and medicating themselves. In the process, they're bypassing the FDA, and the FDA isn't pleased. Nightline recently estimated that 100,000 Americans now use smart drugs. Currently, over 100 cognitive enhancers are under development by major pharmaceutical firms worldwide, including SmithKline, Parke-Davis, and Du Pont, to name just a few. Many hope to cash in on the billions an effective treatment for Alzheimer's would net. And new treatments like nerve-growth factor and human-growth hormone present tantalizing possibilities for those seeking the fountain of youth. (Nerve-growth factor stimulates the growth of new neurons and the regeneration of old ones. Human-growth hormone is being studied for its rejuvenative effects. See the sidebar "New Treatments.")

Smart pharmaceuticals such as Hydergine are often found to have therapeutic uses other than those for which they were originally approved. Though the FDA approval process mandates that drugs can only be marketed for their "approved uses," physicians do prescribe them for unapproved uses. Other pharmaceuticals which have been studied for their efficacy in treating memory impairment due to age or disease and touted by advocates for their cognitive-enhancing properties can have serious side effects. They include those originally approved in the United States to treat diabetes (vasopressin), Parkinson's disease (Deprenyl), and epilepsy (Dilantin).

Ward Dean, a gerontologist in Pensacola, Florida, is something of a renegade in the medical establishment. He believes the aging process is the one chronic, universally fatal disease that everyone over the age of 30 catches. Dean discovered nootropics in Korea where he attended medical school. Like Japan, Korea has a high incidence of stroke. "The first thing we did for stroke victims was give Hydergine and piracetam," he says, "and they reported increases in memory and other cognitive functions."

In his practice, Dean prescribes a variety of nutrients and smart drugs—primarily Hydergine and piracetam—to his patients. "They depend on the functioning of their brains," he says. "As they get older, they realize their memory isn't as sharp as it used to be—the ability to process new material is declining." To Dean, the FDA is the single greatest impediment to medical research in the country.

Take Mark Rennie's anecdotal testimony. Club impresario, lawyer, man-about-San Francisco, Rennie is also a partner in Smart Products along with Jim English. "When I started using SDs, it was like I had a dirty windshield and somebody finally cleaned it," he says. "We live in a toxic-waste dump, and we're not as clear as we used to be." Rennie's been using smart drugs—particularly Hydergine and vasopressin—for more than a decade. "I'm stacking up amino acids, I'm taking Deprenyl, and when I can get it, piracetam. I'll never quit taking Hydergine."

So what are the actual findings? Drugs like Hydergine and piracetam have been extensively researched and tested. The results are more encouraging in rats than in humans. Gary Wenk is a professor of neurology and psychology at the University of Arizona. He also performs independent testing for drug companies. Wenk claims that he's never tested a pharmaceutical for cognitive enhancement that he's found to be effective. The results, he says, are minimal at best, though some drugs like piracetam are "pretty innocuous."

Dean, however, argues that for those suffering from memory impairment, a small benefit is better than none. "It may be a 5-percent improvement," Dean says, "but if you're talking about competitive athletics or business a 5-percent edge can make the difference. If you're 5-percent sharper than the other guy, you're going to win.

Chief scientific officer at Cortex Pharmaceuticals in Irvine, California, Raymond Batus has been investigating cognitive enhancers for nearly 20 years. He believes, based on his testing, that for those with a deficit, particularly in the early stages of Alzheimer's, nootropics can help improve memory and attention. "Though the effects are fairly subtle and variable between patients," he says, "they're relatively safe, and they may improve quality of life." Bartus compares today's cognitive enhances to Piper Cubs before the age of jumbo jets. "They're doing something, but not in all patients. What the FDA's requiring is a drug that works so well that it goes beyond our ability to imagine it because the technology hasn't been invented yet." Besides, he adds, there are no other treatments on the market.

Researchers face a number of problems in testing a cognitive enhancer. Not only is there no consensus about what intelligence is, there are still no objective measurements to diagnose Alzheimer's. And the FDA hasn't been able to come up with a set of guidelines to help researchers meet approval standards for the scores of "antidementia" drugs now in development.

"We just don't know enough about the brain," Wenk says. But when we do, we will be able to design really effective cognitive enhancers. He predicts that Alzheimer's patients will benefit first, and then "we'll work our way back through the decades, and children will be popping enhancers from day one." Until that day, however, Wenk believes people need the protection of the FDA from the "charlatans out there looking to make bucks."

At the center of the controversy are the FDA regulations. The approval process, the most stringent in the world, requires that a drug be proven not only safe, but (since 1962) effective at treating a particular condition. Getting a new drug approved costs an estimated $231 million and requires about 12 years of controlled clinical trials. Once a drug is approved, manufacturers have patent as well as marketing protection for its approved use. Since the approval process only allows for drugs which treat a known condition, enhancement of normal function is not a category that is officially recognized. Furthermore, drug officials say, claims made by smart-drug producers and users are anecdotal and can be attributed to placebo effects.

Last January, the FDA announced an Import Alert, instructing FDA field officers to "automatically detain all imported drugs by six overseas companies" (including InHome Health Services and Interlab) who promote and distribute their products in the United States. Citing safety concerns as well as illegal promotions, the "embargo" includes both non-FDA-approved drugs (piracetam and Lucidril) and foreign versions of drugs approved here—Hydergine and vasopressin. Many of the targeted companies happen to be those that include smart drugs on their rosters.

The smart-drug community says the FDA policy is an attack on their civil liberties by an overweening bureaucracy. Steve Fowkes, editor of numerous newsletters, including *Smart Drug News*, is a Libertarian who's sure we'd be better off without the FDA. "The most common mental-deficit problem in the world is age-related mental decline. This is something the FDA says is not a disease." Fowkes believes that if there's a treatment for it, the FDA has no business telling people they can't have it. "The bottom line is, do these drugs work?" he says. "The FDA wants to be the only arbiter of that."

According to Fowkes, smart drugs are among the safest substances there are; they deal with performance enhancement, improving personal power, bringing more control into the hands of the individual. "And most economic institutions are fundamentally opposed to that," he says.

But does the public need protection from what the FDA cites are unsafe manufacturing conditions, unregulated directions for use, and unscrupulous "snake-oil salesmen" (as one FDA official described them) who make bogus claims and big bucks? Or is the FDA holding up the works, creating an unnecessary burden for the consumer, and, as some claim, protecting the market for American drug companies?

Don Leggett of the FDA's compliance office contends, "The firms listed in the Import Alert are promoting and selling drugs which are already approved here, which is in violation of FDA policy. And suppliers are advertising lower prices for them. The marketing of unapproved versions of these drugs is illegal." The FDA maintains that offshore companies offer drugs often of "unknown quality with inadequate directions for use and which may pose medical risks." Legget adds, "These people are self-medicating with prescription drugs, which can be dangerous."

Wenk, concerned about the potential profits involved, sees FDA protection as a benefit. "People are so eager, particularly with Alzheimer's, that they'll spend money on anything," he says. "We have to weigh the costs against the benefits. If something isn't efficacious, why spend money on it?"

What particularly irks the FDA are the claims made by marketers and promoters of SDs. Promoters of unapproved uses for prescription drugs, as well as nutrient marketers who make health claims, have been blatantly challenging the FDA's authority. Last March, in response to the amount of publicity smart drugs have generated, the FDA finally drafted a policy statement intended to clarify its position. Citing television appearances—for example, Jim English and John Norgenthaler on Nightline telling people about nutrients and smart drugs— and "word of mouth" promotion, the FDA targeted marketers and promoters of both pharmaceuticals and nutrients said to have cognitive-enhancing effects. Though stating that no injuries had been reported, the FDA warned; "Any product, regardless of its composition, that is clearly associated with SD claims, is illegal and subject to seizure or other actions."

The FDA also said it is evaluating strategies for regulating dietary supplements (including amino acids), a 2-billion-dollar-a-year industry.

While the FDA's task force evaluates the status and safety of nutrients and the claims being made about them, it has already sent investigators to Rennie and English's Smart Products' San Francisco offices. The Durk and Sandy products they license and distribute—Fast Blast, Memory Fuel, Power Maker—flirt with regulations against claim making. What's more, their newsletter, *IntelliScope*, is chock-full of articles and interviews combining testimonial and fact to promote the cognitive-enhancing and antioxidant benefits of their products. Whether or not their claims are true, the FDA is within their jurisdiction.

By law, a substance can be regulated as a drug when it claims a therapeutic effect and is therefore subject to FDA action. This applies to any written, oral, or visual promotion that implies an intended use for a product. Manufacturers and distributors often attempt to circumvent restrictions with brochures, newsletters, and ambiguous labeling that advertise healthful effects for their products. Though technically most vitamin and nutrient manufacturers could be considered in violation of the law, a gray area exists regarding what constitutes a health claim. Research reporting new therapeutic possibilities for substances from vitamin E to beta-carotene threatens to blur the distinction between drugs and nutrients and erode FDA regulations.

The FDA doesn't have to show that substances are dangerous to remove them from the marketplace; it has seized both beta-carotene and vitamin C for claims made about their effects. According to Leggett, the association of therapeutic claims with "nutrient" products is led by overzealous marketers. "They're not willing to put up the money, to perform the clinical investigations," he says. "It may be because they can't patent them; but if they make drug claims, they have to play by the rules."

Rennie thinks the real conspiracy is between the FDA and the pharmaceutical manufacturers. "The next American pharmaceutical firm to come up with a nootropic will make billions, and the drug companies want the market swept before that happens," he says. "The FDA testing process is a big con. And this happens in the home of the free. Hey, give us a break."

Whether or not playing by the rules is in the public's best interest is an open question. The publicity surrounding amino acids and their cognitive-enhancing potential may be so much hype, but a growing body of research suggests that amino acids may provide a nontoxic, nonaddictive alternative for those trying to overcome addiction to cocaine and amphetamines. Researchers at MIT and Harvard Medical School have also found that the amino acid tyrosine may be effective in treating depression. And Sigma-Tau, a pharmaceutical firm, is currently developing an acetylized version of L-Carnitine, an amino acid, to treat Alzheimer's.

For more than a year, research pharmacist at the Haight-Ashbury Drug Detox Clinic, Gantt Galloway, has been using combinations of amino acids—mainly tyrosine and phenylalanine—in open trials to treat cocaine and amphetamine dependence. "People report they have more energy, less craving; they feel better and they come back for more," he says. Galloway's impression is that the amino-acid combinations keep drug addicts in treatment longer than if they get no medication. He's currently trying to organize a double-blind study using tyrosine. "It's not on patent, though, so it's hard to get funding," Galloway says.

Our view of what we see as drugs and what we see as nutrients changes, and the way we use them is changing. As Wenk puts it, "Fifty years ago, people thought of vitamins as drugs. Now we pop them every day with our breakfast. A nutrient is a drug, and some drugs are actually nutrients. They're all chemicals." In order to keep pace with research findings, SD users contend, the FDA will have to alter its position and rewrite current regulations to reflect the changing technology—which could make marketing restrictions, the approval process, and placebo effects obsolete.

While advocates and users insist on the right to benefit from the smart technologies that exist and others dismiss the claims of cognitive enhancement as anecdotal, everyone agrees on one thing: The smart drugs of today are a window on the cognitive enhancers of the future.

"Smart drugs," says Sirius, "are an indication of the evolving knowledge of brain chemistry. Now the chemicals are awkward and crude. One day there will be a brain implant. You'll be able to push a button and release the chemicals you want." □

Answers on page 201

Questions:

1. Do you think that the ingestion of a smart drug can help the brain? Cite some examples of body physiology.

2. Is there a biochemical basis for brain function?

3. Compare the use of smart drugs to the benefits of a well-balanced diet.

Several means of signaling through the brain and other regions of the nervous system have been understood for a long time. The action potential is an electrical change that travels along the length of the nerve cell membrane. This wave of depolarization travels rapidly over space and time. The most common mode of signaling between neurons occurs at the synapse. Its nature is biochemical, existing in the form of a neurotransmitter that diffuses from the presynaptic membrane to the postsynaptic membrane. Aside from exciting the membrane of a postsynaptic cell biochemically, there is another from of signal transmission between neurons. It occurs at an electrical synapse or gap junction. However, it now appears that neurons can communicate by other means. Through volume transmission, chemical and electrical signals leave the usual track between cells and travel through the large fluid - filled spaces between nerve cells. Experimental evidence reveals that this mode of communication can travel over large distances through the brain when moving from cell to cell. Apparently volume transmission is another fundamental mode of communication between cells of the brain.

Volume Transmission in the Brain

by Luigi F. Agnati, Börje Bjekle and Kjell Fuxe

How does the brain process information? the common answer to the question relies on the same model that engineers have used to design electrical circuits and rail networks. Just as electrons flow along wires in a circuit, the neurons in the brain relay information along structured pathways, passing messages across specific points of contact called synapses. Information can no more leave the neuronal circuitry than a train can safely leave its tracks.

But there is increasing evidence that neurons can communicate without making intimate contact. The relaying of messages across synapses may be the fastest means of processing information, but it is quite likely that information often leaves the track. Some communication in the brain might be like a radio broadcast, where signals travel through the air and can be picked up by any properly equipped receiver. This mode of communication is called volume transmission. The medium of communication is the great fluid-filled space between the cells of the brain, and the carriers of the messages are chemical and electrical signals that travel through the space and can be detected by any cell with the appropriate receptor.

Over the past several years we have developed a theory of volume transmission in the brain. It is not an alternative to the traditional view, commonly called wiring or synaptic transmission, but a complement to that established theory. Neurons clearly relay signals efficiently and quickly across synapses; the processing of the words you are reading now is undoubtedly being handled in that way. But our experiments have shown that neurons also release chemical signals into the extracellular space that are not necessarily detected by neighboring cells but by cells far away, in the same way that hormones released by a gland into the bloodstream can have effects on cells far away. These processes occur on much longer time scales than does synaptic transmission, and they probably play a distinct role, perhaps regulating or modulating the brain's responses to synaptic signals.

In this article we shall trace the development of the traditional view of intercellular communication in the brain, and of our theory of volume transmission. By blending the two views neuroscientist may someday have a comprehensive view of how the brain processes information.

Wiring in the Brain

To a certain extent, the inside of the brain looks like a tangle of wires pulled from an old electronic computer. Nerve-cell processes wend their way along tortuous paths, weaving a three-dimensional fabric whose order is not immediately apparent. On closer inspection it becomes evident that the mission of the nerve processes is to connect one part of the brain with another. These connections form the basis for the current view of how information is relayed between neurons.

The connecting wires in the brain are axons, extensions from the cell's body that may range from a few micrometers to many centimeters in length. Electrical signals, or action potentials, traveling down the axon encode information by the frequency and the duration of their transmission. At its end, the axon swells to form a terminal bouton,

Reprinted by permission from AMERICAN SCIENTIST, Journal of Sigmi Xi, The Scientific Research Society.

where the electrical signal is converted into a chemical signal. The chemical signal, or neurotransmitter, is released from the neuron into a narrow (synaptic) cleft, where it diffuses to contact specialized receptor molecules embedded within the membrane of the target, or postsynaptic, neuron. Activation of the receptors in the postsynaptic neuron can then transduce the chemical signal by opening channels that admit ions, changing the electrical potential of the cell's membrane. This transduction process may ultimately result in exciting the postsynaptic neuron to send action potentials along it axon, or it may inhibit the neuron from sending such signals down the axon.

A key element in the transfer of information between neurons is the chemical synapse. The synapse is a highly specialized interface where the distance between the presynaptic and the postsynaptic membrane may be as little as 30 nanometers. It is this distance that a neurotransmitter must cross in the standard synaptic interaction. The molecular events involved in the release and the diffusion of the neurotransmitter across this synaptic cleft impart a delay of about five milliseconds before the postsynaptic neuron responds. This may be a brief period on the scale of daily events, but it is about 10,000 times slower than the speed at which the action potential travels down the axon.

There is another common form of signal transmission between neurons, where the delay is negligible: at an electrical synapse, or gap junction. In these synapses, the membranes of two neurons are applied directly against each other, only a few nanometers apart. Specialized channels in the membranes of the two neurons form a link across this gap, so that the cytoplasmic contents of the two neurons are confluent, but still excluded from the extracellular space. The size of the channels in a gap junction limits the size of the molecules that may pass across an electrical synapse. Consequently the signal typically consists of ions that flow through the channels, from one cell to another. In the chemical synapse the neurotransmitter always flows from the presynaptic to the post synaptic neuron, but the transmission in an electrical synapse is usually bidirectional.

The action potential, the chemical synapse and the electrical synapse form the basis for cellular communication in the wiring-transmission scheme of the brain. Information flows from one point to another through specialized structures that are designed solely for signal transmission. This view of the brain as a wired structure is widely accepted among neuroscientist. Indeed, there is little doubt that these "wiring" mechanisms play a fundamental role in the processing of information in the brain. But it is significant to our story to note that this was not always the dominant view.

In the late 1800s, a passionate debate arose concerning the integral structure of the brain. Earlier in the century, the German scientists Matthias Jakob Schleiden and Theodor Schwann had proposed that all living things were made of distinct units called cells. According to their theory, each cell was essentially a membrane-bounded bag containing a nucleus. Although 19th-century biologists widely accepted cell theory, most students of the nervous system failed to appreciate that the brain was also made of cells. Instead the brain was imagined to be a large confluent network of protoplasm in which each nerve cell body was merely a nodal point.

This reticular doctrine, as it came to be called, was actually a reasonable scheme in the 19th century. After all, the microscopic techniques of the era could not resolve the presence of the cell's delicate plasma membrane — scientists had no means of espying the synapse. (The membrane itself remained invisible until the advent of electron microscopy in the 1950s.) But the reticular doctrine was not merely a consequence of the anatomist's techniques; there was also a strong conceptual foundation at the heart of the theory. Many neuroanatomists thought it more likely that a signal could be conveyed by a continuous process, than interrupted and somehow regenerated between cells. The Italian anatomist Camillo Golgi espoused the reticular doctrine because it suggested that the nervous system operated in a holistic manner, enabling neurons to communicate over relatively large distances.

An opposing view — the neuron doctrine — held that the brain was made of discrete cellular entities that only communicated with one another at specific points. Although the German anatomist Wilhelm Waldeyer was the first to apply cell theory to the brain, in 1891 (he suggested the term neuron), it was the Spanish neurohistologist Santiago Ramón y Cajal who aggressively championed the neuron doctrine. Through many years of intense labor, and a series of brilliant deductions, Cajal amassed volumes of circumstantial evidence in favor of the neuron doctrine. Ironically, much of the evidence brought forth by Cajal was based on a histological technique developed by Golgi. The Golgi method revolutionized neuroanatomy by allowing a microscopist to see an unobscured brain cell in its entirety. Cajal's work, using the Golgi method, lies at the root of the wiring-transmission view of the brain.

It is in this historical context that we might consider the notion of volume transmission. For all of

Camillo Golgi's contributions to cellular biology (including the discovery of the Golgi apparatus), he is often chastised for his steadfast support of the reticular doctrine. But in some respects, the volume-transmission mode of neural communication is a vindication of Golgi's adherence to the reticular doctrine. In both views there is a holistic conception of the brain: Neural information flows along confluent pathways between relatively large, cellular territories, and not only at specific points between individual cells. With respect to the function (rather than the structure) of the brain, Golgi's view may prevail after all.

Leaks in the Brain

In the wired transmission of information, a signal flows along well-defined pathways. The method is reliable and rapid. In contrast, the volume-transmission mode is predicted on the passage of information in the "cracks" between neurons. Although a signal may be released from a particular site, and affect a specific target, there is no linear route for its passage; rather, there is a space for the diffusion of the message.

This space is the extracellular fluid that bathes the neurons. By one estimate, the extracellular space occupies about 20 percent of the brain's volume (Van Harreveld 1972). Of course the extracellular space is not empty; it is filled with many types of molecules — ions, proteins, carbohydrates and so on. This neuronal bath has been envisioned as playing a generally passive role in neural function, the molecules merely a maintenance crew that provides a milieu conducive to the proper functioning of the cells. This is not the case in volume transmission. Here these extracellular molecules are active players taking part in the relay of signals.

Information can be carried by extracellular molecules in either of two ways: by electrical currents or by chemical signals. In general, electrical effects in the extracellular space are thought to be a consequence of the movement of ions — such as potassium, calcium and sodium — across the neuronal membrane. The action potential is one example of ion movement. As the action potential travels down the axon, large numbers of ions cross the axon's membrane, affecting neighboring neurons. When many neurons exhibit action potentials at the same time, it can give rise to relatively large currents that can produce detectable signals. The electroencephalogram (EEG) is a record of the waves of extracellular currents that arise from the simultaneous discharge of many neurons in the cerebral cortex.

One advantage of this passive flow of ions is that it requires no energy beyond that expended during the neuron's action potential (Shepherd 1988, Agnati et al. 1988). Although their effects are generally diffuse, surrounding cells can sometimes dictate the path of the current. One method is supplied by the supporting cells in the brain, the neuroglia. These cells can make boundaries, with high electrical resistance, by repeatedly layering their plasma membranes. As a consequence ionic current is forced to cross the membranes of neighboring neurons. Other factors that can determine the flow of current include the ionic content of the extracellular fluid, and the nature and number of the ion channels in the neuronal membranes. The flow of extracellular ionic current generated by nerve cells has been called volume conduction (Martin 1985), a term that suggested to us the name volume transmission (Agnati et al. 1985, 1986) for the more general phenomenon that also includes the movement of chemical signals through the extracellular space.

Just as the ionic current can flow into the extracellular space, the chemical mode of volume transmission involves the release of a neuroactive substance from a neuron into the extracellular fluid, where it can diffuse to other neurons. The relay of chemical signals in the volume-transmission mode has some similarities to the release of a hormone from a gland into the bloodstream. In both instances, cells can communicate with each other without making intimate contact by releasing molecular signals into a fluid medium. As in the hormonal system, the release of a humoral signal in the brain can be used to regulate the cell that releases the signal (autocrine control), a neighboring cell or a distant cell. Humoral signals in the brain may also reach a distant target — in the spinal cord, for example — by diffusing through the cerebrospinal fluid, the clear liquid in which the brain floats.

There are a number of factors that can affect the distance a substance travels in the extracellular space. Charles Nicholson and Margaret Rice of the New York University Medical Center have made explicit some of the general principles that could affect the diffusion of a molecule in the extracellular space. One of these is the tortuosity of the brain tissue. For example, if a molecule frequently encounters obstructions, such as a cell's membrane, the course of its travels will be significantly slowed. The molecular weight of a substance also appears to affect its diffusion, but not in any simple way. The potassium ion and the calcium ion, for example, have similar weights but different rates of diffusion. The other significant factor is the relative proportion of space occupied by the extracellular fluid compared to that occupied by the cells of the brain. Although the extracellular space is thought to occupy about 20 percent of the brain's volume, this may not be a constant value in all regions of the brain. We would also add to this list the presence of extracellular molecules that can bind, destroy or otherwise limit the diffusion of a signal molecule.

The functional link between the nervous system and the endocrine system is especially important to our argument. For one thing, a well-established precedent for the presence of humoral signals in the brain already exists. Several steroidal hormones, such as testosterone and estradiol, enter the central nervous system, where they exert their effects on neurons in various parts of the brain. There is little question that these hormones reach their targets by diffusing considerable distances through the brain. But neurons are not merely targets of hormones; they also release substances into the bloodstream. For example, neurons in the posterior pituitary release ten humoral factors oxytocin and vasopressin, which affect target cells in distant parts of the body. Since the mechanisms for the release and the response to humoral factors are already established in the brain, one might suspect that neurons also interact with one another in this manner.

Many molecules that act as humoral signals in other parts of the body are also found in the brain. More than 50 such molecules — short chains of amino acids called neuroactive peptides — have been found in the brain. Neuroactive peptides, such as beta-endorphin, substance P and neuropeptide Y, do not affect their target neurons in the same way as the classical neurotransmitters do. Instead, neuroactive peptides are thought to have a modulating role, affecting not only the activity of ion channels in the plasma membrane, but also enzymes in the cytoplasm and gene expression in the nucleus. These processes are typically enacted through a class of receptors coupled to G-proteins, which themselves activate second messengers inside the cell. A neuroactive peptide may thus have long-term effects on the relative excitability of the target neurons, or it may affect the cell's metabolism and growth.

It is often the case that one or more neuroactive peptides and a small neurotransmitter coexist within the same nerve terminal (Hökfelt et al. 1982.) The small neurotransmitters are usually stored within small synaptic vesicles near the active site of a nerve terminal: neuroactive peptides, on the other hand, are usually found within large granular vesicles, away from the active site. In these instances the small neurotransmitter may be acting primarily in the standard synaptic interaction, whereas the peptide may be acting mainly as a humoral signal. The coexistence of multiple neuroactive peptides within a nerve terminal may reflect the need for creating multiple channels for volume transmission. Since each co-transmitter has specific receptors, only those neurons that carry appropriate receptors can detect the peptide signal. As a result, a single neuron can have an effect on several types of neurons without making contact with them.

But the process goes further. Once in the extracellular space, neuroactive peptides can be accosted by enzymes specifically designed to split the signal molecules into smaller fragments. In some cases the enzymes limit the region of the signal's effectiveness by reducing the peptides to their inactive components. In others the peptide fragments are themselves active, stimulating or inhibiting the receptor for the original peptide. The peptide fragments may even have their own receptors, which can affect the receptor of the parent peptide through receptor-to-receptor interactions.

The coexistence of neuroactive substances, and the fragmentation of peptides into a cascade of active molecules, together suggest that the release of signals from a few neurons could result in an elaborate spatial and temporal pattern of neuronal activation over relatively large regions of the brain. In total, there may be hundreds of neuroactive substances and active fragments in the extracellular fluid. In conjunction with the flow of extracellular ions, the impact of volume transmission may be extremely complex.

Experimental Evidence
We have briefly reviewed the chemical and electrical mechanisms that might be involved in volume transmission. But what evidence is there that the phenomenon actually takes place i the brain? Are neuroactive substances found outside the synapse? Can these molecules travel appreciable distances in the extracellular space without being inactivated or destroyed? Over the past two decades, a number of experiments have been performed that support the hypothesis of volume transmission.

The best evidence for the diffusion of ions in the brain comes from recent work by Nicholson and Rice (1991). Nicholson and Rice used microelectrodes that were specifically sensitive to calcium and potassium ions to show that these ions could diffuse several hundred micrometers in about 100 seconds. Their results argue that some of the effects of ionic currents are confined to local groups of neurons, and that they take place over relatively long periods of time. Because of these limitations, extracellular ionic currents may have their greatest effects on cellular microcircuits — parts of two or more neurons in close contact tat integrate synaptic information (Rakic 1875, 1979; Agnati et al. 1981). Since these structures appear to be especially sensitive to low-intensity signals (Metzler 1977), they may be capable of detecting and integrating electrical signals in the extracellular space.

The first indications that neuroactive substances could be found in the extracellular spaces of the brain came in the 1960s. It was found that amphetamine drugs can promote the release of catecholamines, a class of neurotransmitters that includes norepinephrine and dopamine, into the extracellular space (Fuxe and Ungerstedt 1968). Moreover, when catecholamines are injected into the cerebrospinal fluid within the ventricles of the brain, the molecules can diffuse into the surrounding brain tissue. (Glowinski and Axelrod 1966, Fuxe and Ungerstedt 1968). Another neurotransmitter, serotonin, can also prevail in the extracellular fluid when it is prevented from returning to the neuron after its release (Carlsson et al. 1969). these early studies hinted that the catecholamines and serotonin (collectively called monoamines) can exist for at least brief periods in the extracellular space without being destroyed.

The electron microscope has also provided anatomical evidence suggesting that monoamines act as humoral signals in the brain. Laurent Descarries and his colleagues at the University of Montreal have shown that many of the monoamine nerve terminals in the cerebral cortex lack the specializations characteristic of a chemical synapse. If these terminals are not releasing the neurotransmitter at a synapse, is the substance being released as a humoral signal? Intriguingly, the binding sites for these monoamines are located several micrometers from the terminals that release them (Descarries, Séguéla and Watakins 1991). In conjunction with physiological evidence that norepinephrine has widespread effects on the metabolism of terminal boutons

(Mobley and Greengard 1985), this anatomical evidence strongly suggests that the chemical signal may travel beyond the confines of a traditional synapse.

Electron micrographs of nerve terminals that contain neuroactive peptides also hint that these molecules may act as humoral signals. In many instances, the large granular vesicles containing these peptides appear to fuse with the plasma membrane, away from the active site of the synapse, and secrete their contents into the extracellular space (Thureson-Klein and Klein 1990, Matteoli, Reetz an DeCamilli 1991). Frequently, the peptide's receptor on the postsynaptic neuron is located at some distance from the synaptic site; in other cases the postsynaptic neuron may not carry a high-affinity receptor for the peptide at all (Dana et al. 1989).

We have also investigated the distance between the location of the neuropeptide-bearing terminal and the peptide's receptor with the light microscope. Nerve terminals that contain a particular neuroactive peptide can be identified with fluorescently labeled antibodies that bind specifically to the signaling molecule. The location of the peptide-sensitive receptors can be determined by labeling the peptide with a radioactive isotope such as iodine-125, applying the labeled peptide to brain slices, and allowing it to bind to its receptors in the tissue. By comparing the locations of the terminal boutons containing the labeled peptide, and the receptors labeled with the radioactive isotope, the distance between the terminals and the receptors can be determined.

It turns out that the mismatch between the location of the transmitters and their receptors as visualized in the light microscope is even more dramatic than the disparity suggested by the electron microscope. We have used this technique to

determine that relative locations of neuropeptide Y and its receptors in various regions of the brain. We found that many instance were labeled terminals were unaccompanied by a receptor, as well as cases where labeled receptors were distant from labeled terminals. In many cases the receptors were millimeters away from terminals that contained the peptide. The distance between the site where a peptide is released and its target receptor can, therefore, be about one million times the distance across the synaptic cleft. Remarkably, such distances were the rule rather than the exception.

Can neuropeptide Y travel in the extracellular fluid from the site of its release to the receptor, several millimeters away? To test this possibility, we injected small amounts of neuropeptide Y that had been labeled with iodine-125 into various regions of the brain. We found that the labeled molecule diffuses at the rate of about 1 to 1.5 millimeters per hour, depending on the direction of its travel. Clearly, if neuropeptide Y does act as a humoral signal in the brain, the course of its action must be considered to be very slow, on the order of hours rather than the few milliseconds it takes for a classical neurotransmitter to cross the synaptic cleft.

We also performed experiments using unlabeled peptides, such as the opioid beta-endorphin and the pituitary hormone prolactin. When these molecules were injected into that part of the brain called the striatum, they not only diffused long distances, but they also appeared to be taken up by nerve processes and transported back to the nerve-cell body several millimeters away. It seems likely that the cells ingest these peptides through some form of uptake mechanism. The transport of these substances back to the cell body offers the possibility that these molecules have some effect on gene expression in the nucleus. These observations are intriguing because they suggest that peptides in the extracellular fluid need not rely solely on diffusion to act on distant neurons.

Other methods of delivering prolactin into the brain also result in its diffusion through the extracellular space. Surgical implants of the anterior pituitary into the striatum release prolactin into the surrounding tissue. When we examined the tissue with the aid of a confocal laser-scanning microscope, we found that prolactin was indeed diffusing through the extracellular space of the striatum. We observed, however, that the diffusion of prolactin can be reduced with the drug bromocriptine, which specifically binds to certain dopamine receptors (of the D2 type) on the transplanted pituitary cells. Activation of these receptors is known to inhibit the release of prolactin from the pituitary cells.

Could the nerve terminals in the striatum that naturally release dopamine be inhibiting the amount of prolactin released from the transplanted pituitary cells? One way of testing this possibility would be to remove the dopamine terminals in the striatum and observe the effects on the diffusion of prolactin from the transplant. It is possible to do this by injecting a neurotoxin into the substantia nigra, the region that supplies the dopamine-releasing terminals in the striatum. When we performed this experiment we found that the absence of the dopamine terminals increased the amount of prolactin that is released from the transplant, allowing the molecule to diffuse greater distances. Since the dopamine terminals were not making synaptic contacts on the transplanted pituitary cells, the neurotransmitter must have reached these cells by diffusing in the extracellular space. It may be that dopamine is released naturally as a

humoral signal in the striatum.

These experiments hint at some of the mechanisms involved in the course and treatment of Parkinson's disease. One of the most persistent findings in this ailment is the loss of neurons in the substantia nigra, the same region of the brain that supplies the dopamine-releasing terminals to the striatum. the onset of Parkinsonian symptoms, such as tremor and slowness of movement, does not occur until about 80 percent of the dopamine has been lost from the striatum. One possible explanation for this observation is that the remaining dopamine terminals are releasing the neurotransmitter as a humoral signal in the striatum. The humoral release of dopamine would also explain how surgical treatments involving transplanted dopamine neurons in the striatum can alleviate the symptoms of Parkinsonism (Bjekle et al. 1991).

Conclusion

We propose that volume transmission is a fundamental mode of communication between cells in the brain It is a slow form of communication in which cells influence each other overlong periods of time, rather than the fleeting effects typically associated with standard synaptic transmission. Information is distributed to a general region of sensitive cells, rather than a specific set of neurons that make up a circuit. As a result, volume transmission may be involved more in regulating the general level of activity of a circuit than in determining what that activity is.

We might speculate that volume transmission is involved in the neuroendocrine system and the central autonomic system. Changes in the activity of the brain during sleep and wakefulness, relative levels of alertness, mood and sensitivity to pain may be highly

dependent on volume transmission. Thus, although information regarding the location of pain is carried by the circuitry of the nervous system, the intensity and the duration of the pay may be somewhat modulated by the ambient humoral signals. In this respect, acupuncture may also be a phenomenon that is dependent on volume transmission.

The volume-transmission mode of communication may affect cells other than neurons in the brain. IN a region where neurons are releasing peptides or other types of neurotransmitters, a wide-spread signal would allow for a coordinated response from neuroglial cells and cells that regulate the dilation of blood vessels, some of which have receptors for neuroactive substances. In this way neuronal activity can be associated with an appropriate change in energy metabolism and blood flow.

There are other hypotheses for non-synaptic communication in the brain that are consistent with our theory of volume transmission. In 1984 Francis Schmitt of the Massachusetts Institute of Technology proposed a larger view of neuroactive substances. he suggested that the behavior of a neuron is regulated by several classes of informational molecules — including transmitters, hormones and growth factors. Some of these substances could be relayed between neurons through the standard circuitry, but others may be relayed in a parasynaptic system, in parallel with the synapse. In this way, neurons not only modulate one another's electrical and metabolic activity, but also one another's growth and survival.

Also in 1984, E. Sylvester Vizi of the Institute of Experimental Medicine in Budapest proposed that a molecular signal released into the extracellular space could act on the receptor of another neuron about one micrometer away. Vizi suggested that the humoral signal acts on a terminal bouton to inhibit the release of a classical neurotransmitter at a synapse. Such an effect is important because it may override the signal to release the neurotransmitter that is delivered by the action potential. Although such presynaptic inhibition has long been known to take place between two neurons through a standard synaptic contact, such an immediate effect for a humoral signal suggests that in certain instances, volume transmission may also have brief and rapid effects on a circuit's activity.

Our work also raises the question of how volume transmission and wiring transmission interact as modes of intercellular communication. What does volume transmission signify for network models, which assume that the brain is merely a series of wired circuits? How can these models take into account the action of a diffuse humoral signal? These questions are especially critical because network models are typically based on an idealized neuron that has specific properties governing its behavior, including its tendency to respond and its connections to other neurons. Ultimately, it may not be possible for rigid neuronal networks to model the brain if volume transmission plays a significant role.

Finally, although most neuroanatomical studies aim to describe the circuitry of the brain, it is the structure of the extracellular space that is important for understanding the flow of information in volume transmission. The space is defined not only by the boundaries of the neuronal and glial membranes, but by the molecular composition of the space. For example, ions in the space may act as buffers, and large molecules may be limited in their movements. New research efforts might be directed toward understanding the extracellular space i the

brain and how it may limit the diffusion of electrical and chemical signals. In this regard, a number of question remain to be answered. Given a medium with certain properties, what is the maximum distance a specified number of neuroactive molecules can travel to activate particular receptors? (Nicholson and Rice 1991). Is the diffusion of a certain neuroactive substance the same in all directions? Do active processes facilitate or inhibit the movement of extracellular molecules? We hope that our work will stimulate further experiments and theoretical considerations. □

In Memoriam
This article is dedicated to the memory of Professor N.Å. Hillarp (1916-1965) and Professor Francesco Infantellina (1916-1991) in honor of their scientific achievements, deep humanity and great culture.

Acknowledgement
This work has been supported by a grant (04X-715) from the Swedish medical research council.

Bibliography
Agnati, L.F., K. Fuxe, M. Ferri, F. Benfenati, S.O. Ögren. 1981. A new hypothesis on memory. A possible role of local circuits in the formation of memory trace. Medical Biology 59:224-229.

Agnati, L.F., K. Fuxe and M. Zoli. 1985. Considerazioni sulle relazioni tra sistema mervoso centrale e sistema endocrino. Crescita 14:34-38.

Agnati, L.F., K. Fuxe, M. Zoli, I. Zini, G. Toffano and F. Ferraguti. 1986. A correlation analysis of the regional distribution of central enkephalin and b-endorphin immunoreactive terminals and of opiate receptor in adult and old male rats. Evidence of the existence of two main types of communication in the central nervous system: the volume transmission and wiring transmission. Acta Physiologica Scandinavica 128:201-207.

Agnati, L.F., M. Zoli, E. Merlo Pich, M. Ruggeri, K. Fuxe. 1988. The Emerging Complexity of the Brain: Limits of Brain-Computer Analogy. In Traffic Engineering for ISDN Design and Planning. ed. M. Bonatti and M. Decina, 209-230, Amsterdam:Elsevier Science Publishers B.V.

Bjelke, B., L.F. Agnati and K. Fuxe. 1991. Experimental evidence for volume transmission by analysis of host-graft interaction using intrastriatal adenohypophyseal transplant in the rat in combination with three-dimensional reconstruction. In Volume Transmission in the Brain, eds. K. Fuxe and L.F. Agnati, 463-478. New York: Raven Press.

Carlsson, A., J. Jonason, M. Lindqvist and K. Fuxe. 1969. Demonstration of extraneuronal 5-hydroxytryptamine accumulation in brain following membrane-pump blockage by chlorimipramine. Brain Research 12:456-460.

Dana, C., M. Vial, K. Leonard, A. Beauregard,P. Kitabgi, J.P. Vincent, W. Rostene and A. Beaudet. 1989. Electron microscopic localization of neurotensin binding sites in the midbrain tegmentum of the rat. I. Ventral tegmental area and the interfascicular nucleu. Journal of Neuroscience 9:2247-2257.

Descarries, L., P. Séguéla and K.C. Watkins. 1991. Nonjunctional relationships of monoamine axon terminals in the cerebral cortex of adult rat. In Volume Transmission in the Brain, eds. K. Fuxe and L.F. Agnati, 53-62. New York: Raven Press.

Duggan , A.W., P.J. Hope, B. Jarrott, H.-G. Schaible and S.M. Fleetwood-Walker. 1990. Release, spread and persistence of immunoreactive neurokinin A in the dorsal horn of the cat following noxious cutaneous stimulation. Studies with antibody microprobes. Neuroscience 35:195-202.

Fuxe K., and L.F. Agnati. 1991. Two principal modes of electrochemical communication in the brain: volume versus wiring transmission. In Volume Transmission in the Brain: Novel Mechanisms for Neural Transmission, eds. K. Fuxe and L.F. Agnati, pp. 1-9. New York: Raven Press.

Fuxe, K., L.F. Agnati, M. Zoli, B, Bjekle and I. Zini. 1989. Some aspects of the communicational and computational organization of the brain. Acta Physiologica Scandinavica 135:203-216.

Fuxe, K. and U. Ungerstedt. 1968. Histochemical studies on the distribution of catecholamines and 5-hydroxytryptamine after intraventrciular injections. Histochemie 13:16-28.

Glowinsky, J., and J. Axelrod. 1966. Effects of drugs on the disposition of ³H-norepinephrine in the rat brain. Pharmacological Reviews. 18:775-785.

Hökfelt. T., I.M. Lundberg, L. Skirboll, O. Johansson. M. Schultzberg and S.R. Vincent. 1982. Coexistence of classical transmitters and peptides in neurons. In Co-transmission, ed. A.C. Cuello, 77-126. London and Basingstoke: MacMillan.

Martin. J.H. 1985. Cortical neurons, the EEG and the mechanisms of epilepsy. In Principles of Neural Science, eds. E.R. Kandel and J.H. Schwartz, 636-647. Amsterdam:Elsevier.

Matteoli, M., A.T.Reetz and P. DeCamilli. 1991. Small synaptic vesicles and large dense-core vesicles: Secretory organelles involved in two modes of neuronal signaling. In Volume T Transmission in the Brain, Eds. K. Fuxe and L.F. Agnati, 180-194 NY:Raven Press.

Metzler, J. 1977. Mental transformations. A top-down analysis. In Systems Neuroscience, Ed. J. Metzler, 1-24. New York:Academic Press.

Mobley, P., and P. Greengard. 1985. Evidence for widespread effects of noradrenalin on axon terminals in the rat frontal cortex. Proceedings of the Natural Academy of Sciences 83:945-947.

Nicholson, C. and M.E. Rice. 1991. Diffusion of Ions and Transmitters in the Brain Cell Microenvironment. In Volume Transmission in the Brain: Novel Mechanisms for Neural Transmission, eds. K. Fuxe and L.F. Agnati, 279-294. New York: Raven Press.

Rakic, P. 1975. Local circuit neurons. Neuroscience Research Program Bulletin 13:3.

Rakic, P. 1979. Genetic and epigenetic determinants of local neuronal circuits in the mammalian central nervous system. In The neurosciences: Fourth Study Program, eds. F.O. Schmitt and F.G. Worden. 109-127. Cambridge, Mass.:The MIT Press.

Shepherd, G.M. 1988. Neurobiology. New York:Oxford University Press.

Thureson-Klein, Å.K. and R.L. Klein. 1990. Exocytosis from neuronal large dense-cored vesicles. International Review of Cytology 121:67-126.

Van Harreveld, A. 1972. The extracellular space i nthe vertebrate central nervous system. In: The Structure and Function of Nervous Tissue, ed. G.H. Bourne, pp. 447-511. Academic Press, New York.

Vizi, E.S. 1984. Non-synaptic Transmission Between Neurons: Modulation of Neurochemical Transmission. New York:John Wiley.

Answers on page 201

Questions:

1. How does signaling along a neuron differ from signaling between neurons? Use specific reference to neuron anatomy and physiology that you have learned.

2. How does neuronal signaling differ from signaling by the endocrine system?

3. What is the advantage of evolving volume transmission in the brain?

24

Consciousness is a product of the human brain, related to its role of processing information. Yet, very little is known about how this organ receives stimuli, recodes this input, and synthesizes it into the form of human thought. Within milliseconds, numerous synaptic circuits engage to launch a conscious experience in the human brain. This neuronal activity can be detected through an electrical recording, the EEG (electroencephalogram). However, how does a recording of alpha waves, beta waves, delta waves and theta waves correspond to the real events unfolding in the brain? In the last decade of the twentieth century the origin of a thought and the mechanism of human consciousness remain great mysteries, questions that are virtually unanswered through modern scientific techniques.

Consciousness on the Scientific Agenda
by Jeffrey Gray

How does the brain generate conscious experience? Twenty years ago, most scientists thought this a question for the philosophers; and most philosophers regarded it as a symptom of some kind (though what kind never became clear) of deep linguistic confusion. In sharp contrast to this picture, there was a large measure of agreement among both scientists and philosophers at a recent symposium* that not only is there a real problem of consciousness but that it is a scientific problem and that the time has come for scientists to tackle it.

Hardly anybody today doubts that consciousness is in some way a product of the brain, a product that is intimately connected with the brain's role in behavior and the processing of information. Cartesian dualism — the notion that brain-stuff and mind-stuff are essentially separate, though able to communicate with each other — has virtually no contemporary followers. Dualism has not, however, vanished without trace. It has left behind the notion of a 'Cartesian theatre' (located by

Ciba Foundation Symposium No. 174 on Experimental and Theoretical Studies of Consciousness, London, 7-9 July 1992.

Descartes himself in the pineal gland, the place where brain-stuff and mind-stuff were supposed to communicate): a privileged site at which brain events enter into conscious awareness, those happening elsewhere remaining in a kind of outer darkness.

This notion came under strong attack at this month's meeting. Instead, it was proposed (M. Kinsbourne, Sargent College, and D.C. Dennett, Tufts University, Massachusetts) that consciousness is a property of whatever pattern of distributed neural firing is (in some still obscure sense) 'dominant' at any particular time. This proposal, however, raises the so-called 'binding' problem: what feature of neuronal activity unifies the ramified patterns of individual neuronal events that take place in millions of separate cell-bodies and in even more millions of synaptic junctions at any given instant? And there is a temporal as well as a spatial aspect to the binding problem: neuronal events take place on a timescale measured in milliseconds, whereas the contents of consciousness have an apparent duration of at least two orders of magnitude greater.

(Perhaps the only important gap in the coverage of topics at the symposium was a consideration of recent proposals[1,2] that a solution to the binding problem may unexpectedly be provided by quantum mechanics.)

Also more or less absent from the contemporary scene are behaviorism and its philosophical equivalent, positivism; the notion that, because only behavior, not conscious experience, is amenable to direct observation, talk about conscious events has not place in the world of science. There has also been a more radical behaviorist view that holds, on the same grounds, that consciousness does not really exist. (One is tempted to add: it is 'just a figment of the imagination'; but that leads to some rather obvious problems.)

I once asked such a radical behaviorist what, on this view, is the difference between two awake individuals, one of them stone deaf, who are both sitting immobile in a room in which a record-player is playing a Mozart string quartet? His answer: their subsequent verbal behavior. Mercifully, there were no radical behaviorists at the symposium.

But behaviorism too has not

vanished without trace. Its modern offspring is a vigorous form of functionalism, which enshrines as the touchstone of the presence of consciousness the Turing test: if you feed symbols (for example, a string of English words) to a machine (or, more particularly, to a digital computer) and a human being and get symbols back from each of them in reply, and if you cannot distinguish between the machine and the person on the basis of their replies, then, if the human being is conscious, so is the machine. On this view, when the machine correctly uses the symbols for 'red', 'pain', 'itch' and so on, it makes no sense to ask further whether the machine has the sensory experiences ('qualia') that belong to these terms.

When this view is held strongly, it is known in the trade as 'strong AI' (artificial intelligence). A famous attack on strong AI — the 'Chinese room argument' — was made some years ago by John Searle[3] (University of California, Berkeley). Essentially, the argument demonstrated that computers have only syntax, not semantics; that consciousness is permeated with semantics (one is typically conscious of things such as 'a red rose situated on a table over there'; indeterminate itches and the like are the rarity); so computers cannot, just by computing transformations of symbol strings, achieve consciousness.

On the whole, Searle seems to have won his argument: at any rate, nobody at the meeting made a serious attempt to refute it. Instead, the current form of functionalism goes beyond the mere digital computer, allowing it now to have limbs and sense organs (that is, the computer becomes a robot). Moreover, proponents of the contemporary functionalist approach, unlike the earlier radical behaviorists, are willing — indeed, eager — to take account of what is known about events in the real brain that lie between input (stimulus) and output (response).

Dennett, for example, argued that one can already take what is known about the way the brain codes and recodes visual stimuli, put it together with existing data from psychophysics and use the knowledge that results to predict, successfully, new visual phenomena, such as illusory experiences. One can even 'inject' such experiences into the brains of experimental animals, as demonstrated in elegant experiments (W.T. Newsome, Stanford University) in which monkeys responded (behaviorally) to microstimulation of a circuit encoding a particular direction of motion in the same way that they had been trained to respond to an external stimulus having the same directional value processed by the retina and what lies behind it.

What more than this, Dennett asked, do we need? There is no theoretical problem about consciousness, we just have to go on gathering new data, and when we have all the details of what goes on in the brain and how this brain activity interfaces with the environment, that will be in and of itself a complete scientific account of consciousness.

Some, however — probably the majority at the meeting — remain unconvinced that it is going to be so simple. For them, more than just dedicated gathering of experimental data will to be needed. What is needed is, rather, a new theory (one comparable with, say, the theory of heat for the relation between the gas flame and the boiling of the kettle) that will render the relations between brain events and conscious experiences 'transparent', to use T. Nagel's (New York University) felicitous term. A series of brute correlations would not suffice. This, after all, is the standard set in all other domains of science, so why not here? This new theory is at present unimaginable, but only in the sense that no-one could have imagined relativity or quantum mechanics before *they* were invented, not because we are dealing with the unknowable or a bad language habit. ☐

Jeffrey Gray is in the Department of Psychology, Institute of Psychiatry, De Crespigny Park, Denmark Hill, London SE5 8AF, UK.

1. Penrose, R. The Emperor's New Mind (Oxford University Press, 1989).
2. Marshall, I.N. New Ideas in Psychology 7, 73-83 (1989).
3. Searle, J. Behavl. Brain Sci. 3, 417-457 (1980).

Question: *Answers on page 202*

1. Where do you think that the greatest discoveries will be made in neurophysiology in the next twenty years?

2. How do the association areas of the cerebral cortex relate to human consciousness and thought?

3. Why are behaviors such as reasoning, memory, and problem-solving so difficult to understand and study through research?

25

Until recently the measurements of muscle performance originated from isolated muscles that were surgically removed from the organism and studied in an artificial setting. Such experiments did not reveal the pattern of muscle activity naturally as an intact part of the living organism. New studies on bivalve mollusks, however, have supplied data on the action of muscles when the organism is swimming freely. By studying the action of the muscle that closes the two valves of a scallop, scientists have learned about the natural force and speed of muscle contraction. The force depended on the cross- sectional area of the muscle. The work of this slow - acting, powerful muscle was also calculated. It was found to be proportional to the volume of the organ. Muscle activity was investigated down to the level of its myofilaments and their response. Other data from the muscle performance of birds has added to information on the topic.

The Work That Muscles Can Do
by R. McNeill Alexander

By choosing a particularly suitable subject, Marsh and his colleagues (page 411 of this issue[1]) have been able to measure a muscle's natural performance. That is something of an achievement — experiments with isolated muscles have told us how strongly and how quickly many different muscles can contract, but it is much more difficult to discover how muscles perform in natural movements of living animals.

The subject used by Marsh *et al.* was the scallop, a bivalve mollusc that swims by clapping the two valves of its shell open and shut: a single joint is worked by a single muscle so that there is no ambiguity about the muscle's contribution. Marsh and his colleagues attached transducers to free-swimming scallops, to record the movements of their valves and the pressure inside them. From the records they calculated the length changes of the muscle, the force it exerted and the work it did in each contraction. The significance of this type of work was highlighted by A.V. Hill in a classic paper[2]. The forces that different muscles can exert are proportional to

their cross sectional areas, and the distances they can shorten are proportional to their lengths. Thus the work (force multiplied by shortening distance) done in a contraction is proportional to muscle volume. Different muscles have different intrinsic speeds and are capable of very different power outputs, but the work done per unit volume (or per unit mass) of muscle, in a single contraction, should be about the same for all muscles of similar composition.

How much should this work be? Vertebrate striated muscles, shortening very slowly, exert stresses up to about 3×10^5 Nm^{-2} (ref. 3). Other muscles with myofilaments of similar length can be expected to do about the same (but some invertebrate muscles with longer myofilaments can exert larger stresses). Muscles seem typically to shorten by 25% or less in normal movements (if they shortened much more, the stresses they could exert near the extremes of the length range would be greatly reduced). The density of muscle is about 1,060 kgm^{-3}. Thus muscles can be expected to do about $3 \times 10^5 \times 0.25 / 1,060 = 70 \mathrm{J}\ kg^{-1}$ in a

strong, slow contraction.

To maximize power output, the muscle must contract faster, at a rate at which the stress is reduced by a factor of 0.3 (ref. 3), making the work output only 20 J kg^{-1}. Thus we may expect vertebrate and other similar muscles to do 20 J kg^{-1} in each contraction when maximum power output is required, and up to 70 J kg^{-1} when the quantity of work from a single contraction is more important than the rate at which the work is done. Work-loop experiments with isolated fibre bundles have usually given values lower than 20 J kg^{-1}, although R.S. James (personal communication) finds values of up to 23 J kg^{-1} for the mouse extensor digitorum longus.

The scallop muscle investigated by Marsh *et al.* has myofilament lengths similar to those of vertebrate muscle and had a work output of up to 20 J kg^{-1} in each contraction. That is about what we should expect of a muscle contracting at the rate that maximizes power output. The best other data for muscle performance in intact animals come from wallabies and starlings with surgically implanted transducers. Griffiths[4] found that when a wallaby accelerated from rest the medial gastrocnemius muscle did about 30 J kg[-1] (from his Fig. 6D, using my own measurement of the mass of the muscle in a closely related species). Biewener and colleagues[5] attached strain gauges to the humerus of starlings which they then flew in a wind tunnel. They used films to record wing movements and were able to calculate that the work done by the muscle in each wing beat was only 7 J kg[-1].

These values contrast with much higher estimated work outputs in some more strenuous activities in which (unfortunately) the contributions of individual muscles have not been measured. The highest recorded jump of a bushbaby (Galago)[6], to an amazing 2.3 m, must have been powered by the leg and back muscles which are 24% of body mass[7] and must have done 88 J kg[-1]. Sprint cyclists reach up to 560 J

for each revolution of the pedals[8] using leg muscles totalling (probably) about 14 kg: this is 40 J kg[-1]. Hawk wing muscles have been estimated at 50 J kg[-1] for each wing beat of a fast climb[9]. The squid Illex jet propels with pressures up to 50 kN m[-2], driving out water amounting to 45% of body volume[10]. To achieve this, its mantle muscles (40% of body mass) must put out more than 50J kg[-1] of work. This performance can reasonably be compared with that of the vertebrate and scallop muscles, as all these muscles have similar myofilament lengths, but not with the well-known data for insect jumping (the insect muscles have longer myofilament).

We might expect to find muscles doing 20J kg[-1] in activities that require maximum power (for example, swimming or cycling) and 70 J kg[-1] when maximum work is required from a single contraction (such as jumping). By these stan-

dards, some of the work outputs recorded so far seem remarkably high, but none of the high values came from measurements of the output of the individual muscles. We need more measurements like those of Marsh and his colleagues to give us a general picture of how muscles perform in strenuous activities.

R. McNeill Alexander is in the Department of Pure and Applied Biology, University of Leeds, Leeds LS2 9JT, OK

1. Marsh, R.L., Olson, J.M. & Guzik, S.K. Nature 357, 411-413 (1992).
2. Hill, A.V. Sci. Progr. 38, 209-230 (1950).
3. Woledge, R.C. Curtin, N.A. & Homsher, E. Energetic Aspects of Muscle Contraction (Academic, London, 1985).
4. Griffiths, R.I. J. Exp. Biol. 147, 439-456 (1989).
5. Biewener, A.A., Dial. K.P. & Goslow, G.E. J. exp. Biol. 164, 1-18 (1992).
6. Hall-Craggs, E.C.B. J. Zool., Lond. 147, 20-29 (1965).
7. Grand, T.I. Am. J. phys. Anthropol. 47, 211-240 (1977).
8. Davies, C.T.M. & Sandstrom, E.R. Eur. J. appl. Physiol. 58, 838-844 (1989).
9. Pennycuick, C.J. Fuller, M.R. & McAllister, L. J. exp. Biol. 142, 17-29 (1989).
10. O'Dor, R.K. & Webber, D.M. J. exp. Biol. 160, 93-112 (1991).

Questions: *Answers on page 202*

1. How is the action of a muscle the result of the collective action of its parts? Refer to levels of organization of muscle structure.

2. How does the subject of physics pertain to this biology-based study?

3. Why do you think there is a lack of data on muscle activity in living human organisms?

26

Fish have the ability to see ultraviolet light, a part of the electromagnetic spectrum that humans cannot detect. Among other members of the animal kingdom, the honeybee and other invertebrates have this ability. Apparently birds can also see ultraviolet light. By perceiving ultraviolet light and using it to detect the polarization of light, these species are also aware of the changing position of the sun, using it as a cue for navigation. What allows fish to have these abilities? Also, why are these abilities important in their underwater world? Unlike the retina of the human eye, this innermost layer of the fish eye contains cone cells that are sensitive specifically to ultraviolet light. How this ability is useful remains unknown. Perhaps it is also a spatial cue used in fish migration. Further research is needed to solve this mystery.

Polarization Vision in Fish

by Craig W. Hawryshyn

Looking into the stolid eyes of a fish, one might think that the animal has only a limited view of its surroundings. But the phlegmatic countenance of a carp belies it perceptual abilities. The common goldfish can see a world of light that we can only attempt to imagine.

As a graduate student I was confronted with the challenge of understanding what a fish might see, and how it sees it. Some experiments I performed seems to suggest that fish are sensitive to ultraviolet light, a part of the spectrum that people cannot detect. Can a fish actually see ultraviolet light? If so, why would a fish need to see it?

I was not the first to be plagued by such questions and another species' exotic senses. We know well that other animals' experience of the world differs from our own. Bats can hear high frequencies of sound, which they use to locate objects in their environment. Snakes have a refined ability to sense infrared radiation, with which they detect the body heat of another animal. And pigeons use the earth's magnetic field as a navigational aid.

Each of these senses contributes to the mélange of sensations that the Dutch ethologist Nikolaas Tinbergen described as an animal's Merkvelt, or perceptual world. A perceptual world that includes ultraviolet light had been described in other species, most notably the honeybee and other invertebrates. Among vertebrates, only birds appeared to have the ability to see ultraviolet light. It seemed that a clue to understanding what it is that a fish sees might be found by turning to studies of these other species.

The earliest observations of an animal's ability to detect ultraviolet light were those of the Austrian ethologist Karl von Frisch in the 1940s. In his attempt to understand how honeybees navigate when foraging, von Frisch noticed that the bees communicated the location of a source of food by means of a dance that pointed in the direction of the goal. The considerable distance to the food suggested that the bees were not fixing their bearings on the local terrain. Indeed, by blocking the bees' view of the sun and the sky, von Frisch was able to prevent the proper orientation of the dance. One the other hand, if the bees were given a view of as little as one percent of the sky, they were able to communicate the location of the food accurately.

What cues in the sky did the bees use? Von Frisch placed a series of optical filters between the bees and the sky to determine whether certain wavelengths of light are more important than others. He found that filters that blocked ultraviolet wavelengths eliminated the bee dance. Filters that allowed ultraviolet wavelengths to pass, but blocked all others, did not affect the dance at all. Somehow the bees were using ultraviolet light. At the suggestion of a colleague, he then tried a polarizing filter. The results were striking: The orientation of the bee dance changed as von Frisch rotated the polarizing filter. Bees could detect the pattern of ultraviolet, polarized light in the sky.

Since the work of von Frisch, other investigators have shown that many invertebrate species and birds use the pattern of polarization, revealed by their perception of ultraviolet light, as an aid in navigation. Could it be that fish also use ultraviolet light as a means to detect polarization?

Initially, certain logic barriers prevented my colleagues and me

Reprinted by permission from AMERICAN SCIENTIST, Journal of Sigmi Xi, The Scientific Research Society.

from pursuing the idea. For one thing, it was widely believed that ultraviolet light could not penetrate water. Other studies suggested that the lens of the fish's eye absorbs all the ultraviolet light before it can reach the light-sensitive receptors. It turned out, however, that these impediments were nothing more than the human eye predicting what the fish eye could see.

It is now apparent that fish do perceive ultraviolet light, and use this ability to detect the pattern of polarization underwater. In the past few years my colleagues and I have conducted a series of investigations to understand how the fish's eye detects ultraviolet, polarized light, and how the brain encodes this information and translates it into the fish's behavior. Although we have only just begun, the results suggest that polarization vision is an important part of the fish's perceptual world.

Light in the Sky

When you or I look up on a sunny day, the only pattern we may observe is the smattering of clouds drifting across the blue sky. There is, however, an order to the light in the sky that the human eye fails to perceive. It is a phenomenon that was discovered accidentally by the French physicist Etienne Malus in 1809. Malus had looked up at the sky through a special type of crystal and noticed that the light was polarized. Although he did not fully appreciate what he saw, the phenomenon is now well understood.

The polarization of light is best explained by acquiring some appreciation of the physical properties of light. As a form of electromagnetic radiation, light can be produced when there is a change in an electric or magnetic field. For example, the electromagnetic field between a pair of oppositely charged particles, or dipole, will remain constant if the particles do not move

relative to each other. However, if the two particles begin to oscillate, moving together and apart repeatedly, a wave of electromagnetic energy emanates from the pair in all directions perpendicular to the line connecting the particles. Such a wave consists of an electric field and a magnetic field.

The electric field oscillates in a direction parallel to the axis of the oscillating particles, whereas the magnetic field oscillates at right angles to the electric force. The orientation of the electric field is described by a vector that is perpendicular to the direction in which the wave propagates and is called the electric vector or E-vector. It is the E-vector orientation that appears to be important in the detection of polarization by fish.

There are countless numbers of oscillating dipoles in the sun. The complex mixture of dipoles guarantees that light emitted from the sun is oriented in all direction, so that many different E-vector orientations strike the earth's atmosphere. Some of this light is able to pass directly through the atmosphere to the surface of the planet. Much of the light that reaches the earth's atmosphere, however, strikes it at an oblique angle. Some of this light is directed out into space again, but some of it is redirected to the earth's surface by particles in the atmosphere.

The light that is scattered to earth from the atmosphere does not contain a random assortment of E-vector orientations. This is because the light's path has changed direction: The only E-vector orientations that reach the surface are those that are perpendicular to both the path of the light from the sun to the atmospheric particle, and also perpendicular to the path of the light from the particle to the earth's surface. Other E-vector orientations are directed out to space. Since only certain orientations are able to reach

the surface, scattered light is said to be polarized.

The earth's atmosphere not only polarizes the light from the sun; it also permits the passage of certain wavelengths to the surface in preference to others. Whereas the sun emits radiation that ranges in wavelength from less than one millionth of a nanometer to over 1,000 kilometers, most of the solar energy arriving at the earth's surface has wavelengths in the limited range of 300 nanometers to 1,100 nanometers.

Within this narrow range, the human eye perceives the wavelengths between 400 nanometers and 700 nanometers as visible light. Here the different wavelengths correspond to the familiar colors of the rainbow, ranging from the short wavelengths of less than 500 nanometers that we perceive as blue and violet, to the longer wavelengths above 600 nanometers that we see as orange and red. Above 700 nanometers, lies the infrared part of the spectrum, which we can detect as heat (but not light) if the intensity of the radiation is sufficient. Our senses are also neglectful of wavelengths less than 400 nanometers, in the ultraviolet part of the spectrum.

Our inability to detect ultraviolet wavelengths is regrettable since so much of the sun's energy that reaches the earth is in the form of ultraviolet radiation. Although only about 8 percent of the photons that strike the atmosphere have ultraviolet wavelengths, fully 20 to 30 percent of the radiation that reaches the planet's surface is in the ultraviolet part of the spectrum. Of course, part of our inability to perceive ultraviolet light is probably related to the fact that ultraviolet radiation can be harmful — shorter wavelengths have higher energy that can damage cells. Fortunately, proteins in the lens of the human eye absorb much of the ultraviolet radiation before it can reach the retina. (People are able to see ultraviolet light only when the lens of the eye is removed.)

Even though it is potentially harmful, the relatively large proportion of ultraviolet radiation in scattered light at least partially accounts for the use of these wavelengths for the detection of polarization. There are, however, other reasons why animals might exploit ultraviolet wavelengths to detect the polarization of light. For one thing, atmospheric disturbances can affect the orientation of the E-vector. Such disturbances have a greater effect on the longer wavelengths of light, and the least effect on the ultraviolet part of the spectrum. Moreover, longer wavelengths are used by the animals to detect motion and form, and it may be more efficient to keep this information separate from information about the polarization of light. Finally, light that is polarized as a result of reflection from a surface is usually richer in longer wavelengths; by using shorter wavelengths an animal is more likely to be analyzing the sky and not some other polarization pattern.

Just what does an animal see when it perceives the pattern of polarized light in the sky? We cannot see the world through other animals' eyes, but we do know a little about the pattern of polarization that could be perceived. The pattern is actually quite simple, and it is best understood by imagining oneself looking at the dome of the celestial hemisphere from a distance outside the hemisphere. Viewed from this perspective, the E-vector of the scattered light would be parallel to a series of concentric ring arcs emanating from the position of the sun. The greatest amount of polarization, as much as 70 to 80 percent of the scattered light, occurs in the arc 90 degrees from the sun's position. The amount of polarization

is symmetrical about this radial arc, so that the amount of polarized light gradually decreases at angles less than and greater than 90 degrees from the sun.

Since the pattern of polarization in the sky depends on the position of the sun, the E-vector map of the sky changes throughout the day. At sunrise and at dusk, the band of greatest polarization passes through the zenith point of the celestial hemisphere. When the sun is at the zenith point, however, the arc of greatest polarization appears as a circular band at the horizon. Although it is not clear precisely how terrestrial animals use the E-vector pattern in the sky, there is some evidence that they use the pattern to locate the apparent position of the sun, if it is obscured by clouds, and then use this location as a compass for orientation and navigation.

The underwater E-vector pattern differs markedly from the celestial pattern and is much less well understood. In the early 1950s, Talbot Waterman of Yale University showed that the underwater pattern of polarization no longer takes the shape of a dome, but rather forms a sphere that surrounds the animal. He also showed that on cloudy days most of the underwater polarization can be attributed to the scattering of light by water molecules, with little contribution from the polarization pattern in the sky.

Unfortunately, even less is known about the relationship between the wavelength of the light and its polarization underwater. Only recently was it shown that ultraviolet radiation is present in the underwater light field. In general, the relative intensity and spectral composition of light underwater depends on a number of factors, including the turbidity of the water, the amount of particulate matter, the depth of the time of day. We do know that the smallest degree of

polarization occurs at approximately 470 nanometers, in the blue part of the spectrum. Unfortunately, the degree of underwater polarization at the ultraviolet wavelengths has yet to be investigated.

The Angling of Fish

Although polarization vision has been studied in the honeybee since the work of von Frisch in the 1940s, the ability was not observed in fish until relatively recently. Establishing that a fish can see polarized light requires special care in the experimental procedure. It is important to make sure that the fish is not responding to some other aspect of the light, such as its intensity, which may vary with the degree polarization.

One of the first observations that fish could discern the polarization of light was made in 1971 by Peter Dill, then at the University of British Columbia. Using food as a reward, Dill found that sockeye salmon could be trained to distinguish horizontal and vertical planes of polarized light. A little later, Talbot Waterman at yale and Richard Forward of Duke University showed that some species of fish orient themselves at a particular angle in response to the polarization pattern of the light.

These early studies were followed about a decade later by the research of Gunzo Kawamura and his colleagues at Kagoshima University, who showed that several species of fish, including trout, have innate physiological responses to changes in the polarization of light. By monitoring a fish's heart beat, Kawamura found that whenever the plane of the polarized light changes, the fish's heart rate decreases. The significance of the physiological change is not clear, but the response has proved useful to investigators.

My colleagues and I have used similar experimental techniques to investigate behavioral and physiological aspects of polarization sensitivity. We soon learned that one of the main problems in this field of research is the maintenance of healthy fish throughout the experimental procedure. It has posed a considerable challenge throughout the course of our research. After experimenting with many different parameters, we found that fish are extremely sensitive to the quality of the water in their tanks. A change in the concentration of chlorine in the water from 3 parts per billion to 5 to 10 parts per billion was sufficient to affect the animal's recovery from an experiment. It was a valuable lesson that revealed how some very subtle factors could affect the outcome of a line of research. Having solved the problem, we could continue with the course of our investigations.

One of our first goals was to determine if the fish's ability to detect the E-vector pattern is a purely visual phenomenon or if the pineal gland, a light-sensitive organ at the top of the fish's head, mediates some of the response. By placing a depolarizing patch over the pineal, we were able to show that the fish's ability to orient to the E-vector pattern remains intact when it is deprived of the use of the pineal. It seemed likely, then, that the fish uses only its eyes for the detection of polarized light.

Next we went about testing whether, like the honeybee, fish use ultraviolet wavelengths to detect the polarization of light. In one set of experiments we trained rainbow trout to swim under a polarized-light field to a target, and provided a food reward for successful orientation. We then observed the behavior of the fish when the polarized light included ultraviolet wavelengths, and when it did not include ultraviolet wavelengths. When the polarized light included ultraviolet wavelengths, the fish were able to orient relative to the polarized-light field accurately. In contrast, the fish could

not orient to the E-vector when ultraviolet wavelengths were not included in the polarized light stimulus. There is little doubt that ultraviolet wavelengths are necessary for the perception of polarization by fish.

Light in the Eye

How is the fish able to perceive the pattern of ultraviolet, polarized light? A number of laboratories, including my own, are currently investigating the phenomenon. An explanation of some of the hypotheses and the experiments requires a brief discussion of the mechanisms by which the eye can detect any king of light at all.

Again consider the oscillating dipole. The electromagnetic wave that emanates from the dipole can be absorbed by another dipole that is parallel to the first — in other words, when the E-vector of the wave is parallel to the absorbing dipole. In this way a signal can be sent from a source that generates electromagnetic radiation to a receiver that absorbs some of the radiation.

The transfer of electromagnetic energy from an oscillating dipole to a receiving dipole is exactly the process that allows an animal's eye to see a wave of light emitted from the sun. The receiving dipole in the eye is a molecule called 11-cis-retinal, a derivative of vitamin A that is located in the light-sensitive cells of the retina. The dipole-like behavior of 11-cis-retinal was studied by the American biochemist George Wald, who won the Nobel prize for his work in 1967. Wald showed that when the long axis of the 11-cis-retinal is aligned parallel to the E-vector of the incident light, the molecule is able to absorb some of the energy. In doing so, the 11-cis-retinal changes its conformation to form the isomer all-trans-retinal. The change in conformation from the cis to the trans form of the

molecule is the only light-dependent step, and it initiates the visual process. The mechanism is so exquisitely sensitive that a single quantum of light, or photon, can be detected by an individual photoreceptor cell.

The intensity and the color of the light that an animal can see depends on the different types of photoreceptor cells in the eye. Like the human eye, the retina of a bony fish contains two types of light-sensitive cells: cones, which are used for color vision in bright light, and rods, which are able to detect various shades of gray in dim light. In the human retina there are three types of cones, each of which is primarily sensitive to light in a different range of wavelengths. Cones that respond to short wavelengths of light, centered on about 420 nanometers, are called short cones, and they contribute to the perception of blue. Cones that respond best to intermediate wavelengths of about 530 nanometers are called middle cones, and they make a strong contribution to our perception of green. Cones that are most sensitive to longer wavelengths (about 560 nanometers) are called long cones, and they contribute to our perception of red.

Even though each type of cone responds best to a certain wavelength of light, smaller responses can be elicited by wavelengths on either side of the preferred wavelength. For example, long cones respond best to wavelengths of 560 nanometers, but still respond to a lesser degree to wavelengths of 500 nanometers and 600 nanometers. It is important to note here that a cone also has progressively greater responses to increasing intensities of light (until it saturates). Because a cone has differential responses to both the wavelength and the intensity of light, no single cone can distinguish color from brightness. This means that a single type of cone cannot carry all of the necessary information about color.

Eyes with two or more type of cones can avoid this confusion by comparing the responses of the receptor cells. The fact that each type of cone has a different response to a particular wave length means that the pattern of cellular responses is unique for each wavelength of light. By comparing the activity of three types of cones, human eyes are able to reduce the ambiguity almost entirely.

The retinas of many species of fish differ from ours in that they have four types of cones, each with its own range of spectral sensitivity. Our experimental investigations of the carp retina show that its short cones respond best to wavelengths of 460 nanometers, its middle cones have a peak sensitivity of about 530 nanometers, and its long cones respond best to light with a wavelength of 600 nanometers. The fourth type of cone, which the human retina does not have, is most sensitive to ultraviolet wavelengths — it is called an ultraviolet cone — with a peak response at about 380 nanometers. (The peak sensitivity of the ultraviolet cone varies from one species to another.) It is important to note that the absorption spectra of all the cones overlap in the ultraviolet wavelengths. This means that ultraviolet light not only stimulates the ultraviolet cones, but also the long cones, the middle cones and the short cones.

We have recently shown that certain types of cones in the fish retina do have preferences when it comes to the orientation of the E-vector. In particular ultraviolet cones, middle cones and long cones respond best to certain orientations of the E-vector, whereas short cones show no sensitivity to the polarization of light. (Recall that the blue part of the spectrum has the lowest degree of polarization underwater.) Furthermore, the ultraviolet cones

have a polarization sensitivity whose orientation is exactly opposite (or orthogonal) to that of the middle and long cones. That is, the ultraviolet cones respond best to vertically polarized light, whereas middle and long cones both respond best to horizontally polarized light.

The differential sensitivity of the cones to the polarization of light, in conjunction with the cones' capacity to respond to ultraviolet wavelengths, provides a mechanism for the discrimination of the E-vector. Were it not for the overlapping absorption spectra of the cone systems, the wavelength of the light would be confounded with the E-vector orientation. In this case, the same ultraviolet-light stimulus can elicit different responses from cones that are differentially sensitive to the polarization of the light.

How does an individual cone receptor distinguish between vertically and horizontally polarized light? It must be that the receptor is dichroic — it preferentially absorbs certain E-vector orientations and not others. One way to achieve this would be to ensure that all 11-cis-retinal molecules in the receptor are oriented in only one direction. This way the receptor would respond only if the plane of the polarized light were parallel to the long axis of the molecules.

There are, however, other ways to achieve receptor dichroism. Retinal molecules are not floating freely in the cytoplasm of the receptor cells; instead they are embedded in a series of flat, membranous disks. Viewed from the face of the disk (the usual angle at which light strikes a receptor cell), the retinal molecules are oriented randomly. Viewed from the edge of the disk, however, most of the molecules are preferentially aligned parallel to the face of the disk. Rotating a receptor cell so that light strikes the retinal-bearing disk at its edge would make the cell dichroic.

Such a mechanism, in fact, appears to be present in anchovies. However, not all fish have their receptor cells oriented in such a way.

Another hypothetical mechanism to produce polarization within a receptor was recently proposed by David Cameron and Edward Pugh of the University of Pennsylvania. Cameron and Pugh noted that light must first pass through the part of the cone cell called the inner segment before reaching the light-sensitive retinal molecules in the outer segment. As it happens, the cross-sectional profile of the inner segment of the middle and long cones is roughly elliptical. Cameron and Pugh argue that the elliptical profile of the inner segment acts as a birefringent filter that selectively allows only certain E-vector orientations (parallel to the long axis of the ellipse) to reach the retinal molecules. That is, only certain E-vector orientations would elicit a response from a cone cell.

There are, however, problems with this hypothesis. Currently, there is no physical evidence suggesting that the inner segment shows birefringence or acts as a dichroic filter. Moreover, our own experiments suggest that the middle and long cones show the same type of polarization sensitivity (to horizontal orientations) despite their orthogonal orientation in the retina. If the inner segment of the middle and long cones is acting as a dichroic filer, then these cones should have opposite polarization sensitivities.

Processing Light

How does the fish's nervous system use the information provided by the cones? It has been known since the work of Stephen Kuffler at Johns Hopkins University in the early 1950s that some information processing occurs directly in the retina. This should not be surprising since the retina is a part of the brain.

However, an understanding of the processing that occurs in the retina requires a brief description of its structural organization.

Aside from rods and cones, there are at least four other classes of nerve cells in the retina: horizontal cells, amacrine cells, bipolar cells and ganglion cells. In general, the horizontal cells and the amacrine cells mediate the lateral flow of information in the retina. The bipolar cells mediate the vertical flow of information in the retina, connecting the photoreceptor cells to the ganglion cells. The ganglion cells are the final way station for visual information in the retina; it is here that the horizontal and vertical networks converge. In turn, the axonal processes of ganglion cells come together to form the optic nerve, which conveys visual information from the eye to the rest of the brain.

Since the ganglion cells are the output neurons of the retina, the simplest way to assess how the retina as a whole responds to polarized light is to examine the electrophysiological responses of the ganglion-cell axons in the optic nerve. Fortunately, the size of the optic nerve, and its anatomical location, have made this a relatively task. We decided to follow in the footsteps of many others who have studied the information processing of the retina by examining the responses of ganglion cells to polarized light.

Kuffler's work provides the backdrop to our investigations of ganglion-cell activity. One of the things Kuffler found was that ganglion cells do not have a uniform response to a light stimulus. Some cells would increase their activity in response to light on the photoreceptor that connect to it but would decrease their activity when the photoreceptor around the active area were exposed to light. This meant that the ganglion cell has an on center and an off surround. Kuffler also found that the reverse process could occur: A central spot of light inhibits the ganglion cell, whereas an annulus of light on the surround excites the cell.

The center-surround antagonism of the ganglion cell's receptive field provides part of the mechanism for the coding of color in the visual system. There are, for example, ganglion cells that are excited by a green center spot and inhibited by a red surround; these are known as color-opponent cells. Since color-opponent cells in the fish retina should carry information from the ultraviolet cones and the long or middle cones, they should also carry information about the polarization of light.

We decided to search for color-opponent ganglion cells in the fish's optic nerve. To date, we have found several ganglion cells that have an ultraviolet on center and a red off surround. When we stimulated the ganglion cell's receptive area in the retina with a diffuse light with an ultraviolet wavelength of 380 nanometers, it responded best to vertically polarized light — suggesting that the ultraviolet cones were involved. In contrast, light with a red wavelength of 660 nanometers produced the best response if the light was horizontally polarized — suggesting that the long cones were involved.

The next step was to show that a single ganglion cell could make differential responses to ultraviolet light that had different E-vector orientations. (Recall that long cones and ultraviolet cones both respond to ultraviolet light.) First, we reduced the sensitivity of the ultraviolet cones so that they would not overwhelm the response of the long cones. It was possible to do this by providing a low-level ultraviolet light in the background that raises the response threshold of the ultraviolet cones. We could then record the activity of the ganglion cell in response to a very intense ultraviolet light with a wavelength of 380 nanometers while we modulated the E-vector.

What we found confirmed our suspicions. Light that was vertically polarized produced an on response, suggesting that the ultraviolet cones were active, whereas horizontally polarized light produced an off response, suggesting that the long cones were active. The ganglion cell must integrate information from the two types of cones. This means that the ganglion cell sends different signals to the rest of the brain depending on the E-vector orientation of the ultraviolet light.

It seems that the visual system uses a similar mechanism for the perception of polarization and color. As in color vision, it is only by comparing the outputs of the cones that the visual system can discern the polarization of ultraviolet light. If the mechanism were not arranged this way, the fish would not be able to discriminate the E-vector orientation independent of the brightness or the hue of the light. Of course, the analogy between polarization vision and color vision is limited because polarized light in the atmosphere is a broad field stimulus. This means that the response of a ganglion cell to a change in the orientation of the E-vector is a combination of the center-and surround-generated pattern.

Despite our efforts to understand how a fish might detect the pattern of polarized light, we do not yet fully understand why it needs to do so. As with the birds and the bees, polarization vision might provide fish with spatial cues that can be used in orientation. Migration is common among fishes (a salmon may range over 3,000 kilometers), and it may be that polarization vision plays a role in locating the position of the sun during the days when its obscured.

There are other possible uses for polarization vision in fish that have yet to receive adequate attention. Recent studies in a number of laboratories suggest that a sensitivity to the polarization of light can enhance the contrast of an underwater target. Scuba divers, for example, can detect underwater targets more easily with the aid of a polarizer. Moreover, an underwater target itself may polarize light. Zooplankton, a primary source of food for some fish,, are birefringent — hence they reflect polarized light that differs in its composition from the background. Finally, the scales of a fish, like particles in the atmosphere, can polarize light. It may be that fish recognize predators, prey and mates by the pattern of polarized light they reflect.

Acknowledgments

I would like to express my gratitude to my friend and mentor Professor William N. McFarland, whose enthusiasm and excitement for basic biology continues to bring joy to his colleagues and students. I would also like to thank my graduate students, Luc Beaudet, Inigo Novales Flamarique and Daryl Parkyn, and my postdoctoral fellows, Howard Browman and David Coughlin, for their interest and advice. This research is supported by the Natural Sciences and Engineering Research Council of Canada, and the Department of Fisheries and Oceans, Canada.

Bibliography

Bernard, G.D., and R. Wehner. 1977. Functional similarities between polarization vision and color vision. Vision Research 17:1019-1028.

Bowmaker, J.K., and Y.W. Kunz. 1987. Ultraviolet receptors, tetrachromatic colour vision and retinal mosaics in brown trout (Salmo trutta): Age-dependent changes. Vision Research 27:2101-2108.

Brines, M.L., and J.L. Gould. 1982. Skylight polarization patterns and animal orientation. Journal of Experimental Biology 96:69-91.

Cameron, D.A., and E.N. Pugh. 1991. Double cones as a basis for a new type of polarization vision in vertebrates. nature 353:161-164.

Coemans, M., J. Vos and J. Nuboer. 1990. No evidence for polarization sensitivity in the pigeon. Naturwissenschaften 77:138-142.

Cone, R.A. 1972. Rotational diffusion of rhodopsin in the visual receptor membrane. Nature New Biology 236:39-43.

Delius, J.D., and Emmerton J. Perchard. 1976. Polarized light discrimination by pigeons and an electroretinographic correlate. Journal of Comparative Physiology and Psychology 70:560-571.

Dill, P.A.. 1971. Perception of polarized light by yearling sockeye salmon. Journal of Fisheries Research Board of Canada 28:1319-1322.

Douglas, R.H., and C.W. Hawryshyn. 1990. Behavioral studies of fish vision: an analysis of a visual capabilities. In The Visual System of Fish, pp. 373-418. London: Chapman and Hall. London.

Forward, R.B.. Jr., K.W. Horch and T.H. Waterman. 1972. Visual orientation at the water surface by the teleost Zenarchopterus. Biological Bulletin 143:112-126.

von Frisch, K. 1949. Die polarization des himmelsichtes al orientierender fraktor bei den tanzen der bienen. Experientia 4, 142-148.

Hawryshyn, C.W. 1990. E-vector and color coding in ganglion cells of salmonids. Society for Neuroscience Abstracts 16:1217.

Hawryshyn, C.W., M.G. Arnold, E. Bowering and R.I. Cole. 1990. Spatial orientation of rainbow trout to plane-polarized light. The ontogeny of E-vector discrimination and spectral sensitivity characteristics. Journal of Comparative Physiology A 166:565-574.

Hawryshyn, C. W., 1991. Light-adaptation properties of the ultraviolet-sensitive cone mechanism in comparison to the other receptor mechanisms of gold fish. Visual Neuroscience 6:293-301.

Hawryshyn, C.W., M.G. Arnold, D. Chiasson and P.C. Martin. 1989. The ontogeny of ultraviolet photosensitivity in rainbow trout (Salmo gairdneri). Visual Neuroscience 2:247-254.

Hawryshyn, C.W., and A.E. Bolger. 1990. Spatial orientation of trout to partially polarized light. Journal of Comparative Physiology A 167:691-697.

Hawryshyn, C.W.. and F.I. Harosi. 1991. Ultraviolet photoreception in carp: Microspectrophotometry and behavioral determined action spectra. Vision Research 31:567-576.

Hawryshyn, C.W. and W.N. McFarland. 1987. Cone photoreceptor mechanisms and the detection of polarized light in fish. Journal of Comparative Physiology A:459-465.

Ivanoff, A. and T.H. Waterman. 1958. Factors, mainly depth and wavelength, affecting the degree of underwater light polarization. Journal of Marine Research 16:283-307.

Kandell, E.R., J.H. Schwartz and T.M. Jessel. 1991. Principles of Neural Science. New York: Elsevier.

Kawamura, G., A. Shigata and T. Yonemori. 1981. Response of teleost to the plane of polarized light as determined by the heart-beat rate. Bulletin of the Japanese Society for the Scientific Study of Fish 47:727-729.

Kleerekoper, H., J.H. Matis, A.M. Timms and P. Gensler. 1972. Locomotor response of the goldfish to polarized light and its E-vector. Journal of Comparative Physiology

Loew, E.R., and W.N. McFarland. 1990. The underwater visual environment in The Visual System in Fish (ed) R.H. Douglas and M.B.A. Djamgoz, pp. 1-43 Chapman and Hall.

Loew, E.R. and C.M. Wahl. 1991. A short wavelength sensitive cone mechanism in the venile yellow perch. Perca flavescens. Vision Research 31:353-360.

Novales Flamarique, I., A. Hendry, and C.W. Hawryshyn. In press. The photic environment of a salmonid nursery lake. Journal of Experimental Biology.

Rossell, S., and R. Wehner. 1986. Polarization vision in bees. Nature 323:128-132.

Waterman, T.H. 1981. Polarization sensitivity. In Comparative physiology and evolution of vision in invertebrates b: Invertebrate visual centers and behavior I, ed. H. Autrum (Handbook of Sensory Physiology, Vol. VII/6B). pp.281-469. Berlin: Springer Verlag.

Waterman, T.H. 1984. Natural polarized light and vision. In Photoreceptor and vision in invertebrates, ed. M.A. Ali, pp. 63-114. New York: Plenum Press.

Waterman, T.H. 1989. Animal Navigation. New York: W.H. Freeman and Company. Waterman, T.H., and K. Aoki. 1974. E-vector sensitivity patterns in the goldfish optic tectum. Journal of Comparative Physiology 95:13-27.

Waterman, T.H., and R.B. Forward. 1970. Field evidence for polarized light sensitivity of fish Zenarchopterus. Nature 220.06 07.

Waterman, T.H., and H. Hashimoto. 1974. E-vector discrimination by goldfish optic tectum. Journal of Comparative Physiology 95:1-12.

Wehner, R. 1989. Neurobiology of polarization sensitivity. Trends in Neurosciences 12:353-359.

Wehner, R., G.D. Bernard and E. Geiger. 1975. Twisted and non-twisted rhabdoms and their significance for polarization detection in the bee. Journal of Comparative Physiology 104:225-245.

Questions: *Answers on page 202*

1.   Why is the inability of humans to detect ultraviolet light a possible disadvantage?

2.   Is it false to state that fish, or other animal species, see with their eyes?

3.   Do you think that light is matter or energy or both?

27

A recent flight of Spacelab J subjected a variety of animal species to the conditions of space flight. Members included fish (two carp), pregnant frogs, fertilized chicken eggs, white rats, hornets, jellyfish, and fruit flies. The results on the physiological processes of these species from the eight-day biomedical mission were monitored and studied.

One major discovery was the change in energy processing in the muscle cells of rats while in flight. As an energy source, they relied more on carbohydrates than fatty acids. Reasons for the switch are not clear. Skeletal muscle mass, also witnessed in the astronauts, decreased. This atrophy was probably in part due to inactivity. This change occurred faster in mice, probably due to their faster metabolic rate.

Other topics for study in space included neuromuscular changes in jellyfish and the reproductive rates in frogs. Other species will be studied for space-flight response in future missions.

Biomedicine Blasts to New Heights
by Kathryn Hoppe

Some fundamental questions about life on Earth may best be answered in space, as NASA researchers continue to expand our medical knowledge with a broad range of life sciences experiments shuttled into orbit.

The most recent biomedical mission, Spacelab J, launched 20 studies into space for eight days last month. Astronauts kept busy conducting experiments on themselves and caring for a menagerie that included two carp, four pregnant frogs, 30 fertilized chicken eggs, 180 hornets, and more than 7,600 fruit flies.

While investigators still need time to analyze the data from this latest venture, results from earlier flights have already raised new questions, prompting scientists to reexamine some traditional theories about how the human body functions, says Ronald J. White of NASA's life sciences division in Washington, D.C.

NASA's first Spacelab Life Sciences mission (SLS-1) carried 18 investigations into orbit for a nine-day flight in June 1991. After more than a year of analysis, researchers gathered this August at the World Space Congress in Washington, D.C., to discuss their results.

Unlike previous biomedical studies, SLS-1 was designed so that scientists could monitor and compare the long-term health of their astronaut subjects for several months before and after, as well as during, the actual flight. These studies revealed new details about the physiological changes that occur as the human body adapts to different gravitational environments (SN:8/1/92,p.70).

Other investigations focused on the 29 white rats flown aboard the SLS-1 mission. Many of these experiments probed adaptations of the so-called antigravity muscles, which support the body and maintain posture in Earth's gravity.

One of the most unexpected results revealed that muscle cells in space do not process energy in the same way they do on the ground, White says.

In Earth's gravity, muscle cells use both fatty acids and carbohydrates for fuel, explains Kenneth M. Baldwin, a physiologist at the University of California, Irvine. Using biochemical tests, Baldwin and his co-workers determined that, in space, muscles appear to loose their ability to process fatty acids and thus rely more on carbohydrates for fuel.

This change in energy processing came as a "big surprise" to the researchers, says Baldwin, who adds that the study provided only a few hints as to the causes of the switch. He believes further studies may reveal details that could lead to alterations in astronauts' diets.

Related investigations provided new details about previously studied effects such as muscle atrophy — the loss of mass and strength that occurs when muscles remain inactive. In the past, astronauts have lost muscle mass in response to the absence of gravitational stress in space, but such changes remain difficult to measure in humans.

The rat's higher metabolic rate causes these effects to occur faster, making short-term changes easier to study. During the nine days of SLS-1, Baldwin and his co-workers found that the rats lost as much as 25 percent of the mass of their antigravity muscles. Muscles used primarily for locomotion atrophied less. After returning to Earth, the rats regained muscle mass slowly: It took twice as long to regain the mass as to lose it.

These and related studies highlight the need to counteract the effects of space travel as that astronauts can readapt more easily to life on the ground. Baldwin believes future astronauts may prevent atrophy by adding new exercises to their daily routine. Astronauts should "do the equivalent of pumping iron" while in space, he says.

Baldwin plans future studies to investigate the relationship between lost muscle mass and the loss of muscle function in space, as well as the biochemical signals that control atrophy.

The human and rat passengers aboard SLS-1 also kept company with more than 2,400 jellyfish polyps. These primitive animals prove ideal for space-based experiments because of their small size and quick development, says Dorothy B. Spangenberg, a developmental biologist with Eastern Virginia Medical School in Norfolk. During metamorphosis, jellyfish undergo cellular processes similar to those seen in more complex animals and thus may reveal details about the effects of microgravity on organic development, Spangenberg says.

In the early stages of their life cycle, jellyfish exist as stationary polyps less than 3 millimeters long.

Under the proper conditions, these polyps metamorphose into a free-swimming stage, known as ephyrae, in just six days.

After SLS-1 returned to Earth, Spangenberg and her co-workers examined the jellyfish ephyrae that had developed in space. While most of these animals appeared normal, the group as a whole displayed a significantly higher rate of locomotor abnormalities — such as trouble coordinating their muscular pulsing movements — compared with a control group of ephyrae raised entirely on Earth. Roughly 18 percent of the space-developed ephyrae showed such effects, as opposed to less than 3 percent of the Earth-developed controls.

This suggests that microgravity interferes with the development of the neuromuscular system, the nervous system, or the gravity receptors in certain jellyfish, says Spangenberg, who plans further space-based studies to investigate the causes of these effects. She notes, however, that the normal metamorphosis observed in the majority of the experimental jellyfish indicates that cellular development does not depend on gravity.

Jellyfish will again blast into space in 1994 for a series of experiments designed to investigate their exact tolerance for different levels of gravity, Spangenberg says.

While planning for future missions, scientists await for the first results of Spacelab J, which should be announced toward the end of this year. All of the experiments aboard yielded data, says Mission

Manager Aubry King of NASA's Marshall Space Flight Center in Huntsville, Ala. However, not every investigation proceeded as planned. High humidity plagued an orbiting horde of hornets, which did not construct the combs scientists has planned to examine for structural differences.

The high-flying frogs produced more solid results, laying hundreds of eggs that astronauts artificially fertilized in space. Hundreds of tadpoles from these eggs, along with embryos from the 30 chicken eggs flown on the mission, should provide investigators with information on vertebrate development and bone formation in the microgravitational conditions.

Scientists will examine the effects of another environmental change experienced during space travel — higher levels of radiation — by searching for mutations among the mission's fruit flies. Human astronauts and fish studied in other Spacelab J experiments will provide additional clues to physiological adaptations to space.

As researchers analyze the data from these past experiments, they look forward to new studies planned for the future. NASA's schedule for the next few years includes three shuttle missions that will carry biomedical experiments, including a second Spacelab Life Sciences (SLS-2).

Such investigations are "only the first in a series of shuttle missions dedicated to life sciences," says White, who predicts that space-based experiments will continue to produce significant results. □

Questions: *Answers on page 202*

1. What is the value for learning about space-flight responses in different animal species?

2. Why do you think a preference for carbohydrates developed for a body fuel in space?

3. What specifically changes in a skeletal muscle to reduce its mass?

Part Five

The Continuation of Life - Reproduction and Development

28

Stick insects have been studied for their mimetic ability. Their resemblance to sticks and leaves makes their detection difficult for predators. However these unusual species also offer excellent models for studying some unusual patterns of reproduction. Their most unique form of reproduction is the development of the egg cell without the activation by a sperm cell or the union with the male gamete. Although these reproductive phenomena have been induced artificially, they are not common in nature. Apparently there are many unusual forms of sexual reproduction found in stick insects. In another surprising form, the genetic material of the female is eliminated after fertilization. As a result the egg develops entirely under directions from the inherited male chromosomes.

Sex and the Male Stick Insect
by Piero P. Giorgi

Stick insects are better known for their mimetic ability than for their contribution to our understanding of reproductive biology. During the day they remain immobile on branches, where they generally escape detection because their bodies bear an uncanny resemblance to sticks or leaves (Fig. 1). During the night they slowly move about feeding, and reproducing — often, as a group at the University of Bologna has found[1-4], in decidedly peculiar ways.

There are many variations on the theme of sexual reproduction. An egg can develop without activation by a sperm (parthenogenesis), or it can eliminate the male pronucleus after fertilization (Gynogenesis). The male genetic material, however, is much less versatile and development in the absence of the female pronucleus (androgenesis) has so far been described only in plants. Since 1887, when the Hertwigs first demonstrated experimental androgenesis in echinoderms[5], students of animal biology have been taught that this phenomenon occurs only when induced artificially. But now the

group at Bologna has discovered natural androgenesis, as well as other unexpected forms of reproduction, in certain species of stick insect.

Using gel electrophoretic separation of alloenzymes and chromosomal analysis to reconstruct the evolutionary history of the stick insect population of Sicily, Scali, Mantovani and Tinti identified two hybridogenetic strains which have arisen naturally from hybridization between Bacillus rossius females and B. grandii males of different subspecies[1,2].

Hybridogenesis is a type of reproduction particular to interspecific hybrids, and involves the maintenance of one parental genome in the gametes and the elimination of the other. The discarded genome is provided anew through fertilization by the host fathering species (Fig. 2c); thus the hybridogenetic form is a perpetual F[1] hybrid, which can evolve to a full self-sustaining species if it shifts to parthenogenesis. Males of the two hybridogenic strains discovered by Scali and co-workers (B. rossius-grandii benazzii and B. rossius-grandii) are infertile,

whereas females, besides maintaining the hybridogenetic strain, can also reproduce by gynogenesis. Hybrodogenesis has often been advocated as a possible mechanism for speciation in fish and amphibians[6], but no case had yet been found in invertebrates.

The female stick insects had more surprises in store. When a female of B. rossius-grandii benazzii is fertilized by B. rossius male, up to 20 percent of the offspring have the genetic make-up of the father alone (Fig. 2e). This was first demonstrated by use of enzyme electrophoresis, and has now been confirmed by chromosomal analysis[4]. The genetic material of the mother is eliminated so that her eggs develop entirely under the instruction of paternal chromosomes. The resulting androgenetic males and females are fertile, and diploid through duplication of the active chromosomes, or by fusion of two sperm pronuclei. Females of the hybrid B. rossius-grandii benazzii cannot reproduce by parthenogenesis, suggesting that perhaps androgenesis could be a strategy to cope with complex requirements for nucleocytoplasmic interactions in hybrid development.

The discovery that androgenesis can occur naturally in animals opens new prospects in evolutionary and developmental biology. It has been pointed out that hybridogenesis is, practically, a case of sexual parasitism, where females kidnap and exploit sperm from other species. In fact the term 'klepton' (from the Greek for 'thief') has been proposed to denote a systematic group of hybridogenetic origin[7]. But Scali *et al.* identify cases of androgenesis when B. rossius-grandii benazzii hybrid females mated with males of B. rossius and, in some instances even with B. grandii benazzii. The production of offspring corresponding to the original male species through androgenesis may now rehabilitate the hybridogenetic female from the status of a thief's daughter to that of a potential genetic store for the fathering taxon.

Finally, the existence of these complex modes of reproduction casts doubt on Mayr's 'isolation concept' of speciation[8] while supporting the more recent 'recognition concept' proposed by Paterson[9]. This second concept views speciation in terms of specific mate recognition rather then reproductive isolation. Both hypotheses are restricted to sexual organisms. But in the recognition concept, parthenogenesis is considered as an escape from hybridity[10], as originally suggested by C.D. Darlington[11], whereas Mayr and others see problems of hybridity in terms of isolation mechanisms.

Androgenesis, which also amounts to an escape from hybridity, would be expected to occur under the recognition concept, but is contrary to the expectations of the isolation concept. So the reproduc- tion of stick insects can be explained better in terms of specific behavioural or molecular signals mediating recognition between individuals or gametes, rather than in terms of teleological barriers to mating. Piero P. Giorgi is in the School of Anatomy, University of Queensland, Brisbane, 4072 Australia. □

1. Mantovani, B. & Scali, V. Invert Reprod. Develop. 18, 185-188 (1990).
2. Mantovani, B., Scali, V. & Tinti, F. J. Evol. Biol. 4, 279-290 (1991).
3. Scali, V., Mantovani, B. & Tinti, F. Frustuia Entomoi NS, 12, 103-108 (1989).
4. Mantovani, B. & Scali, V. Evolution 46, 783-796 (1992).
5. Hertwig, O. & Hertwig, R. Jen. Zeitschr. Med. Natur. 120 (1887).
6. Uzzel, T., Holtz, H. & Berger, L. J. exp. Zool. 214, 251-259 (1980).
7. Dubois, A. & Gunther, R. Zool. Jahrb. Syst. 109, 290-293 (1982).
8. Mayr, E. Animal Species and Evolution (Harvard University Press, Cambridge, Massachusetts, 1963).
9. Paterson, H.E.H. i nSpecies and Speciation (ed Vrba E.S.) 21-29 (Transvaal Museum Monograph No. 4, Pretoria, 1985).
10. Paterson, H.E.H. Pacific Sci. 42, 65-71 (1988).
11. Darlington, C.D. Evolution of Genetic Systems (Oliver & Boyd, Edinburgh, 1958).

Questions: *Answers on page 202*

1.   How would an egg develop through parthenogenesis?

2.   How would destruction of female chromosomes in a fertilized eggs affect the heredity of the offspring?

3.   What biotechnology was used to reconstruct the evolutionary history of the stick insect?

Does the consumption of cholesterol affect the level of this steroid in a person's blood? It depends on the amount already present in the bloodstream as well as a person's previous dietary habits. An increase in consumption will not affect the blood concentration significantly if the level of cholesterol in the blood is already high or if previous consumption has been great. In these cases the body's capacity for cholesterol is already saturated. Therefore adding more through the diet will not change the level in the blood. However, if this capacity is not saturated then a dietary increase will significantly change the concentration in the blood. Genetics plays a role in how the pattern of baseline cholesterol consumption affects a person's response to dietary cholesterol. Among the organs in the body the liver is easily saturated with the substance, requiring only 400 to 500 mg. daily.

Teasing Out Dietary Cholesterol's Impact
by J. Raloff

To defend their predilection for meat and dairy products, many people cite nutrition studies indicating that cholesterol consumption has little, if any, effect on cholesterol levels in the blood. Cardiologists counter by citing other studies linking dietary cholesterol to significant elevations in serum cholesterol and an increased risk of heart disease.

Who's right? Both, according to a preventative cardiologist from the University of Utah in Salt Lake City.

Paul N. Hopkins pooled data from 27 studies comparing dietary and serum cholesterol levels. Such meta-analyses hunt for trends statistically masked within the smaller, component studies. His findings, detailed in the June AMERICAN JOURNAL OF CLINICAL NUTRITION, appear to reconcile the seemingly divergent data on cholesterol. As a rule, the new analysis shows, the higher an individual's initial serum and dietary cholesterol levels, the less likely that raising or lowering dietary cholesterol will alter serum cholesterol.

So for most people, another egg or two per day should produce little change in blood cholesterol, Hopkins says. Why? Since the 400 milligrams of cholesterol contained in the typical U.S. daily diet have already saturated the body's need for the fat-like substance, any additional cholesterol accumulates in the liver rather than the blood. On the other hand, for people who typically eat less than 160 mg. of cholesterol daily, such as Mexico's Tarahumara Indians, an extra egg or two might seriously spike their naturally low serum cholesterol.

Hopkins notes that genetics plays a far larger role than baseline cholesterol consumption patters in determining a person's responsiveness to dietary cholesterol—accounting for about 50 percent of the variability seen in the populations he studied.

Within the diet, saturated fat remains the most important single influence on serum cholesterol levels. Hopkins says for most of the remaining variability seen—perhaps some 20 percent.

One of the biggest surprises, Hopkins observed, "is how easily the liver is saturated [with cholesterol]," a finding whose significance has not yet been studied. That saturation takes only 400 to 500 mg. per day—the equivalent of two eggs. □

Questions: *Answers on page 203*

1. Do you think that cholesterol is a foreign substance to the body?
2. Would you describe cholesterol as a toxic substance? Why or why not?
3. Why is an elevated level of cholesterol in the blood a threat to the health of the body?

Many of the modern methods of birth control known today were not available to citizens of ancient Greek and Roman societies. Nevertheless effective means of oral contraception were used successfully in ancient times. Many of the antifertility drugs employed were plant - based. Only a faint trail of historical information remains about the exact origin and makeup of these plant - based oral contraceptives. Traces of the legacy of these contraceptives is also part of the historical herbal medicine practiced by the people of China and India. It is also recorded as part of the folklore of communities in the mountain regions of the United States. All of these historical populations probably had knowledge of birth control that we do not know about today. Long before the advent of the "pill," the idea of chemical means of birth control existed. Apparently this idea was practiced and recorded in time periods as ancient as the oldest surviving medical records.

Oral Contraceptives in Ancient and Medieval Times
by John M. Riddle and J. Worth Estes

In the hills near cyrene, an ancient Green city-state in North Africa, there once grew a plant of such economic importance that its image was carried on Cyrenian coins. Bundles of the plant, which was called silphion by the Greeks and silphium by the Romans, were shipped throughout the Mediterranean trading area, where they commanded a price that was said to exceed the plant's weight in silver. Farmers in Greece and Syria tried in vain to cultivate it, buy silphion would grown only in Cyrene, where it was finally harvested to extinction.

The loss of silphion was a blow not only to the economy of Cyrene, but quick possibly to medicine. Its great value apparently derived from a singular use of the plant: Silphion's sap may have been the ancient world's most effective antifertility drug.

Could the Greeks have had an effective oral contraceptive? Modern historians are skeptical. During certain periods many ancient and medieval societies experienced declines in numbers, even in times that were peaceful, prosperous and plague-free. The common wisdom holds that they limited their popula-

tion by non-medicinal means, from abstinence to infanticide. Their antifertility drugs are usually relegated to the realm of magic and superstition. Yet the archaeological and written record is sprinkled with evidence that drugs were a trusted way to prevent conception or induce early-term abortions. These claims, tested against recent studies of plant-based contraceptives, offer one plausible explanation for the fact that ancient and medieval people were able to control their numbers.

Silphion, for example, probably worked. Recent experiments with rodents have shown that extracts taken from some of silphion's surviving relatives can inhibit conception or prevent the implantation of a fertilized egg. Since ancient writers described silphion as far more effective than the other species in its family, perhaps it is not surprising that it was driven to extinction.

Unfortunately, knowledge about agents such as silphion has become nearly extinct as well. The faint trail of historical information about plant-based oral contraception begins in ancient Greece but disappears during the Middle Ages, centuries before the pioneering

endocrinological studies of John Rock and Gregory Pincus in the 1950s led to the development of "the pill." Premodern people probably had knowledge about birth control methods that we do not.

Some of the reasons for the disappearance of such knowledge have to do with the hidden nature of female culture. Although many graphic discussions of sexual practices can be found in ancient literature, the writers had only secondhand knowledge of the birth-control practices of women. All surviving medical works in the West, at least until the 12th century, were written by males. But women were the practitioners or medicinal contraception: Only women knew the secrets of what plants to gather and when to gather them, the part of the plant to use, the method of extracting and preparing the drug, the optimum dose and the best time to take it within the menstrual cycle.

Never an important part of professional medical knowledge, antifertility lore came to belong almost exclusively to midwives as professional medicine developed during the Middle Ages. In time, much of the midwives' knowledge was lost, especially as male physi-

Reprinted by permission from AMERICAN SCIENTIST, Journal of Sigmi Xi, The Scientific Research Society.

cians forced them out of business and gynecology became part of professional medical practice. By the 19th century abortion was a criminal act in many Western countries, and the use of antifertility drugs was excluded from professional medicine. By the time society began to look for medicinal means of birth control again in the 20th century, the historical trail had gone cold.

Traces of the ancient legacy of antifertility drugs are found in folk practices today — in the herbal medicine of china and India, and even in the mountain regions of the United States where midwives still use herbs in their practice. These bits of folklore, combined with the archaeological and written record, hint at what was probably a thriving oral culture of contraception. They offer a trail worth picking up once again, both to re-examine our assumptions about the civilizations that preceded ours and to expand the range of solutions available to a world in need of broader choices.

Population control sometimes seems a modern quandary. Ancient Greeks and Romans did not have to worry about the limits of planetary resources, but they did turn their thoughts to population issues from time to time. "If too many children are being born," said Plato, writing of the ideal city, "there are measure to check propagation."

The Roman Senate of Caesar Augustus's day was alarmed not by population growth but by the economic consequences of a declining birth rate. Abortions were outlawed, and the use of contraceptives forbidden, according to one Roman historian. During the Roman Empire and the early Middle Ages, demographic records show that there were absolute declines in the populations of the Western European continent and the Mediterra-nean area. There are various theories about why populations declined, but it is possible that neither Augustus's subjects nor the Romans of the Empire's waning days paid much heed to the Senate's dictates. The gap between public policy and private behavior is the subject of much speculation.

Changes in sexual activities in and out of wedlock do not provide an adequate explanation. Although many ancient sexual practices are well documented, especially in the Roman literature, the record supplies little evidence that coitus interruptus was widely practiced in the ancient world. In antiquity neither religion nor ethical restrictions regulated sexual activity for males. One French writer, Philippe Ariès, has postulated that both ancient and medieval people limited family size by engaging in other sexual activities (behavior he called perversité sexuelle) that could not lead to childbirth. The documentary evidence reveals that sexual practices were indeed varied during classical antiquity, but this variety does not provide a sufficient explanation for population declines. The rhythm method is unlikely to have had a major impact, since the ancients' understanding of the menstrual cycle would have led those trying this approach to have intercourse during fertile periods.

The condom and the diaphragm are relatively modern inventions, although experiments with barrier devices and vaginal suppositories began deep in antiquity; these are varied and their impact hard to evaluate. Abortions were always available, but it is difficult to find evidence that they were routinely employed for birth control. Both ancient and medieval medical texts recognize surgical or manipulative abortions as dangerous, an agreement that leads modern historians to believe that they were used primarily in desperate situa-tions. Drug-induced abortions are an area of greater debate, and will be discussed in more detail below.

In the Middle Ages all conduct, including that of married couples, was closely regulated in Europe by the church. Procreative intercourse within the bounds of marriage was the only sexual activity condoned, and women in the late Middle Ages and Renaissance tended to marry somewhat later than their classical counterparts. Yet it is interesting that the sexual restraint imposed by Christianity had no obvious impact on population; the demographic ups and downs of medieval times are remarkably similar to the patterns of antiquity. Given the hard-to-explain population declines found throughout the record, it is hard to resist the conclusion that some effective form of family limitation may have been practiced during both periods.

The explanation that has satisfied most historians has been infanticide. The strongest evidence supporting this theory comes from data that appear to show the persistence of a higher ratio of males to females than would be expected in a biologically neutral environment free from interference. Contraception and abortion would have taken place before the sex of the child was known, so they should not have produced the alleged lopsided sex ratio. An interesting body of evidence in this debate comes from the field of paleopathology: Close examination of adult female skeletons — which bear marks that could be used to estimate the number of full-term pregnancies — suggests that there were periods in antiquity during which births declined, a contradiction of the infanticide theory. The precision of the method, however, may not be sufficient to support a comparison between population and birth rate.

Since the hard documentary evidence for widespread infanticide is scant (consisting of occasional references to the abandonment of unwanted babies, who might have died from exposure), the case for this hypothesis rests largely on circumstantial evidence and remains controversial. Some scholars argue that a larger male population is natural in certain conditions. While the debate continues, it is worthwhile to examine an alternative explanation — the use of oral contraceptives.

Ancient medical experts, in both Eastern and Western cultures, regularly prescribed antifertility preparations made from the secretions of plants. But until the 1960s, when studies of such botanical agents began to be reported in Indian and Chinese medical and pharmaceutical journals, Western scientists maintained great skepticism about the possibility that any of these ancient prescriptions worked. A re-examination of these beliefs has been slow to come.

The development of modern oral contraceptives has its roots not in the historical experience of women but in the science of endocrinology, which has supplied an understanding of the human ovulatory cycle that is good enough to point the way to effective birth control. In order to be fertilized, an ovum must first mature while still within its ovarian follicle. Its maturation is prompted by the release of follicle-stimulating hormone from the anterior pituitary gland. The release of this hormone is inhibited when the concentration of estrogenic hormones, which control other parts of the menstrual cycle, rises in the blood. Thus modern oral contraceptive pills inhibit the maturation of follicles and their ova, so that fertilization cannot occur, by keeping estrogen concentrations in the blood at a relatively high level.

Estrogens were long thought to be produced only by animals; it was assumed that plants do not synthesize compounds with estrogenic properties. But this assumption was challenged in the 1930s and 1940s as improved techniques of chemical analysis became available. Two studies published in 1933 offered the first clues. Boleslaw Skarzynski reported to the Polish Science Academy that he had obtained from the willow a substance, trihydrooxyoestrin, that resembled a human female sex hormone in its physico-chemical properties. The same year, Adolf Butenandt and H. Jacobi reported that the date palm and pomegranate produced female sex hormones. Although the latter report could not be duplicated experimentally, the principle that plants can produce substances with estrogenic properties had been established by 1966.

Evidence of how substances in plants might affect animal fertility came, meanwhile, from the observations of Australian sheep breeders who noticed that fertility was sharply reduced in animals that grazed on one species of clover, *Trifolium subterraneum*. A study in the 1940s traced the clover's antifertility effect to isoflavonoids, which stimulate estrogen production when they are metabolized by mammals.

Sheep are not human, and so skepticism about the contraceptive effect of plant extracts in humans persisted following these studies. The question was reopened, however, in 1960, when chemists D.B. Bounds and G.S. Pope followed up on a report by anthropologists that Thai women took an extract of the root of *Pueraria mirifica*, a close relative of kudzu, to induce abortion. They isolated from the plant an estrogenic compound called miroestrol.

The Indian and Chinese scientific literature since that time has reported many studies of crude traditional drugs made from indigenous plants; interestingly, the plants used in the East are often related to species described in the ancient Western literature. Combined with evidence from animal science, anthropology, pharmacy and medicine — and the historical record itself — these studies provide grounds for new investigations of the historical use of plant-based antifertility drugs. The possibilities are many and varied; here we will be able to offer only a few examples. Before we look into the ancient pharmacy, however, it is useful to consider the state of knowledge and belief that guided ancient and medieval approaches to contraception.

The story of ancient oral contraceptives must be woven from threads supplied not only by writers on medicine but also by legend, art and the writings of ancient poets, philosophers, playwrights and satirists. Fortunately, there is a large body of such work in the West, and so we shall focus our attention there; although botanical contraceptives may have been used in China as early as they were in Greece, the nature of Chinese archival evidence makes that record far more obscure. We can press onward from antiquity into the Middle Ages in Western Europe, but it is here that the trail becomes obscured because much of the written record in the record of the church. Eventually church opposition helped to smother birth control in taboo and secrecy.

The ancients did not understand the hormonal basis of the menstrual cycle, but they had some ideas about how conception occurred. Because these ideas sometimes blurred what is now considered the boundary between contraception and early-term abortion, it is not always easy to determine how ancient antifertility drugs acted. Soranus, antiquity's foremost authority on gynecology, did make a distinction in his writings during the second century C.E., explaining: "A contraceptive differs from an abortifacient, for the first does not let conception take place while the latter destroys what has been conceived." Soranus recommended contraception as preferably for safety reasons.

But ancient writers did not, in general, understand conception as we do. They though that is it occurred after sexual union at a time when the male seed began to gown. Indeed, many of the ancients did not believe that the female possessed a "seed." Employing an agricultural analogy, they asserted that the female supplied only nourishment, and that the male seed could live within the female for several days before conception took place. This period corresponds roughly with what we now recognize as the time required for implantation of the fertilized ovum in the uterus. The definition of contraceptives (atokioi, as the Greeks called them) therefore encompassed drugs taken shortly after intercourse. A further confusion arises from the fact that it was not always possible to know whether menstruation had ceased because of pregnancy or for another reason, perhaps fever, chronic disease, malnutrition or depression. Women took a variety of drugs during ancient and medieval times to provoke menstruation. In modern terms, these drugs could be called early-term abortifacients, but a woman of earlier times who took an emmenagogue (as such drugs have been called since the 17th century) to regulate her menses could not know whether she had caused an abortion.

It is important to note also that the ancient ideas about the beginning of life provided a woman a period of time in which she was free to stop a pregnancy without considering the act an abortion. The Stoics believed in the potentiality of life at conception, but maintained that the soul originated at birth.

Most ancient cultures provided a legal and moral zone for action during the period between conception and "animation" or quickening. Under ancient Hebrew law, a woman was regarded as pregnant 40 days after conception. Western society's tacit acceptance of the "morning-after pill" persisted for many centuries; it was not until the 19th century that the taking of a drug to regulate menstruation became widely outlawed as abortion.

There is some evidence of contraception in ancient Egypt, but the first well-documented example of an oral contraceptive is the mysterious silphioin. The historian Theophrastus, who wrote about 310 B.C.E., traced the Cyrenians' discovery of silphion to a date 300 years earlier. Botanical descriptions indicate that it was a species of Ferul, or giant Fennel. A Greek vase of the sixth century B.C.E. shows workers aboard a ship weighing and storing packages under the eye of Arkesilas, king of Cyrene. Although we cannot be sure, the packages have long been believed to contain silphion.

We know that silphion was a valued contraceptive from both objects and writings of the day. On the face of a Cyrenian four-drachma coin, the leaves of a silphion plant just touch the right hand of a woman, who is seated with her left hand pointing to her genital area. The iconography suggests a connection between the plant and reproduction. Both the Greek comedy writer Aristophanes and the naturalist Pliny the Elder mention silphion's high cost, and Hippocrates recorded the failed efforts to cultivate it in Syria and Greece. The reason for the plant's high values was best explained by Soranus, who said that the sap of silphion, taken by mouth, prevented conception. Soranus provided several prescriptions for "Cyrenaic juice," which he said

would either prevent or halt a pregnancy.

The related species still growing in North Africa include Ferula assa-foetida L., commonly called asafoetida and used today as an aromatic ingredient in Worcestershire sauce. F. Assa-foetida was reported to act as a human contraceptive in 1963, and crude alcohol extracts of F. assa-foetida and another relative, F. orientalis, inhibited implantation of fertilized ova in rats at rates of 40 percent and 50 percent, respectively (Prakash et al. 1986). Other Ferula species have produced even more impressive effects: In 1985, organic and aqueous extracts of F. jaeschkaena were reported to be nearly 100 percent effective in preventing pregnancy when administered to adult female rats within three days of coitus (Singh et al. 1985). (The effect was not seen in hamsters, which are far less sensitive than rats to conventional estrogenic inhibitors of fetal implantation.)

The idea of a chemical means of birth control is as old as the oldest surviving medical records. The earliest medical writing from Egypt, the Kahun Medical Papyrus, dating from around 1850 B.C.E., contains fragments of three contraceptive recipes. They call for using crocodile feces, honey and saltpeter to make vaginal suppositories. The Egyptian record is sketchy and full of such questionable (and indecipherable) remedies, but by the time of the Greeks and Romans women had gained more sophistication about contraception. It is no coincidence that two plants used as oral contraceptives in ancient times, and for many centuries since, carry the names of famous women of ancient legend.

Artemis, the goddess of women and the woodland and the protector of childbirth, gave her name to a group of shrub-like plants that can grow to about four feet. Whereas silphion was a rare, exotic plant, Artemisia is the name of a common genus whose members can now be found over much of North America. Their feather-like leaves are gray-green and highly indented. Artemisia is also the name given the drug that can be produced from some of these plants.

In ancient times, artemisia was said to protect against the pangs of childbirth, although no one stated whether it did so by preventing or terminating a pregnancy or by some other means. During the Roman Empire, both Soranus and Dioscorides, a first-century herbalist and writer on pharmacology, reported that physicians prescribed an Artemisia species, wormwood or A. absinthium, as an antifertility drug. Other ancient medical authorities classified wormwood as an abortifacient, a reputation that survived many centuries. Macer, the author of the most popular herbal of the central Middle Ages (1000-1300) and probably a bishop in the 11th century, discussed artemisia. At that time the Christian church opposed abortion, to a limited degree, after the fetus had "formed." Macer related that the plant was given its name by the goddess Artemis herself because she had discovered that it "mainly cures female ailments and, when drunk..., it produces an abortion." (Despite the church's position in favor of procreation, medieval writers — many of whom were churchmen — related practical information about contraceptives and early-term abortifacients. Indeed, one of the prolific writers on birth-control techniques was Peter of Spain, author of the Treasury of the Poor, who became Pope John XXII.)

Artemisia has not been forgotten. One of the most popular herbals in print today, the 1986 Reader's Digest book Magic and Medicine of Plants, says that

artemisia provokes menstruation. There is modest scientific support for the abortion-inducing effects of the drug (Duke 1985), and animal studies over the past 20 years have reported that wormwood has delayed the production of estrus and ovulation and interfered with implantation. One study indicates that wormwood has an effect on males as well — interfering with spermatogenesis. (It is important to note that artemisia, taken internally, has several toxic side effects, it is perhaps because of its toxicity that ancient physicians such as Soranus and Dioscorides recommended that it be taken along with two other drugs, myrrh and rue.)

Another legendary figure gave her name to a less widely used substance — the precious resin myrrh, the sap of shrubs of the genus Comiphora, which grown in East Africa. Although myrrh is most famous for its uses as a fragrant ointment in the Bible, the Roman poet Ovid recounted its origin in legend. Myrrha was the daughter of Theias, also called Cinyras, a legendary king of Assyria. A great misfortune befell the innocent girl because of her father's impiety. When he angered Aphrodite, the goddess of love, Aphrodite caused him to lust after his daughter. As a result of their incest, Myrrha bore a son, Adonis. But Theias continued to assault her, until at length she fled. When Myrrha called upon the gods to help her escape from her father, they transformed her into the plant known by her name. Her tears, the plant's sap, became the rescuer of daughters victimized by a father's lust. "Even the tears have fame," Ovid wrote in his Metamorphoses," and that which distills from the tree trunk keeps the name of its mistress and will be remembered through all the ages."

Myrrh's effectiveness is uncertain; it is listed among the antifertility drugs used in the

traditional medicine of India, but does not seem to have been tested experimentally. Records indicate that it is often a component of antifertility prescriptions in ancient and medieval Western medicine; its exotic nature and high cost may have limited its use.

Artemisia and myrrh are not the only birth-control agents that turn up in ancient legend. Pennyroyal, the aromatic mint plant Mentha pulegium L., plays a role in Aristophanes' play Peace, first produced in Athens in 421 B.C.E. Trigaius is in need of a female consort. The god Hermes provides him with Harvesthome. Trigaius says to her: "Come here and let me kiss you — but, Hermes, won't it hurt me if I make too free with fruits of Harvesthome at first?" Hermes replies: "Not if you add a dose of pennyroyal."

Pennyroyal was known as an antifertility agent throughout classical antiquity and the Middle Ages. Ancient experts and modern science appear to agree that an ingredient of pennyroyal, pulegone, is an abortifacient — and a potentially dangerous one. Quintus Serenus, writing on medicine in the second century C.E., said that, when a pregnancy was less than a month old and the "fetus" (Latin did not provide a meaningful distinction between fetus and embryo) was weak, one should "rush into the bedroom" to administer pennyroyal in tepid water to the woman.

The ingestion of pulegone has been shown in recent studies to induce abortions in women and animals (Thomassen, Slattery and Nelson 1990; Thomassen et al. 1991; Froelich and Shibamato 1990; Gordon et al. 1987). A young woman died in Colorado in 1978 when she took pennyroyal oil to induce an abortion — an incident that may serve to demonstrate both the hazards of herbal abortifacients and the fact that ancient knowledge may have been passed on, but imperfectly. Pulegone is more concentrated in the oil than in the tea taken by premodern women. The oil would be expected to damage the liver of both mice and human beings, if taken in sufficient dose (Sullivan et al. 1979; Gordon et al. 1982).

Pennyroyal and Artemisia grown wild in many parts of the world. They are tow of dozens of plants that women have gathered over the centuries in hopes of preparing a tea, potion, powder or see preparation of prevent or stop pregnancy. Finding a potent plant, it turns out, is easier than you might think.

Queen Anne's lace grows profusely in many areas of North America. Its reddish, slender, strongly aromatic root gave the plant, Daucus carota L., the ancient name of wild carrot. Its stems, two or three feet high and covered with coarse hair, carry the white flowers that are recalled by the plant's modern name. The flowers produce the many small seeds that generate the next growth, and that contain substances with extrogenic activity.

A small number of women in Watauga County, North Carolina, in the Appalachian mountains, gather the seeds of Queen Anne's lace and same them, Each time they have sexual intercourse, those who wish not to be with child drink a glass of water containing a tea-spoonful of seeds saved from the autumn harvest. Rural populations in Rajasthan, India, chew dry seeds of Queen Anne's lace to reduce fertility. Both are practices that were known to women 2,000 years ago.

Dioscorides said that the seeds of the wild carrot brought forth the menses and aborted an embryo. Hippocrates had prescribed the plant in the fifth century B.C.E. for preventing or aborting pregnancy, as did the ancient physicians Scribonius Largus, Galen and Paul of Aegina in the first through the sixth centuries.

Extracts of Queen Anne's lace seeds have been tested on rats, mice, guinea pigs and rabbits. When mice were given the seeds by mouth (in doses of 80 to 120 milligrams) on the fourth to sixth days of pregnancy, the seeds totally pre-vented implantation (Sharma, Lel and Jacob 1976). In other experi-ments with rodents, the seeds were found to inhibit implantation and ovarian growth and to disrupt the estrous cycle (Kaliwal, Ahamed and Rao 1984; Kaliwal and Ahamed 1987). The seeds' antifertility effect is antagonized by progesterone, and recent evidence suggests that terpenoids in the seed block crucial progesterone synthesis in pregnant animals (Kong, Xie and But 1986). A crude boiled preparation of D. carota seeds, on the other hand, was found not to block fertilization in two rat experiments; neither did it stimulate rat uterine contractions in vitro (Lal, Sankaranarayanan and Mathur 1984; Lal et al. 1986). Despite such conflicting evidence, the seeds (or their active ingredient, which has not been isolated) have been viewed as a promising post-coital antifertility agent.

Like Queen Anne's lace, the strong-smelling woody herb rue (Ruta graveolens L.) seems to turn up in many regions and, as an antifertility agent, in many times and cultures. The ancients used rue both as a contraceptive and as an abortifacient. Rather than recom-mending it, Pliny the Elder — who opposed abortion — wrote: "preg-nant women must take care to exclude rue from their diet, for I find that the fetus/embryo is killed by it." Today rue is employed as an abortifacient throughout Latin American and among the Hispanic population of New Mexico as an abortifacient (in the form of a tea), and in the traditional medicine of India.

Chinese investigators have explored the antifertility effects of rue by administering chloroform extracts of the whole rue plant in daily oral doses of 0.8 to 1.2 grams per kilogram of body weight to female rats over the first eight to ten days of pregnancy. The extracts reduced the number of pregnancies in the experimental group by 20 percent to 75 percent (depending on the potency of the extract adminis-tered). As was the case with the Ferula extracts, the same effect was not observed in hamsters. Because the active substance in rue, chalepensin, is highly toxic, its effect on fertility may be simply the result of nonselective toxicity (Kong et al. 1989).

Rue belongs to a botanical family, Turaceae, that includes several plants that produce antifertil-ity agents. Chinese scientists have studied the active substance in another Rutaceae species. The substance, called Yuehchukene and extracted from the plant Murraya paniculata var. M. sapientum L., has been reported to be 100 percent effective in preventing pregnancies in rats when administered orally at doses of 2 milligrams per kilogram during the first six days of preg-nancy, or as a single does of 3 milligrams per kilogram on the first day after coitus (Kong, Xie and But 1986). It is interesting that another Rutaceae species, Pilocarpus jaborandi Homes, produces a cholinergic agent called pilocarpine that has been given to horses to induce abortion. Pilocarpine is not known, however, to have the same effect in women.

The botanist and physician Carolus Linnaeus named a third common plant for its abortion-inducing effect. Squirting cucumber, so called because its fruit squirts it seed out when it dries, carried the Linnean name Ecballium elaterium, from the Greek word ekballion, which means "abortion." Linnaeus was hardly the first to notice its effect; Hippocrates is alleged to be the author of a treatise on women's diseases and problems in which squirting cucumber is recommended for abortion. In fact, the author indicated that sikos agrios, as the Greeks called it, was the aborifacient of choice.

Recent animal tests support the notion that squirting cucumber has a contraceptive effect. When mice were given daily doses of 20 to 100 milligrams per kilogram of extracts from the whole plant or from the flower alone, they failed to ovulate (Farnsworth et al. 1975). The abortifacient effect has not been confirmed, and it is possible that the drug's other known effects, cathar- sis, emesis and diuresis — may have led physicians to connect it with abortion.

It is possible to conclude that women of ancient and medieval times were fooled by physicians, witch doctors, herbalists, witches, midwives, village wise persons and charlatan medicine-show salesmen into taking birth-control potions that did not work. Placebo and psycho- logical effects are bound to be common in experiments with herbal medicines, but in the prevention of pregnancy they work only one way. A woman who takes a potion to prevent conception cannot be sure that it did the trick, but a full-term pregnancy is a sure sign that it doesn't work. Thus any drug that simply does not work should, over time, be discredited.

If the plants we have discussed survived as birth-control drugs on the strength of false reputations, they did so for a very long time. "We've so many sure-fire drugs for inducing sterility!" remarked the Roman satirist Juvenal in the second century C.E., as the emperors continued to struggle with population declines. St. Jerome three centuries later condemned those

who "drink sterility and murder those not yet conceived," as well as people who use poison to destroy those yet to be born. And the eighth century priests asked in confession: "Have you drunk any maleficium, that is, herbs or other agents so that you could not have children?"

If, as we suggest, modern scientists and historians decide to consider the efficacy of ancient oral contraceptives an open and useful question, a number of interesting avenues of inquiry lie ahead. Do some herbs prevent or disrupt pregnancy simply because they are toxic? Probably. The unintended effects of plant substances are many; no one should try ancient Greeks' recipes at home. But other plant substances may have clearly defined actions and controllable toxicity; there is much that is yet unknown about them, and in some circles skepticism still seems to inhibit research. We too quickly draw a hard line that separates us from the premodern period, dismissing the ancients' solutions to problems.

Historians are likely to be intrigued by another question raised by the record: Why do so few now know what so many once did? The disappearance of knowledge — the severing of a trail of learning that extended over many centuries — ought to be of concern to all scholars. In a world in which population control promises to be a growing concern, a closer look at what once was known about contraception might broaden the options that science can offer.

Bibliography

Ariès, Philippe. 1953. Sur les origines de la contraception en France. Population 8:465-72.
Dioscorides. De Materia Medica. 3 vols. Berlin:Weidmann. 1907.
Duke, J.A. 1985. CRC Handbook of Medicinal Herbs. Boca Raton, Louisiana: CRC Press.
Farnsworth, N.A., A.S. Bingel, G.A. Cordell, F.A. Crane and H.H.S. Fong. 1975. Potential value of plants as a source of new antifertility agents. Part 1, Journal of Pharmaceutical Sciences 64(April):535-98. Part 2, 64(May):717-754.
Froehlich, O., and T. Shibamato. 1990. Stability of pulegone and thujone in ethanolic solution. Journal of Agricultural and Food Chemistry 38:2057-2060.
Gordon, W.P., A.J. Forte, R.J. McMurtry, J. Gal and S.D.
Nelson. 1982. Hepatotoxicity and pulmonary toxicity of pennyroyal oil and its constituent terpenes in the mouse. Toxicology and Applies Pharmacology 65:413-424.
Gordon, W.P., A.C. Huitric, C.L. Seth, R.H. McClanahan and S.D. Nelson. 1987. The metabolism of the abortifacient termene, (R)-(+)-pulegon, to a proximate toxin, menthofuran. Drug Metabolism and Disposition 15:589-594.
Kaliwal, B.B., and R.N. Ahamed. 1987. Maintenance of implantation by progesterone in carrot seed Daucus carota extracted treated by albino rats." Indian Journal of Physical and Natural Sciences Section A7:10-14.
Kaliwal, B.B., R. Nazeer Ahamed and M. Appaswomy Rao. 1984. Abortifacient effect of carrot seed (Daucus carota) extract and its reversal by progesterone in albino rats." Comparative Physiology and Ecology 9:70-74.
Kant, A., and N.K. Lohiya. 1986. The estrogenic efficacy of carrot Daucus carota seeds." Journal of Advanced Zoology 7:36-41.
Kong, Yun Cheung, C.P. Lau, K.H. Wat, K.H. Ng, P.P.-H. But, K.F. Cheng and P.G. Waterman. 1989. Antifertility principle of Ruta graveolens. Planta Medica 55:176-178.
Kong, Yun Cheung, J.-X. Xie and P.P.-H. But. 1986. Fertility regulating agents from traditional Chinese medicines. Journal of Ethnopharmacology 15:1-44.
Kong, Yun Cheung, K.H. Ng, K.H. Wat, A. Wong, I.F. Saxenza, K.F. Cheng, P.H. But and H.T. Chang. 1985. Yuehchukene, an novel anti-implantation indole alkaloid from Murraya paniculata." Planta Medica 51:304-307.
Lal, R., M. Gandhi, A. Anakaranarayanan. V.S. Mathur and P.L. Pharma. 1986. Antifertility effect of Daucus carota seeds in female albino rats." Fitoterapia 57:243-246.
Lal, R., A. Sankaranarayanan and V.S. Mathur. 1984. "Antifertility and uterine activity of Daucus carota. A preliminary report." Bulletin of Postgraduate Institute of Medical Education and Research Chandigarh 18:28-31.
Prakash, Anand O., V. Saxena, S. Shukla, R.K. Tewari, S. Mathur, A. Gupta and S. Sharma. 1986. Anti- implantation activity of some indigenous plants in rats. Acta Europaea Fertilitatis 24:19-24.
Riddle, John. 1992 (in press). Contraception and Abortion from the Ancient World to the Renaissance. Boston: Harvard University Press.
Russell, Josiah. 1985. The Control of Late Ancient and Medieval Population. Philadelphia: American Philosophical Society.
Scarborough, John. 1989. Contraception in antiquity: the case of pennyroyal. Wisconsin Academic Review 35:19-25.
Sharma, M.M., G. Lel and D. Jacob. 1976. Estrogenic and pregnancy interceptory effects of carrot Daucus carota seeds. Indian Journal of Experimental Biology 14:506-508.
Singh, M.M., D.N. Gupta, V. Wadhwa, G.K. Jain, N.M. Khanna and V.P. Kamboj. 1985. Contraceptive efficacy and hormonal profile of ferujol: a new coumarin from Ferula jaeschkeana. Planta Medica 51:268-270.
Singh, M.M., A. Agnihotri, S.N. Garg, D.N. Gupta, G. Keshri and V.P. Kamboj. 1988. Antifertility and hormonal properties of certain carotane sesquiterpenes of Ferula jaeschkeana. Planta Medica 54:492-494.
Soranus. Gynaeciorum. J. Ilberg, ed. Leipzig: Teubner. 1927.
Sullivan, John B., Barry H. Rumack, Harold Thomas Jr., Robert G. Peterson and Peter Bryson. 1979. Pennyroyal oil poisoning and hepatotoxicity. Journal of the American Medical Association 242:2873-2874.
Thomassen, D., P.G. Pearson, J.T. Slattery and S.D. Nelson. 1991. Partial characterization of biliary metabolites of pulegone by tandem mass spectrometry detection of glucuronide glutathione and glutathionyl- glucurionide conjugates. Drug Metabolism and Disposition 19:997-1003.
Thomassen, D., J.T. Slattery and S.D. Nelson. 1990. Journal of Pharmacology and Experimental Therapy 253:567- 572.

Questions: *Answers on page 203*

1. How do you think plant-based substances might work in the female body to produce a
    contraceptive effect?

2. If a plant produces estrogens, how could they function as an antifertility drug?

3. How much scientific proof do you think exists for the presence of ancient oral contraceptives?

31

Evolutionary biologists have discarded the theory that females of various animal species are passive creatures that do not expend energy toward mate selection. The female effect on the mating process, however, remains largely unknown. The female invests far more energy than the male toward care of the offspring. This difference suggests that the female selection of a male mate, as well as the genes that he will pass on to future generations of the species, should be a careful choice. The differences in primary characteristics between sexes, such as testes and ovaries, are necessary for reproduction and evolved by natural selection. Secondary sexual characteristics, including body structures or behavior patterns indicative to one sex, are also important to successful reproduction although originally Darwin did not view them as essential for reproduction. This essential difference has led to the recognition of a second type of selection, namely sexual selection. This recognition has promoted further investigation of the female choice of a mating partner. Selection of a more brilliantly-colored peacock or more combative male bighorn sheep, for example, could be meaningful.

Female Choice in Mating
by Meredith F. Small

The large, pink rear end of a female monkey bobbed through the juniper scrub. The skin on her rear end was swollen like a balloon as a result of the hormonal changes that trigger estrus. The swelling showed males, even those at a great distance, that she was ready to mate. She soon approached a young male and swung her hindquarters into his face; he mounted her. Both of them ignored the rattle of paper and scratchy pencil noises as I noted this copulation.

I came to this group of Barbary macaques, housed on 10 hectares of oak forest in southwestern France, to study the mating behavior of females. My job was to follow one of 14 females for half an hour, several times a week, and note her sexual partners. Barbary macaques breed seasonally, from September to January. As a female comes into estrus the swelling on her hindquarters indicates that she is sexually receptive and that ovulation is imminent. A female selects a male by turning her hind end into his face; if he is interested, they copulate. Some males wait for females to

come calling. A more assertive male might approach a female and give her a slight nudge from behind; if she is willing, they copulate. The female typically emits a loud call during the copulation; afterwards, she spends a few minutes grooming her partner. The female always ends the interaction by moving on, usually making a beeline for the next male.

Initially, the data that I gathered on mating by Barbary macaques fit nicely into a relatively recent development in the theory of animal behavior. Over the years, evolutionary biologists have discarded the image of females as passive, coy creatures and have embraced the notion that females, just like males, have been selected to tend to their own reproductive interests during mating. But what are those interests?

For any organism, the fundamental evolutionary interest is passing genes to future generations. Nevertheless, males and females pursue this goal through different strategies because of physiological constraints. One conspicuous

asymmetry between the sexes is that females invest more heavily in offspring than males do; females gestate and lactate while males are free to inseminate other females. This difference in investment suggests that a female should be careful about selecting a potential father for her offspring. Mate selection by females, however, might be a significant evolutionary force acting on males, affecting the passage of particular male genes into future generations. Current evolutionary theory strongly supports a female effect on the mating process.

My observations showed clearly that Barbary females decide which males get to mate and when. These female monkeys are a perfect example of the sexually assertive female primate. But by the end of the breeding season, having observed 506 copulations. I found myself questioning the current consensus about female choice. Yes, female Barbary macaques do make choices, but they seem to choose every male in the group, one after

Reprinted by permission from AMERICAN SCIENTIST, Journal of Sigmi Xi, The Scientific Research Society.

the other. As the breeding season progressed, I recorded a steady increase in the number of different male partners for each female. If Barbary females are supposed to be selective about which males will father the next batch of infants, why were these females moving from male to male with apparently indiscriminate abandon? The day that I watched a female copulate with three different males in the span of six minutes, I knew it was time to reevaluate the current concept of female choice.

Darwin's Feelings on Females

The theoretical investigation of female choice began, as did much evolutionary theory, with Charles Darwin. His theory of evolution by natural selection implies that all individuals are under similar selection pressures. But if this is so, why are there morphological and behavioral differences between males and females? Many of these differences, ones that Darwin called primary sexual characteristics, are required for reproduction. Ovaries, testes, eggs and sperm are all necessary for sexual reproduction, and thus their development can be explained through the conventional mechanisms of natural selection. But Darwin had a more difficult time explaining secondary sexual characteristics — traits and behaviors that appear at puberty and are either limited to one sex or exaggerated in one sex, such as horns, manes and facial hair. Darwin thought that these traits arose through competition for mates — a process that he called sexual selection.

In 1871 Darwin published a further analysis of sexual selection in The Descent of Man and Selection in Relation to Sex. He was explicit about how sexual selection operates, and about which sex is affected more by the process. As far

as evolution is concerned, it is not enough for animals simply to survive; they must also pass on genetic material to win in the game of reproductive success. Darwin wrote: "This depends, not on a struggle for existence, but on a struggle between males for the possession of females; the result is not death to the unsuccessful competitor but few or no offspring."

Darwin described two ways in which sexual selection could operate. First, it could arise through competition between males, as they battle to gain access to females. This male-male competition is now referred to as intra-sexual selection. Many male traits can be explained in this way by the need for weaponry and armor during male-male contests over females. For example, the massive horns of male bighorn sheep grow as males reach sexual maturity, and the horns help males in battles for females. Male lions have huge manes to protect their necks during fights with other males. Male baboons sport giant canine teeth to intimidate opponents who want access to the same females.

In a second aspect of sexual selection, Darwin saw males winning the favor of females by first attraction attention and then waiting to be selected over other males. The peacock's tail is often cited to illustrate this point. The extravagant plumage of peacocks serves no survival purpose, and peacocks do not use their tails for fighting. But peahens are attracted by the large fans, and choose the most demonstrative males as mates. The lavish tail of the peacock has, therefore, evolved because females are attracted to it.

In 1982 Malte Andersson of the University of Göteborg in Sweden empirically demonstrated a direct relationship between female choice and the exaggeration of a male trait in African widowbirds.

Male widowbirds have an extremely long tail, about 50 centimeters in length, whereas the females have a tail more in proportion with their body size. In an ingenious experiment design, Andersson lengthened the tails of some males by adding pieces of tail feathers and shortened the tails of others; for controls he left some males as they were and in other cases clipped a segment of tail and then glued it back onto the same bird. Andersson found more new nests in territories owned by males with artificially lengthened tails. He also showed that males do not use there exaggerated tails in male-male competition. Andersson concluded that the only explanation for the evolution of tail-length exaggeration in widowbirds is female choice. Female widowbirds consistently choose males with the longest tails. Today we call this process inter-sexual selection or female choice.

In both intra-sexual and inter-sexual selection, Darwin emphasized that sexual selection explains how males evolve odd and exaggerated characteristics. Darwin felt that the female role in sexual selection was minor; even when a female chooses a certain male over others the female's behavior only serves to explain male traits. Darwin's emphasis on males over females was based on what he considered the difference in passion between the sexes. Darwin felt that males were the more passionate sex and therefore they would fight to gain females, whereas females were not interested in sex. Darwin wrote of this supposed difference in ardor: 'The female ..., with the rarest exceptions, is less eager than the male. As the illustrious [John] Hunter long ago observed, she generally 'requires to be courted'; she is coy, and may often be seen endeavoring for a long time to escape from the male." Males, with their zealous passion, would be competitive, whereas females, with their passive approach to sex, would be choosey. Darwin clearly thought that female choice, as a force of sexual selection, was secondary to male-male competition, and rather unimportant.

Ruth Hubbard, a historian of science at Harvard university, has suggested that Darwin's view of female sexuality was more influenced by his Victorian times than by the facts of animal behavior. In defense of Darwin, he had few reports of distinctively female animal behavior to guide his conclusions, and as for human models, he had only the behavior of the females around him — his social milieu — to form a picture of femaleness. Darwin's writings about women sound very quaint to modern ears. He saw men as more "courageous, pugnacious, and energetic than women," and thought that a woman differed from a man "in mental disposition, chiefly in her greater tenderness and less selfishness." And it seems he believed the sexual behavior of female animals generally would be like that of "proper" Victorian ladies — passionless and passive. It would take years of field and laboratory research to prove Darwin wrong on this point. Female animals are anything but sexually passive creatures.

Female Choice Affects Females

For almost a century, biologists virtually ignored female choice as a potential evolutionary force. R.A. Fisher a leader in both statistics and genetics during the first half of the 20th century wrote briefly about female choice in his volume The Genetical Theory of Natural Selection, a classic evolutionary text published in 1930. Fisher proposed that mate choice can have an evolutionary effect only if a consistent choice by the female and the favored trait of the male are passed on together. Fisher held that

females choose males because of a particular trait even though the trait probably has little effect on their offspring. The trait in the male and the choice by the female pass through generation after generation, co-evolving in a runaway manner. Natural selection stops the exaggeration only when the trait becomes a hindrance.

In the 1950s John Maynard Smith of the University of Sussex in England began working on the mating process in general, his experiments produced some unexpected information about female mating behavior. In one experiment, Maynard Smith and his colleagues worked with a strain of flies in which many of the males were infertile because of their inbred heritage. The investigators also noticed a ritualized courtship dance of male and female flies. At first, when females bobbed right and left in front of a courting male, Maynard Smith assumed that the females were trying to get away. But other experiments with wax models of female flies showed that female participation in the dance was necessary for mating to proceed smoothly. And in many cases the female started dancing just to see if the male could keep up. When inbred males, who were clumsy dancers with bad timing, tried to mount females, they were often too far forward or too far back, and the female would eventually kick them off altogether. As Maynard Smith put it for these poor males: "The spirit is willing but the flesh is weak."

Maynard Smith proposed that the dance evolved on the female's behalf. In other words, females judge males by their dancing ability, and the dance is a true indicator of male fertility. Remember that the inbred, less-fertile males were poor dancers. Maynard Smith predicted that selfish choices made by a female

might, as a secondary consequence, affect the appearance of male characteristics and behaviors.

This was a new kind of female choice, one based on the reproductive interests of the females rather than on male-male competition or on females mindlessly choosing males with the most impressive displays. Unfortunately, no one listened. Maynard Smith recently wrote: "When, in 1956, I published a paper showing at least to my own satisfaction, how female fruit flies choose males. I do not remember receiving a single reprint request." Two decades passed before anyone seriously considered females as effective factors in sexual selection.

In 1972 Robert Trivers of the University of California at Santa Cruz wrote a paper on parental investment that forever changed the structure and meaning of female-choice theory. Trivers argued that Darwin's dichotomy between mate competition as a male domain and mate choice as a female domain is reasonable, but not because of a difference in passion between the sexes. The same conclusion can be reached because of a difference in reproductive strategies. Males, who produce effectively endless amounts of sperm, and typically spend no time caring for offspring, do best in evolutionary terms when they gain as many mates as possible. Females, on the other hand, produce fewer gametes, and they gestate and usually care for the infants. Females should be choosy, picking the best father for their offspring. This male-female difference is especially critical for mammals because females spend months nursing infants, and each batch of offspring represents a large investment. Trivers suggested that these biological differences produce males who battle for females, and females who should be very selective about their reproductive partners.

Female-choice theory is now somewhat different from what Darwin envisioned. Gone is the supposed difference in passion between males and females. The passive role for females is replaced by a new sexual assertiveness. And females gained a further dimension from Trivers. Females are selected to address their own reproductive requirements, choosing a mate who will pass on the best genes to the next generation. This change in perspective toward female behavior, initiated in the 1970s and still running strong, was socially timely, and it would be naive to ignore the influence of social trends on the acceptance of the theory. the feminist revolution in Western culture paved the way for open-minded thinking about female roles, and more women entered the field of animal behavior. Today animal-behavior journal about with data on female behavior, especially information on female mating behavior. We have learned that females are active participants in mating, but oddly enough we remain unsure about the impact of female choices on inherited traits, behaviors and individual reproductive success.

Fishing for Fitness

Ryne Palombit, a colleague in animal behavior at the Univeristy of California at Davis and I believe that part of the confusion over the effect of female choice on an animal's life is caused by a basic confusion in the theoretical framework. Between the days of Darwin and today, evolutionary theorists have really identified two types of female choice.

Following the traditional theory of Darwin and Fisher, females might choose specific males because the females are attracted to some male trait that appeared in the first place by natural selection. For example, the male widowbird first needed a tail for stability in flight,

not for attracting females. The co-evolution of the male's trait and the female's choice explains the evolution of male feature. This kind of "Fisherian" female choice explains the evolution of male traits and differences between males and females, just as Darwin predicted. To look for this type of female choice, we must ask the question. Are there traits that are exaggerated in or exclusive to one sex, and that cannot be explained by male-male competition."

For the primate order the group of animals that I study the search for traits that might have evolved by Fisherian female choice is difficult. Male primates are usually bigger than females, but this is due to intense pressure on males to compete with other males for matings. Male baboons have huge canines but females have canines too, and the male exaggeration is probably due to fighting among males. There are only a few examples in primates of male traits that could, perhaps, be explained by Fisherian female choice. For example, several lemur species are sexually dichromatic: The males are one color and the females are another. This trait probably did not evolve to help males in fighting. Therefore, it may have evolved by female choice, even if coloration has nothing to do with male fitness. The only other clear example is the brightly painted red and blue face of the male mandrill.

The other kind of female choice is what I call "Triverian" female choice, where females are making choices based on what is best for them. In this type of female choice, the effect on males can be secondary. The only time that a male is affected is when the specific trait on which the females base their choice also has a link to the fitness of the male. Torbjörn van Schantz and his colleagues at the University of Lund in Sweden showed a direct

connection between a male trait (the spurs of pheasants), female choice and improved reproductive success for females. Van Schantz factored out male size, wing span, territory and age by changing the kinds of males available to females. Only spur length was consistently important to females. And females who mated with large-spurred males hatched the largest clutches. If spur length is a heritable trait, then the males from these clutches should also have long spurs. In other words, if a male trait is correlated with health and vigor and females are attracted to that trait, then the females' choice will affect the evolution of the male trait. Traits with this property represent "truth in advertising" on the part of males. But to females, how their choice affects male evolution is not important; the female is only interested, in an evolutionary sense, in how her choice affects her offspring. From this perspective, the mail trait, and how it evolved is secondary to the motivation of the female making the choice.

These two kinds of female choice, Fisherian and Triverian, are lumped together by most biologists, but there is an important distinction between them. Fisherian female choice is always sexual selection in the manner described by Darwin; it explains difference in male and female traits and the exaggerated appearance of those traits. Triverian female choice is really an expression of the ordinary natural selection because it is inconsequential if the female's choice affects male traits. The female has been naturally selected to choose the best male as her fertilizing partner.

Most of the confusion about what females are really doing stems from misunderstanding the differences between these two types of female choice. There is no question that female animals do make choices. At question is the evolutionary effect of a female's choice. Does it simply affect male traits and maintain the difference between males and females, or does it improve the female's fitness? Fisherian female choice gives rise to differences between males and females, without necessarily directly affecting the female's fitness. Triverian female choice may also create differences between males and females, but its primary effect is to enhance the viability of the female's offspring.

Primate Picks

Primates are likely candidates for studies of female choice, because of their big brains, high intelligence and flexible behavior. Twenty years of primatology, including laboratory and long-term field studies, has shown that female primates are active participants in the mating game. They often initiate sexual activity, and they walk away from males as well. Female primates are highly sexual — mating not just for conception at the moment of ovulation but repeatedly during the estrus cycle. Although these females are clearly sexually assertive, we do not know what the evolutionary effect of this assertiveness might be or what it evolved. And we are not sure how a female primate's choice of a partner has shaped the evolution of primate mating systems.

Reproduction by a female primate should be constrained by one critical factor — she has a limited reproductive potential. Female primates tend to reach sexual maturity late (in comparison with other mammals), and they usually bear only one offspring at a time, with long periods between births. Furthermore, the female invests considerable energy in her offspring. Infants are nursed for many months, and mothers usually carry them until the infants initiate independence. One has only to think of the human infant, with its years of

dependence and need, to see the far end of the primate continuum. Considering that a female has a limited amount of energy to expend on the production of offspring, she should carefully select a mate who will enhance the survival of her offspring and, hence, the survival of her genes. Female choice, when it occurs, should reflect the reproductive interests of the female.

But how can the female judge the fitness of a male? Primate females cannot count sperm, do health checks or predict male vigor. Nevertheless, the females do choose on the basis of proximate cues provided by males. These cues are both physiological and behavioral.

In many species of primates, female choice appears to depend at least in part on a male's rank, his place in the troop's hierarchy. Not all primate species have such hierarchies, but when they do, females seem to pay particular attention. In theory, females should prefer males of high status because this is generally correlated with access to resources. In addition, high-status males might provide some protection from predators or aid females during interactions with the troop. In a survey of female response to male rank. I found that females in nine primate species seem to prefer high-status males. For example, female rhesus monkeys, vervet monkeys and savannah baboons choose high-ranking males when given the chance. These males also tend to be the older males in the group, which guarantees mature sexual physiology. But the preference for high-status mates is not absolute of invariant, even in species that have a status hierarchy, some females do not pursue a high-status male.

In addition, female primates may consider the familiarity of the prospective mate. In a few species, females choose familiar males over unknown males. Barbara

Smuts of the University of Michigan at Ann Arbor reports that female olive baboons often have two males friends who gain preferred access during estrus. Correspondingly, orangutan females actively resist nonresident males. But this penchant for the familiar in some species is balanced by a taste for the exotic in others. Many female primates are attracted to unknown males. This selection for novel males may have evolved to avoid inbreeding. Females in nine primate species have been reported to show a preference for "foreign" males. Some females — such as those of patas monkeys and rhesus monkeys — will actively seek males lurking on the periphery of the troop. Japanese macaques, conversely, seem equally willing to mate with either a familiar or an unfamiliar male.

And we must consider the factor of popularity. People base many of their mating decisions on "attractiveness." Perhaps nonhuman primates also have some concept of attractiveness, or beauty. In fact, many of the choices made by nonhuman-primate females that seem inexplicable to human observers might fall into this category. One male in my group of Barbary macaques was repeatedly the target of female attention. To my eyes, he seems a brute — he bit females and never engaged in mutual grooming. But the Barbary females evidently saw in him "a certain something" that only a female macaque could explain. His rate of copulation was higher than that of any other male.

In essence, our knowledge of female choice in primates presents a confusing picture. Sometimes the females choose a high-status male, sometimes they don't. Sometimes the females choose a familiar male, sometimes they don't. Part of this confusion may arise from a lack of data. We do not have a single example of a female primate making a specific mating choice that can be confirmed to increase her reproductive success. Likewise, we lack any direct evidence of a female primate making a choice that affects a male characteristic. Maybe we lack such examples because we have yet to ask the right questions.

There is, however, another potential explanation: Female choice may not be a powerful evolutionary force in primates. For female choice to have any evolutionary impact, females must make consistent choices; yet some female primates do not seem to make any choice. Female chimpanzees seem to have no clear preferences at all; they mate with just about every male. As I have shown, the same is true for Barbary macaque females. However, these females are not passive or indifferent; they do have an agenda of their own, but that agenda is simply to mate with any male in reproductive condition. I believe that some primate females are less

choosy because they are under pressure to conceive. Ovulation is an event with a short window of opportunity, and seasonal breeding means there are few chances to conceive before females fall into an anestrous state. Although all females might prefer the "best" male, best is probably a category that expands and contracts with the timing of ovulation. If there is not time to dawdle, there is no time to be choosy.

The behavior of female Barbary macaques and of other female primates suggests that our current notion of female choice is already antiquated. We have empowered the behavior of females by acknowledging their sexual assertiveness, but we often stop short of accepting that sexual assertive behavior might result in less than choosy behavior. A passionate female with reproduction in mind must do what she must do. It may, in fact, be impossible to be sexually assertive and choosy at the

same time. It is not that the contribution of male genes is unimportant to a female; but she may be more interested in simply conceiving, rather than conceiving with a particular male. Lowering her standards, and taking anyone who is interested, must then be one of the strategies available to a female primate.

Bibliography

Andersson, Malte. 1982. Female choice selects for extreme tail length in a widowbird. Nature 299:183-186.

Darwin, Charles. 1859. The Origin of Species. London. John Murray and Sons.

Darwin, Charles. 1871. The Descent of Man and Selection in Relation to Sex. London. John Murray and Sons.

Fisher, R.A. 1930. The Genetical Theory of Natural Selection. New York Dovet.

Hubbard, Ruth. 1979. Have only men evolved In Women Look at Biology Looking at Women. ed. Ruth Hubbard. Boston and Company.

Maynard Smith, John. 1955. Fertility, mating behavior and sexual selection in Drosophila obscura Genetics 54:261-279.

Small, Meredith F. 1989. Female choice in non-human primates. Yearbook of Physical Anthropology 32:103-127.

Small, Meredith F. 1990. Promiscuity in Barbary macaques (Macaca sylvanus). American Journal of Primatology 20:267-282.

Smuts, Barbara B. 1985. Sex and Friendship in Baboons. New York:Adline.

Trivers, Robert L. 1972. Parental investment and sexual selection. In Sexual Selection and the Descent of Man, ed. B. Campbell. New York:Adline.

Questions: *Answers on page 203*

1. How could the brilliant plumage of a peacock be adaptive?

2. How could the combative behavior of a male bighorn sheep be adaptive?

3. Why do you think biologists ignored female choice as insignificant for such a long period of time?

32

The fern life cycle is an alternation of generations. A dominant diploid sporophyte alternates with a haploid gametophyte. Clusters, each one called a sorus, dot the edges and bottoms of the fern leaflets or fronds. Each sorus contains sporangia where spores are produced meiotically. The haploid spore produced by the sporophyte germinates into the haploid gametophyte. The production and union of gametes by this generation leads to the formation of the diploid sporophyte, thus completing the life cycle. Therefore, the germination of the spore into the gametophyte is a very important link of the plant life cycle. Often a spore lies dormant before germinating. What events trigger its activity? What occurs during the germination process? Light is one signal keying the event. The reaction of phytochrome and other plant pigments to changing conditions of light in the environment produces germination. During germination the spore builds many complex molecules. Many of these are protein molecules. Therefore protein synthesis, perhaps in response to the signal of plant hormones, is another key requirement for germination. All results of research point to a complex series of interactions occurring during the process of germination.

Germination of Fern Spores

by V. Raghaven

Every gardener knows the fern, whether from an apartment flower pot or an informal shade garden. The fern's graceful lines and the soft, feathered triangles formed by its fronds bring the beauty of the forest to the domestic landscape. Botanists, too, have paid much attention to the fern, but they are not interested only in the quality of the plant's form. Rather, in the apparent simplicity of how a fern's life begins — with the germination of a single cell called a spore — lies a story basic to plant biology and yet filled with mystery. A fern spore can rest in a dormant state for years, waiting for environmental conditions to be just right for germination. Then with a little water and either the right sort of light or a minute quantity of a plant hormone, the spore revives, and a complex burst of activity begins the life of a new plant. The fern spore provides a tractable, single-cell model with which to explore many questions about germination, not only of spores themselves but of those more complex reproductive structures called seeds.

Ferns evolved in the Devonian period, nearly 400 million years ago. They radiated dramatically and stood with tree lycopods (club mosses) and horsetails in the Carboniferous forest. Many ferns from the Carboniferous period became extinct, but many other species survived. Most familiar are the small ferns of the temperate and tropical forests. However, one extant tropical fern, Dicksonia, is a tree with stems reaching many meters high. A jungle fern, Gleichenia, has the longest leaves of any plant, up to 50 meters long, and these leaves can cover an entire hillside. Ferns remain the best-represented seedless plant group among modern flora.

Whereas apple trees and avocados make seeds to produce progeny, ferns make spores that, during germination, divide repeatedly to produce multicellular plants. Spores are small and light and can be carried by the wind, allowing ferns to disperse across great distances. During his voyage on the H.M.S. Beagle, Charles Darwin collected air samples that contained fern spores, although the ship was hundreds of miles from shore.

Ferns lead two lives: a sporophytic generation and a gametophytic generation. The familiar lacy, green fern is the sporophytic generation and contains two sets of chromosomes in each cell, making the sporophyte diploid. On the edges or bottoms of the fronds of the sporophyte, small clusters, like microscopic bunches of grapes, dot the surface. Each cluster is called a sorus, or fruit dot, and contains many sporangia, the reproductive structures of the sporophyte. Within the sporangia are found the dust-like spores. the spores are born out of a meiotic division of a sporophytic mother cell; therefore they are haploid, each containing only one set of chromosomes. At maturity, the sporangium bursts open along a line of weak cells and by a spring-like mechanism, catapults the spores into the air. If a spore lands in a favorable location, it germinates to produce a gametophyte, a form of the fern that later develops sperm-producing and egg-producing organs. (Two orders of water ferns are heterosporous — that is, they produce two kinds of spores, one giving rise to female gametophytes and the other to make

Reprinted by permission from AMERICAN SCIENTIST, Journal of Sigmi Xi, The Scientific Research Society.

gametophytes.) In the presence of water, a sperm can fertilize an egg to produce a new sporophyte thereby completing the life cycle.

In this reproductive cycle, the germination of the spore deserves particular attention. The spore's germination initiates the gametophytic stage which later allows genetic recombination and the colonization of new habitats. But what triggers germination in an apparently lifeless cell, launching a flurry of activity that will result in the spore dividing and differentiating into a living, complex plant? What is the source of the genetic signals that guide this process? The purpose of this article is to provide an overview of recent findings in the control of fern-spore germination.

Much of what we know about the germination of fern spores comes from the study of those spores that are normally dormant. These spores are somewhat peculiar and fail to germinate even when environmental conditions appear appropriate. Their germination can be started or stopped by a single essential factor making it possible to identify triggering mechanisms with relative ease. (Less useful germination studies but more common are quiescent spores which germinate readily if placed in an environment that offers ample moisture, light,oxygen and an appropriate temperature.)

The spores of most species undergo a common series of cytological changes during germination. the nucleus of a dormant spore lacks granularity, and its chromosomes are packed tightly together. A few hours after the spore is provided with water, the nucleus looks normal. When the spore receives light or a plant hormone, the nucleus moves from its central position to one end of the spore known as the proximal pole. By convention the proximal pole is the region of the spore in common contact with the

other members of the tetrad _ the group of four spores that arise from meiosis of one sporophytic mother cell and remain bound together. The other end is the distal pole. The equatorial plant lies at right angle to the proximal-distal axis.) When the nucleus reaches the proximal pole, mitosis divides the cell in a line parallel to the equatorial plane. However, the cell is not divided equally. A small lens-shaped cell is cut off near the proximal pole. This cell becomes the rhizoid — a narrow, colorless projection that anchors the delicate, young plant and absorbs water. Later the large cell divides again, along a line perpendicular to the first division. One of these cells becomes the protonema, the forerunner of the gametophyte. At this three-celled stage, germination is considered complete.

Let There Be Light

Light is one signal known to awaken dormant fern spores. Studies of the effects of light had their beginnings in the 19th century, when investigators discovered that different wavelengths of light have different effects on the germination of seeds. Experiments in 1937 by Lewis Flint and Edward McAlister of the Smithsonian Institution showed that red light prompts lettuce seeds (scientifically knows as fruits called achenes) to germinate. In the 1950s, Harry A. Borthwick and his colleagues at the U.S. Department of Agriculture research station at Beltsville, Maryland, Further investigated the effects of light on the germination of lettuce seeds. These investigators found that red light induces germination; but seeds exposed to red light and then far-red light (the wavelengths between red and infrared light) fail to germinate even when all other environmental conditions are proper for germinating. This suggested that red light turns germination on and

that far-red light turns it off.

The presence of a light-sensitive molecule, phytochrome, explains this phenomenon. Phytochrome is composed of a large protein that is covalently linked to a chromophore, a light-absorbing pigment. The absorption of light changes the structure of the chromophore and, thereby, the structure of the phytochrome as a whole; likewise, the chromophore's structure determines what kind of light can be best absorbed. When left in the dark, phytochrome exists in a form that best absorbs red light; this form is called the Pr form (P for phytochrome and r for absorbing red light). Pr phytochrome is inactive; it does not trigger germination. Once Pr phytochrome is exposed to red light, however, the chromophore's absorption of the light changes the phytochrome's structure to a second form, called Pfr phytochrome (the fr indicating it absorbs far-red light). This is the active form whose biochemical characteristics trigger the next steps in germination. However, the Pfr form is unstable; over time it reverts to the Pr form, but only after germination is complete. In essence, phytochrome acts like a simple germination switch, and the wavelength of light flips the switch back and forth between off (the Pr form) and on (the Pfr form).

In 1955, Erwin B_unning and Hans Mohr of the University of T_bingen exposed spores of the male fern (Dryopteris filix-mas) to red and far-red light. B_nning and Mohr discovered that illuminating well-watered spores with red light induces germination. But the same spores are inhibited from germinating if they are later exposed to far-red light. This suggested that fern spore germination is also under the control of phytochrome.

Beginning in the late 1960s, Michizo Sugai and his colleagues at the University of

Tokyo showed that the germination of fern spores depends on three photoreactions. Working with Pteris vittata, they could repeated turn germination on and off with red and far-red light, respectively. But blue light seemed to irreversibly inhibit germination. Once spores are illuminated with blue light, no amount of red light makes them germinate. However, the investigators found that spores inhibited by blue light can be revived by being returned to the dark and then illuminated with red light. Presumably, the phytochrome interacts with another pigment that is sensitive to blue light.

Work with spores of a wide variety of ferns showed that germination is generally controlled by phytochrome or by a combination of phytochrome and a blue-light-absorbing pigment. The key factor in inducing germination is adequate production of the active form of phytochrome (Pfr) before the blue-light-absorbing pigment stops the process. The ratio of the active Pfr phytochrome to the inactive Pr phytochrome in a spore determines whether or not it will germinate. However, Pfr phytochrome is not the ultimate control.

What step does Pfr phytochrome initiate? In the early 1980s Randy Wayne and Peter K. Hepler of the University of Massachusetts at Amherst examined the effect of calcium on the germination of spores of the sensitive fern (Onoclea sensibilis). The investigators eliminated calcium from outside the spore by applying EGTA (ethyleneglycol-bis (_-aminoethylether)-N,N,_,_-tetra-acetic acid), a substance that binds calcium. They found that germination of the spores could be induced only if both red light and a calcium-rich medium were provided. In another set of experiments, calcium inside the spore was increased by bathing the spore in a solution that contained the ionophore A23187, which helps carry calcium into the spore through its membrane. Even without light, the increase in internal calcium triggered germination. Later, Wayne and Hepler examined spores by atomic-absorption spectroscopy and saw that exposure of the spores to red light was followed by an increase in internal calcium. Likewise, far-red light decreased internal calcium.

This suggested a more complete pathway to germination in fern spores. Illuminating spores with red light increases the proportion of phytochrome molecules that are in the Pfr state. Apparently Pfr phytochrome affects the membrane of the spore, which allows an increase in the internal calcium. Then the internal calcium triggers the next step in germination. On the other hand, if spores are first illuminated with red light, later illumination with far-red light increases the level of Pr phytochrome, which causes the spore to expel calcium, and germination is inhibited.

The role of calcium in coupling a stimulus to a response is well known in animal systems. For example, electrical activity in a neuron leads to an increase in calcium inside the cell, which is necessary for the release of a chemical messenger — a neurotransmitter — that carries the signal to the next neuron. So finding a similar scenario in fern spores is not surprising. It looks as though evolution settled on this ubiquitous metallic ion as a second messenger in diverse living systems.

The experiments with calcium revealed a useful approach to tracing the course of germination — experimenting with substances known to mimic the effects of light. If the added substance induces spores to germinate in the dark, maybe a substance with similar properties is produced by the spore as it responds to light and begins the process of germination.

The plant hormone gibberellic acid is another substance that can induce fern-spore germination in the absence of light. In 1962 Helmut Schraudolf, then working at the Iustus-Liebig University of Giessen and now at the University of Ulm, discovered that light-requiring spores of the fern Anemnia phyllitidis germinate in the dark if a small amount of gibberellic acid is added to the medium. Other investigators confirmed the effects of gibberellic acid on the germination of spores of Anemia phyllitidis; but, there is general agreement that gibberellic acid is not a universal trigger for fern germination, as light is. Gibberellic acid causes a high proportion of spores to germinate only in species of the genus Anemia and in the Japanese climbing fern (Lygodium japomam). Other ferns germinate sparsely or not at all in the presence of gibberellic acid. Moreover, it is not clear that light-induced germination in spores of the genus Anemia involves the production of gibberellic acid.

Pointing to Proteins

Stimulated by red light the spore tunnels its resources into a rapid burst of metabolic activity. Neither the choreography nor the chronology of this activity is yet fully understood. Nevertheless, some basic characteristics are clear.

Early in germination, the spore "digests" many of the complex compounds that have been kept in storage. These compounds are mostly macromolecules such as starch, proteins and lipids. They are converted into small molecules such as simple sugars, amino acids and fatty acids. Spores of some ferns, such as Anemia phyllitidis and Pteris vittata, contain granules that appear to be filled with proteins. My colleagues and I found that these granules are empty within about 36 hours after a spore begins to germinate. The degradation of these proteins as well as that of starch and lipids may raise the osmotic pressure within the spore. Splitting the larger compounds into smaller packages raises the concentration within the spore. More water enters the spore to dilute the molecules. The water stretches the spore like an over-inflated balloon until the wall ruptures and the rhizoid escapes. This may be one of the final acts in germination.

Germination, however, is a building process, and the spore must at the same time construct many complex molecules. The synthesis of various proteins appears to be initiated within the first few hours after germination begins. But the rule of these molecules in the early stages of germination remains something of a puzzle. In several experiments, germination has been induced in the presence of substances that inhibit the synthesis of the nucleic acids DNA and RNA, which control protein synthesis. These experiments have produced varied pieces of information but not a full picture of how proteins necessary for germination are made.

Protein synthesis is the final step in a complex cascade that carries information through a cell. One stage of this cascade involves messenger RNA (mRNA), the nucleic acid that carries the blueprint for making proteins out of the nucleus and into the cell's cytoplasm. The function of mRNA raises an interesting question for those studying the early stages of germination: What is the source of the particular mRNA that serves as a template for the synthesis of the first proteins? Finding an answer to this question has been a preoccupation in my laboratory for 20 years.

Seeds and fungal spores contain stored mRNA that codes for the first proteins of germination. My colleagues and I thought that fern spores might do the same thing. To study this possibility, we used the antibiotic actinomycin D, which blocks mRNA synthesis in a diverse group of plants and animals. For the experiments with actinomycin D, we developed a simple ground rule: If a stage of germination proceeds in the presence of this antibiotic, that stages does not require synthesis of mRNA — a result that might suggest that stored mRNA is being used.

We began by investigating spores of the bracken fern (Pteridium aquilinum). When induced to germinate in the presence of actinomycin D, the spore coat ruptures as it would during the normal process, but the rhizoid and protonema fail to elongate as they would under normal conditions. this shows that the fern spore does not need to synthesize new mRNA for germination; but the spore does require mRNA synthesis for the elongation of the rhizoid and protonema.

Next, we added a radioactive precursor of protein synthesis (^{14}C-leucine) to spores both with and without actinomycin D. In both cases and at various stages of germination, we found radioactivity in proteins extracted from the spores. This meant that the spores built new proteins even without the synthesis of new mRNA. In a similar experiment, we supplied a radioactive precursor of RNA synthesis (^{3}H-uridine) to the spores. When the labeled RNA was fractionated on a sucrose gradient, no germinating spores showed any radioactivity in their mRNA. These experiments suggested that germination of the bracken-fern spores proceeds normally, at least through the early stages, without the production of any new mRNA. We concluded that, like seeds and fungal spores, fern spores have mRNA in storage, waiting for the right conditions for germination.

Other experiments confirmed that dormant fern spores contain stored mRNA. Annette Fechner and Schraudolf isolated the mRNA from dormant spores of Anemia phyllitidis and used this mRNA as a template for protein synthesis in vitro. This experiment produced proteins, proving that the dormant spores contain functional mRNA.

In my laboratory, we labeled mRNA that was stored in the spore. We used a radioactive probe (^{3}H-polyuridylic acid) that binds to all the mRNA in the spore. Once the mRNA was tagged with the radioactive label, we located it with autoradiography. In this process, this sections of the spores are placed on a glass slide and then covered with a photographic emulsion. Particles emitted from any radioactive source (in this case, ^{3}H-polyuridylic acid bound to mRNA) activate the emulsion, much like light or x-rays, and leave tiny, black grains of metallic silver. These tiny grains mark the location of the labeled mRNA. The mRNA is found throughout the dormant spore. As germination advances, there is progressively less mRNA in the cytoplasm, but mRNA remains in the nucleus. At the end of germination, mRNA is present in the rhizoid and protonema. These experiments suggest the following scenario. During the early stages of germination, stored mRNA codes for protein. As germination progresses, the stored mRNA is gradually depleted. Later, new mRNA is synthesized and appears in the newly formed cells, such as the rhizoid and the protonema.

The final question was: Does the stored mRNA produce the same proteins that a germinating spore produces? To approach this question, we first isolated the stored mRNA from dormant spores of the sensitive fern and used this mRNA as a template to synthesize protein.

This gave us the proteins that the stored mRNA could produce. Next, we placed spores in a medium that contained ^{35}S-methionine, a radioactive amino acid that would label newly formed proteins. At various intervals, we isolated the proteins formed in the spores and compared them with those produced from the stored mRNA. As early as four hours after illumination, a spore begins to synthesize small amounts of three proteins that are encoded by the stored mRNA. Within 24 hours after illumination, the spore produces 20 percent of the proteins made by the stored mRNA. In the end, the germinating spore produces all of the proteins that the stored mRNA can produce, and three proteins that are not produced by the spore's stored mRNA. This suggests that the spore needs to make only minor adjustments in its genetic program to go from dormancy to germination.

We thought this was the answer: Stored mRNA codes for proteins that trigger germination. It seemed as if these proteins must be specific to germination. But it is possible that they are not. These proteins from the stored mRNA are also synthesized although in sightly smaller quantities by spores provided with water and kept in complete darkness — spores that are not germinating. So these proteins do not necessarily trigger germination. Maybe some of the rare proteins formed by the stored mRNA are specifically for germination. But such rare proteins produced in extremely small amounts are not detectable by our methods.

Some investigators then challenged the hypothesis that proteins from stored mRNA are specific to germination. Fechner and Schraudolf argued that the stored mRNA codes for proteins that act in the ripening of the spore, not in the germination of the spore. Nevertheless, my colleagues and I began to search for rare proteins produced from the stored mRNA.

We approached this problem by constructing a complementary-DNA (cDNA) library of stored mRNA from spores of the sensitive fern. Using protocols from genetic engineering, my colleagues and I isolated a highly purified mRNA fraction and, through the action of reverse transcriptase, used that mRNA to make single-stranded and double-stranded DNA. The cDNA was then ligated into plasmids, and these plasmids were used to transform bacterial cells. We eliminated any sequences of DNA that are common to spores, leaves and mature gametophytes in the sensitive fern. Thereby we obtained a small number of bacterial clones that contained spore-specific messages. From these messages we selected one important sequence of DNA that complements mRNA stored by fern spores and that codes for proteins that appear specific to germination. The related mRNA begins to decrease within 12 hours after one exposes the spore to light, and it is completely gone within 48 hours. However, the same mRNA does not decrease, even after 96 hours, in spores that remain in the dark and, therefore, do not germinate. Presumably this mRNA produces proteins that participate in the process of germination. Now we know at least one sequence of DNA that is important for germination. However, the sequence does not correspond to any known proteins. In other words we do not know what this sequence of DNA produces, and therefore do not yet know the identity of the first proteins of germination.

We have not reached our desired destination — understanding the pathway from a dormant spore to a growing gametophyte. Yet we have answered some difficult questions. Different wavelengths of light can either trigger or inhibit the process of germination. If the light is red, germination proceeds. Inside the spore, a second messenger, calcium, appears to transmit the external message. Then internal biochemistry takes control. Complex molecules degrade into simpler elements, and protein metabolism commences. And indeed, some of the stored mRNA in the spore does produce proteins that affect germination. We have yet to define the precise nature of those proteins, but we continue to look.

Bibliography

Borthwick, Harry A., Sterling B. Hendricks, Marion W. Parker, Eben H. Toole and Vivian K. Toole. 1954. A reversible photoreaction controlling seed germination. Proceedings of the National Academy of Sciences of the USA 38:622-666.

Bünning, Erwin, and Hans Mohr. 1955. Das Aktionsspektrum des Lichteinfiusses aut die Keimung von Farnsporen Naturwissenschaften 42:212.

Fechner, Annette and Helmut Schraudelf. 1982. Isolation and characterization of poly adenylic acid - containing ribonucleic acid from dry spores of Anemia phyllitidis L. Sw. Zeitschrift für Pflanzenphysiologie 108:419-428.

Fechner, Annette and Helmut Schraudolf. 1984. Translator and transcription in inbibed and germinating spores of Anemia phyllitidis L Sw Planta 161:451-458.

Raghavan, V. 1970. Germination of bracken fern spores. Regulation of protein and RNA synthesis during initiation and growth of the rhizoid. Experimental Cell Research 63:341-352.

Raghavan, V. 1974. Control of differential in the fern gametophyte. American Scientist 62:465-475.

Raghavan, V. 1977. Cell morphogenesis and macromolecule synthesis during phytochrome-controlled germination of spores of the fern, Pteris vittata. Journal of Experimental Botany 28:439-436.

Raghavan, V. 1980. Cytology, physiology, and biochemistry of germination of fern spores. International Review of Cytology 62:69-118.

Raghavan, V. 1989. Developmental Biology of Fern Gametophyte. Cambridge University Press, Cambridge.

Raghavan, V. 1991. Gene activity during germination of spores of the fern, Onoclea sensibilis: RNA and protein synthesis and the role of stored mRNA. Journal of Experimental Botany 42:251-260.

Raghavan, V., and Joseph C. Kamalav. In preparation Isolation and characterization of germination-specific cDNA clones from dormant spores of the sensitive fern (Onoclea sensibilis).

Sugar Michizer and Masaki Furuva. 1968. Photomorphogen-esis in Pteris vittata. Recovery from blue-light induced inhibition of spore germination. Plant and Cell Physiology 9:671-680.

Vogelmann, Thomas C., and John H. Miller. 1980. Nuclear migration in germinating spores of Onoclea sensibilis: The path and kinetics of movement. American Journal of Botany 67:648-652.

Wayne, Randy, and Peter K. Hepler. 1984. The role of calcium ions in phytochrome-mediated germination of spores of Onoclea sensibilis L Planta 160:12-20.

Wayne, Randy, and Peter K. Hepler. 1985. Red light stimulates an increase in intracellular calcium in the spores of Onoclea sensibilis. Plant Physiology 77:8-11.

Questions: *Answers on page 203*

1. Why is a knowledge of the environment necessary to understand germination?

2. Why is a knowledge of physics necessary to understand germination?

3. Why is a knowledge of biochemistry necessary to understand germination?

Part Six

The Evolutionary Legacy - Unity and Diversity of Living Things

33

Were humans responsible for the extinction of many late-Quaternary mammals or did climatic changes cause their demise? Paleontologists have studied the fossil bones of pygmy hippos and small elephants, mainly discovered in the Americas and Australia. Usually they have concluded that climatic changes caused the extinction of these diminutive species, just before the arrival of humans into their habitat. However, recent discoveries in other parts of the world, particularly from the Mediterranean islands, have established that humans and the small hippos may have coexisted. There is evidence that the adjacent human societies were hunters-gatherers. They would have relied on the small mammals for food. The controversy of small-mammal extinction is still unresolved, as the study in the Mediterranean sites may not explain the evolutionary result worldwide.

Twilight of the Pygmy Hippos

by Jared M. Diamond

Who or what caused the late-Quaternary extinctions of many large mammals — were humans responsible, or climate change[1]? For a long time, the best-known examples were provided by the Americas and Australia, to which more recently has been added every palaeontologically explored island of the Pacific. Surprisingly, European biologists and archaeologists have tended to ignore examples much closer to home, on the islands of the Mediterranean[2] and western Indian Ocean[3]. Discoveries about the extinct pygmy hippos of Cyprus[4-7], and Madagascar[8] are now yielding tantalizing new insights into the date, speed and cause of the extinctions.

Mediterranean islands such as Cyprus, Crete, Corsica and Sardinia are surrounded by such deep water that they remained unconnected to Europe even at Pleistocene times of low sea level. All were reached (by swimming or rafting) by flightless mammals, which then underwent striking evolutionary changes in body size[9-12]. In general, big mammals became smaller, while little ones became larger. The islands developed pygmy elephants, hippos and deer, and even a paradoxically named pygmy giant deer (Megaceros). Leg bones indicate that these dwarfs had become slower-moving than their mainland ancestors, as would be expected given the absence of large predators[9]. Side by side with the dwarfs lived monsters: big shrews and voles, as well as giant dormice, owls, swans, lizards and tortoises.

All of these insular dwarfs are known only from subfossil bones, some of which were formerly believed to be remains of early Christian martyrs. The bones date to no more recently than early or middle Holocene, which began about 10,000 years ago. Human settlement of Mediterranean islands did not begin until the early Holocene, and (as elsewhere in the world) this approximate coincidence between megafaunal extinction, human settlement and the end of the Pleistocene cries out for clarification. Were the extinctions due to climate and habitat changes at the Pleistocene Holocene boundary, or to a Blitzkrieg of human hunting or habitat changes caused by humans and their domestic animals?

Until recently, the earliest known human occupation of Cyprus was by Neolithic herders and farmers[11]. Their sites are marked by remains of stone houses, stone bowls and the bones of the sheep, goats and pigs that they brought with them. The sole remains of the extinct megafauna at these sites are two debated scraps of hippo bones. Conversely, none of the previously known Cypriot sites with abundant bones of pygmy hippos and elephants yielded signs of human presence. That would seem to mean that the dwarfs succumbed to climate changes just before human arrival.

Discoveries[4-7], at the Cypriot site of Akrotiri-Aetokremnos[13] reverse this conclusion, however, and at last support the view that hippos and humans coexisted (even if only briefly). Simmons and Reese have unequivocally established the presence of humans at the site by identifying hearths, shell beads and stone artifacts very distinct from the Neolithic implements found on the island. There is no trace of domestic animals, suggesting that the people were hunter-gathers.

The presence of the pig-sized hippos is also unequivocally

established — their bones occur in prodigious quantities, and over 240,000 of them are known from at least 200 individuals. Along with the hippo bones are remains of lots of shell fish, crabs and sea urchins, many big birds, and at lease three pygmy elephants, plus turtles and snakes. The bones are disarticulated and many are burnt, indicating that the people who left the stone tools cooked the animals that left the bones.

The age of the site is well supported by carbon-14 dates on various materials clustering around 8500 BC (uncalibrated). This makes the site 1,500-2,000 years older than Cyprus's Neolithic horizon, and possibly the oldest established site for human occupation of any remote Mediterranean island[14]. Interestingly, the proportion of hippo bones decreases, and the proportions of bird and mollusc bones increase, from deeper to shallower levels at the site, but both levels have indistinguishable carbon-14 dates. This implies that these first hunters to reach Cyprus quickly exterminated the pygmy hippos, which would have been conspicuous prey, easy to locate and kill down to the last individual.

It remains to be seen whether this human role in the extinction of Cyprus's pygmy hippos will apply to the miniaturized megafaunas of other Mediterranean islands. For Corsica and Sardinia, Vigne[10] has summarized evidence that

megafaunal species were alive when humans arrived, and vanished soon afterwards. My own guess is that all megafaunal species, on all remote Mediterranean islands, that survived into the Holocene, will prove to have become extinct through human agency.

Parallels and contrasts are presented by Madagascar's pygmy hippos, on which the Sun set long after their Cyproit brethren had vanished. Madagascar has yielded subfossil bones of giant lemurs and elephant birds, and of semi-pygmy (cow-sized) hippos. The island has generally been believed to have been settled by humans around A.D 500, and all these species were apparently extinct by the time that Europeans began describing Madagascar's animals in the seventeenth century. Very few human-modified bones of extinct megafauna are known from Madagascar but MacPhee and Burney[8] have now identified seven Madagascar hippo femurs bearing marks of having been cut and then snapped by humans. The cuts were apparently made by iron tools at a time when the bones were still fresh. Accelerator mass spectrometry carbon-14 dates on collagen from the bones fall between AD 0 and 300, pushing back the date of human occupation of Madagascar by a few centuries.

A possible contrast between Madagascar and Cyprus is the hint that the Cyprus megafauna succumbed quickly, where some of the

Madagascar's giant lemurs were still alive in the thirteenth century AD. The possibly slower pace of extinction on Madagascar may reflect its much greater area, and also a difference between the economies of the first Malagasy and Cypriot settlers[15]. The discoverers of Cyprus may have been full-time hunters-gathers relying heavily on hippos for food, whereas Madagascar's first settlers were probably herders, farmers and fishermen for whom hunting was no more than a pastime. However that may be, the message of the new work is that zoology departments in Europe can dig into the reasons for extinctions on islands close to home.

Jared M. Diamond is in the Department of Physiology, University of California Medical School, Los Angeles, California 90024, USA. □

1. Martin, P.S. & Klein, R.G. (eds) Quaternary Extinctions (University of Arizona Press, Tucson, 1984).
2. Martin, P.S. in Quaternary Extinctions (eds Martin, P.S. & Klein, R.G.) 354-403 (University of Arizona Press, Tucson, 1984).
3. Dewar, R.E. in Quaternary Extinctions (eds Martin, P.S. & Klein, R.G.) 574-593 (University of Arizona Press, Tucson, 1984).
4. Simmons, A.H. Antiquity 65, 857-869 (1991).
5. Simmons, A.H. in Perspectives on the Past (ed. Clark, G.) 282-304 (University of Pennsylvania Press, Philadelphia, 1991).
6. Reese, D.S. Cyprus View 6, 50-53 (1992).
7. Simmons, A.H. & Reese, D.S. Archaeology (in the press).
8. MacPhee, R.D.E. & Burney, D.A. J. arch. Sci. 18, 695-706 (1991).
9. Reese, D.S. Field Mus. nat. Hist. Bull. 60, 22-29 (1989).
10. Vigne, J.D. Atti Convegni Lincei 85, 369-392 (1990).
11. LeBrun, A. et al. L'Anthropologie 91, 283-316 (1987).
12 Sondaar, P.Y. Nat. Hist. 95, 50-57 (1986).
13. Simmons, A.H. Nature 333, 554-557 (1988).
14. Cherry, J.F. J. Mediterranean Arch. 3, 145-221 (1990).
15. Wright, H.T. et al. Taloha 11, 121-145 (1992).

Questions: *Answers on page 203*

1. How do you think that the controversy can be resolved regarding the extinction of the small mammals?

2. How could climate affect the extinction of mammals?

3. How can human intrusion affect the extinction of other mammals?

34

Aside from all of the impressive advances in the biological sciences, a very fundamental question about life remains unanswered. How many different kinds of organisms inhabit the Earth? The truth is that no one knows the answer. Currently, not all species on the planet have been identified. In addition, the archives recording total numbers over the last 250 years are incomplete. Estimates for this total picture range from three million to thirty million species. Answering the question about the current number of living species is important in order to understand the degree of biological diversity existing throughout the habitats of the world. As these habitats are being destroyed at an alarming rate, species are being threatened with extinction. The necessary biological diversity for an ecologically-healthy planet is therefore also being threatened.

How Many Species Inhabit the Earth

by Robert M. May

If an extraterrestrial explorer were to land on the earth, what is the first question it would ask? The alien would, I think, inquire about the number and variety of living organisms on this planet. Given that the earth's physical attributes derive from universal and essentially deterministic laws, the presumably well-traveled visitor would probably have seen countless similar worlds throughout the universe. But the warp of evolutionary forces and the weft of chance that crafted the rich tapestry of life on the earth are almost certainly unique.

Surprisingly, we humans cannot even approximately answer the alien's query. Despite more than 250 years of systematic research, estimates of the total number of plant, animal and other species vary widely, all the way from three million to 30 million or more. Because no central archives exist, no one even knows how many species have already been named and recorded.

I find the current state of affairs particularly distressing in light of the rapid rate at which wild habitats are being destroyed. A knowledge of the total number and distribution of species is fundamental for developing a rational program to conserve as much as possible of the remaining biological diversity. The British government expressed the argument with admirable clarity in a 1990 white paper: "The starting point for this Government is the ethical imperative of stewardship which must underlie all environmental policies.... We have a moral duty to look after our planet and to hand it on in good order to future generations."

Political leaders face a baffling assortment of environmental concerns that cry out for a better understanding of how ecosystems change as constituent species die out and are succeeded by other forms of life. Such an inquiry explores the fundamental relation between diversity and the stability of a biological community. It also touches on other compelling issues, such as predicting changes in the earth's climate. After all the earth's oxygen-rich atmosphere was originally produced by living organisms, a fact that underlines the extent to which ecosystems and the atmosphere are intertwined.

More immediately, utilitarian reasons for counting and cataloguing species are also noteworthy. A considerable fraction of modern medicines has been developed from biological compounds found in plants. Society would be well advised to keep looking at other shelves in the larder rather than destroying them. Many nutritious fruits and root crops remain largely unexploited: cultivating them could expand and improve the global food supply.

Even within familiar genera of crop plants, researchers continue to uncover new geographic varieties. Such variants contain the raw material from which more productive and disease-resistant strains can be created by selective interbreeding or by genetic engineering. The triumphs of modern intensive agriculture have been accompanied by a dangerous narrowing of the diversity of the plants that farmers use which increases the susceptibility of crops to disease and climate variability. The likelihood of shifts in the global environment gives fresh emphasis to the desirability of conserving the gene pool and to exploring the use of different plants.

Efforts to view the organic world as an ordered system date back at least to Aristotle. The task of systematically naming and recording species, known as taxonomy, was begun by the great natural scientist Carolus Linnaeus, in Uppsala, Sweden. The canonical 10th edition of his book Systema Naturae, which recorded some 9,000 species of plants and animals, was published in 1758 — a full century after Isaac Newton had arrived at an analytic and predictive understanding of the laws of gravity, based on centuries of detailed astronomical observations.

Since then taxonomists have added species to Linnaeus's list at highly disparate rates in different diological categories. By far the most attention has been lavished on animals endowed with the charm of feathers or fur. Scientists have nearly completed cataloguing such species. For example, less than a century after Linnaeus's work, half of the 9,000 known species of birds had been recorded. Researchers currently uncover only about three to five new bird species a year. A similar situation holds for the 4,000 or so species of mammals although on average about 20 species and one genus are discovered annually. About half of those are truly undiscovered species (mostly rodents, bats or shrews), whilst the others result from reclassifications of old species based on updated biochemical findings.

Quite a different pattern emerges for organisms other than birds and mammals. A record of the growing number of known species of arachnids and crustaceans (essentially, arthropods other than insects) shows that relatively high rates of discovery prevailed in Victorian times, followed by a prolonged lull. More than half of the current total has been added within just the past few decades. In a recent study, Peter M. Hammond of the Natural History Museum in London demonstrated that from 1978 to 1987 the number of known species of birds increased an average of just 0.05 percent a year. Over the same period, the recorded number of species of insects, arachnids, fungi and nematodes expanded by 0.8, 1.8, 2.4 and 2.4 percent a year, respectively.

The varying rates of discovery to some extent reflect different sizes in the work force of taxonomists dedicated to studying each group of organisms. Accurate statistics on the work force are hard to obtain, but a cursory survey of researchers in Australia, the U.S. and Britain by Kevin J. Gaston of the Natural History Museum and me offers some perspective. We found that if N represents the average number of taxonomists studying each species of tetrapod (all vertebrates other than fish), then there are approximately 0.3N taxonomists for each species of fish and only from 0.02N to 0.04N for each invertebrate. Some 10,000 taxonomists work in North America; the global total is perhaps three times greater.

Overall, there are twice as many taxonomists for each recorded plant species as for each animal. Within the animal kingdom, the average vertebrate species receives about one order of magnitude, or factor of 10, more taxonomic attention than does the average plant, whereas the average invertebrate receives about an order of magnitude less. The distribution of taxonomists is manifestly ill-matched to the species richness of the various taxa. Moreover, only about 4 percent of taxonomists work in Latin America and sub-Saharan Africa, where much of the earth's biological diversity resides.

The lack of a central repository of information on species poses an even more basic obstacle to the compilation of a complete taxonomic list. The records mainly consist of old-fashioned file cards poorly coordinated among scattered institutions. There is no official count of the species that have already been named. Scientists know far more about (and spend vastly more money studying) the systematics of stars than the systematics of earthly organisms. Consequently, they have as good a knowledge of the number of atoms in the universe — an unimaginable abstraction — as they do of the number of species of plants and animals.

According to the best estimates taxonomists have identified 1.5 to 1.8 million species, but the list is obviously far from complete. An assortment of approaches — some empirical others theoretical — has been used to determine, at least approximately, the real total. Even the lowest estimates indicate the existence of three million species, so many that present methods will not suffice to discover and catalogue them all within a reasonable time.

Several authors have estimated the global species total by extrapolating from trends in recording species. Such work is prone to divergent interpretations depending on one's statistical procedures. A recent study, conducted using separate statistical projections from trends of discovery for each major biological group, concluded that there are six to seven million species. Another approach, which drew on the opinions of experts in each group, implied the existence of upward of five million species.

Many projections rely on a simple intuitive argument based on the relative abundances of species in different taxonomic categories. For well-studies groups such as birds and mammals, tropical species are about twice as common as temperate or boreal ones. But among insects which account for the majority of all recorded species, northern faunas are much better known than are tropical ones; approximately two thirds of all named insect species live outside the tropics. If the ratio of numbers of tropical species to temperate and boreal ones is the same for insects as it is for mammals or birds (a far from certain assumption), there should be two yet unnamed species of tropical insects for each named temperate or boreal species. Applying the above logic expands the 1.5 to 1.8 million recorded species to from three to five million.

A more direct way of estimating the global species total — especially the number of tropical insect species — involves thoroughly sampling the organisms living in some region that has been relatively unstudied and then determining what fraction of those plants and animals has already been described. That method presents quite a challenge; even within a limited area it is difficult to sample all of the tropical insects. Identifying and classifying them presents an even more onerous task. And one must worry whether the site or group is typical of the general patterns of species distribution.

Ian D. Hodkinson and David S. Casson of the Liverpool Polytechnic in England assessed the hemipterans, or true bugs, in a moderately large, topographically diverse region of tropical rain forest in Sulawesi, Indonesia. They found a total of 1,690 species of terrestrial bugs, 63 percent of which were previously unknown. If this fraction is representative of insects generally, then one could infer from the 900,000 recorded species of insects an actual total of two to three million species.

Hammond carried out a different version of this local-population technique. He observed that the 22,000 insect species found among the well-inventoried fauna of Britain (where generations of clergymen have given more attention to their six-legged flocks than to

their two-legged ones) include 67 species of butterflies. Many naturalists accord butterflies the honorary status of birds, and hence the 17,500 known species represent a nearly complete inventory. The true number almost surely is no more than 20,000.

If the taxonomic composition of the insect fauna of Britain resembles the global composition, the world should contain a total of about six million (22,000 times 20,000, divided by 67 insect species). Such scaling-up estimates suffer from a built-in uncertainty. One cannot be certain that any individual group of insects or any particular place is typical of the insect population in general.

Terry L. Erwin of the Smithsonian Institution directed a particularly intriguing study in which he and his co-workers scrutinized beetle faunas in the canopies of tropical trees. Beetles are distinguished from other insects by their hard, wing-like plates that cover functional inner wings. Approximately one in five of all recorded species are beetles, a fact that prompted the British geneticist J.B.S. Haldane to joke that his studies of the natural world had taught him that the Creator has "an inordinate fondness for beetles." Furthermore, tropical forests are rich sites of biodiversity. Although they cover only about one sixteenth of the earth's land area, tropical forests may harbor at least as many species as the rest of the world's regions added together. Even so, few researchers anticipated the stunning implications of Erwin's work.

Using an insecticidal fog, Erwin collected canopy beetles in Panama from Luehea semannii trees, tropical cousins of the linden. Over three seasons, Erwin found a remarkable total of some 1,200 beetle species. He has not yet sorted them out to determine how many of the species are uncatalogued, so the extrapola-

tion method described above cannot be applied. Instead Erwin used the following chain argument.

First, he needed to know how many of the beetle species he collected live specifically on L. seemannii, as opposed to being distributed across many kinds of trees. Erwin guessed that around 2 percent of the herbivorous beetles, the largest group in his sample are specialized to each tropical tree species. On that basis, he estimated that one kind of tree holds an average of 160 species of canopy beetle. Second, Erwin inferred the plenitude of all insect species from the species density of canopy beetles. Forty percent of known insects are beetles; if this proportion applies in tropical tree canopies, then 400 kinds of canopy insects occupy each tree species.

Third, Erwin supposed that the canopy contains two thirds of the insect species on the tree, implying a total of 600 insect species on every variety of tropical tree. Finally, he cited a widely accepted estimate that the earth supports 50,000 species of tropical trees. Multiplying 600 insect species times 50,000 tree species yields 30 million kinds of insects. The number of all species worldwide would obviously have to be much larger still.

Each link in Erwin's chain of logic involves extreme uncertainty. In my opinion tropical beetles as a group are likely to be significantly less specialized than are temperate ones, in which case Erwin's figure of 20 percent should be closer to 2 or 3 percent. On the other hand, Erwin probably underestimated the species density on the noncanopy part of the tree, which I suspect may hold as much as two thirds of the insect species.

If one incorporates what I consider to be more realistic numbers, the calculations imply the existence of three to six million insect species. I nevertheless believe that Erwin's work is important, both in providing a focused approach to the problem of estimating how many species exist and in emphasizing how taxonomic questions inter weave with ecological ones. Erwin's efforts do not settle the question of how many species exist so much as they define an agenda for research.

The ultimate aim in recording biological diversity is to build a factual foundation for answering basic questions about evolution and ecology. Taxonomic lists serve as points of departure for studying the structure of food chains, the relative abundance of species, the number of species or total number of organisms of various physical sizes, and general trends in the numbers and distribution of living things. Some of those studies suggest broad rules that make it possible to estimate the number of species in other independent ways.

One such technique derives from observing patterns of how many species of land-dwelling animals fall into various body-size categories. In general taxonomists find that for each 10-fold reduction in length the number of species fitting into that size category increases 100-fold. The pattern holds for animals ranging in length from a few meters down to around one centimeter.

Below one centimeter the relation begins to fall apart, possibly because of the highly incomplete record of such tiny terrestrial animals. If the size-versus-species density pattern were arbitrarily extrapolated down to animals having a characteristic length of around one millimeter — an arbitrary dividing line between macroscopic and microscopic life — it would imply a global total of some 10 million species of land-dwelling animals. Such a purely phenomenological assessment would be more persuasive if taxonomists confidently understood the physiological,

ecological and evolutionary factors responsible for body-size distributions.

(The size distribution rule sheds light on Noah's problem with the size of the Ark. Pundits have often suggested that Noah would have faced impossible obstacles getting all the myiad insect species on board. But if a 10-fold decrease in size — equivalent to a 1,000 fold decrease in volume — results in only 100 times as many species, then the difficulty really lay in getting the largest creatures on board.)

General patterns in the structure of the food chain provide the basis for another kind of species tally. Photosynthetic plants create the raw organic material that forms the first link in the food chain. If one could obtain detailed statistics on how many other life-forms each kind of plant can support then the total number of species could be derived from the comparatively complete counts of plant species.

Although scientists remain a long way from this grail, Gaston has extended earlier research on food chains by collecting available evidence on the average numbers of insect species associated with each plant species in communities of highly disparate sizes and locations. On average, he finds approximately 10 insect species for each plant species. Given the fairly secure estimates of about 270,000 species of vascular plants overall, Gaston crudely derives a total of three million insect species.

The discussion up to this point has dwelt mainly on calculating the numbers of terrestrial insect species, and for good reason: those creatures already constitute more than half of all recorded species, and yet they are far from being completely catalogued. Several other taxonomic groups may rival the diversity of the insects, however. In particular, organisms that are small or

unglamorous, or both, have probably been disproportionately overlooked.

David L. Hawksworth of the International Mycological Institute in Kew, England, has proposed an upward revision of the total number of fungal species at least as striking as the one suggested by Erwin's work on insects. Hawksworth begins by noting that taxonomists have recorded some 69,000 species of fungi. In Britain and other well-scrutinized northern European regions, fungal species outnumber vascular plant species by about six to one. If that ratio holds true around the globe, then the 270,000 cataloged vascular plants coexist with 1.6 million species of fungi, more than 20 times the number now known.

Biological patterns seen in temperate zones might not apply to tropical communities, of course. Tropical fungal species may each associate with a broader range of plants than do those in the temperate zones: such a discrepancy would lower the ratio of fungal to plant species. Conversely, Hawksworth ignores fungi associated with insects rather than with plants, which steers him toward a low estimate. In recent studies of particular tropical sites, the proportion of previously undescribed species of fungi hovers around 15 to 30 percent, significantly below the 95 percent one might expect if Hawksworth's calculation were correct. On the other hand, the studies are not even close to being synoptic, so it may be unfair to expect them to uncover the full panoply of uncatalogued species.

The sketchy knowledge of the number of fungal species belies the fact that they are vital elements of most ecosystems, aiding the decomposition of organic material and the formation of new soil. Fungi have undoubtedly shaped the development of biological diversity, first by helping plants colonize dry land and then, especially through symbiotic relations, by assisting in the spread and diversification of vascular plants, insects and other organisms. Such an important part of the earth's biota deserves more attention.

Nematodes are probably the least well inventoried of all animals visible to the naked eye. These tiny worms live as parasites in plants and animals and as free-living creatures in freshwater and marine settings. As of 1860, only 80 species of nematodes had been recognized. Today's total stands at about 15,000. Recent investigations of nematodes on land and in fresh water suggest that recorded species account for only a tiny fraction of the total population. Other studies indicate even greater diversity in marine environments. Few taxonomists would quarrel with Hammond's assessment that the number of nematode species is at least some hundreds of thousands.

The smallest living creatures — those invisible to the unaided eye — probably also contribute greatly to the global cache of species. Micro-organisms such as protozoa, bacteria and viruses account for only about 5 percent of recorded living species. Recent work has revealed, however, that the diversity among natural populations of microbial organisms is far greater than that seen in studies of familiar laboratory cultures. One study of RNA from a mat of photosynthetic bacteria in a hot spring in Yellowstone National Park found eight distinct genetic sequence types, none of which matched the 12 strains of laboratory-cultured bacteria that had been believed to be characteristic of such mats. Only one of the eight sequences bears any close resemblance to those belonging to a recognized bacterial phylum.

Biologists who have examined ribosomal RNA sequences in natural marine microbe populations have arrived at broadly similar results. Those studies are, in some ways, even more remarkable than the revelations about tropical canopy faunas because they demonstrate how astonishingly little taxonomists know about the simplest, most common forms of life.

The task of categorizing bacteria and viruses is complicated by the fact that different strains readily exchange genetic material and that a single parent can clone itself to create an entire population. Also, some viruses mutate notice-ably from year to year. Basic notions about what constitutes a species are therefore more vague for microbes than for vertebrates. Manfred Eigen of the Max Planck Institute for Biophysical Chemistry in Gottingen and Peter Schuster of the University of Vienna argue that the basic unit of classification of many viruses should be the quasispecies, a well-defined set of distinctive RNA sequences. Natural selection acts not on viral species as such but on quasispecies swarm.

The contribution of microorgan-isms and nematodes to the overall gene pool may be surprisingly large. One rather flippant, but not entirely unreasonable, view is that every species of arthropod and vascular plant (which together account for the vast majority of all recorded species) has at least one parasitic nematode, one protozoan, one bacterium and one virus specialized to it. If so, the previously described species counts should automatically be scaled up by a factor of five. In that case, the total could potentially exceed 100 million, although I doubt the number is so large.

Species counts are useful because the species is a fairly concrete and readily defined measure of genetic uniqueness. But many evolutionary questions or practical purposes direct a researcher's interest toward diversity at higher or lower taxonomic levels.

Moving up the taxonomic hierarchy, from species through genera, families, orders and classes to phyla, emphasizes even more fundamental genetic variations. For instance, less than 13 percent of all recorded species dwells in the ocean. But more than 90 percent of all classes of organisms and essen-tially all phyla are represented in the sea. Indeed, two thirds of all phyla is found only in the sea. Hence, in terms of the range of body plans of organisms, the ocean is much more diverse than the land.

At the other end of the hierar-chy, conservationists are often interested in the genetic differences among individuals within a single species. When a species nears extinction, much of its genetic diversity disappears, a loss that cannot be reversed by breeding programs. Investigations of diversity within a species therefore offer a measure of the degree to which endangered species have already been irretrievably depleted. Genetic studies can also suggest whether a species may have approached extinction in the not too distant past.

Improving the catalogue of life will require a huge coordinated and sustained effort. I believe such dedication is warranted. Conserva-tion will increasingly necessitate intervening in and managing ecosystems, along with making agonizing choices of where to concentrate such efforts. Those actions will demand better informa-tion than is currently available.

The human race also needs to understand the diversity of living things — how much there is and why — for the same reasons that compel people to grope for an understanding of the origin and fate of the universe or of the sequence of molecules in the human genome that code for our own self-assembly. Unlike those other quests, the task of studying and conserving biological diversity has a strict time limit. Each year from 1 to 2 percent of the world's tropical forests are destroyed to create farmland and provide fuel and raw material. At that rate, the tropical forests will be gone in 50 years or so.

The time constraints mean that taxonomists must develop imaginative approaches for cataloguing diversity. One method is to enlist the help of nonprofessionals (so-called barefoot taxonomists) to collect and classify specimens using rough-and-ready techniques. They can gather most of the wanted information in a fraction of the time that would be needed to satisfy the canons of traditional scholarship. Simplified taxonomic research programs are being pioneered in Costa Rica by Rodrigo Gamez and others at the National Institute of Biodiversity; Australia is sponsoring similar work as part of an effort to codify the national collection of species.

Other approaches make use of high technology. Researchers could collect, store and distribute species information on computer disks; efficient programs could be developed to cross-check a species against an organized data base, thereby revolutionizing the time-consuming task of separating and classifying new species. Reference images could be made into holograms that would be readily available. The centralized Australian data base already incorporates some of those features. If widely and creatively applied, modern information technology could make extensive European collections — the legacy of empire — accessible to underequipped researchers in the tropics.

Future generations will, I believe, find it incomprehensible that in the late 20th century, Linnaeus continued to lag far behind Newton. They may also be saddened — as people should be today — that our society has devoted so little money and effort toward quantifying and conserving the forms of life that define the earth's unique glory. ☐

Questions: *Answers on page 204*

1. Why do you think it is so difficult to record the total numbers of species on the Earth?

2. How can the controversy of the species concept cloud the picture of tabulating the total number of species?

3. Why is the maintenance of biological diversity important for the healthy survival of the habitats and ecosystems on the Earth?

35

Human beings belong to the order of primates, one of many orders in the class of mammals. Unlike other animal groups, such as the dinosaurs, the early primates did not leave a large number of fossils for study. As a result the primate fossil record is incomplete. Therefore, sparse amounts of evidence have been used to describe the succession of primates leading to the origin of the human species. As an alternative, Ian Tattersall, a paleontologist and primatologist, describes a technique that will shed more light on the unanswered questions concerning human evolution. Working with skimpy examples of human fossils, mainly replicas of fragmented bones, scientists can sculpt a complete human skull and make a mold from it. A new skull is then cast from the mold, serving as a foundation onto which skeletal muscles, fat deposits, and external facial features are carefully added to model a life-like head of a human ancestor. This far-reaching approach surpasses the traditional approach of interpreting human evolution from bones and teeth alone. The life-like models may help to fill in some of the gaps about the progression of primate forms leading to humans.

Evolution Comes to Life

by Ian Tattersall

Ancient bones are the objective evidence of biological history. From my standpoint as a paleontologist, they are vastly more informative about extinct creatures than reconstructions or models, in whose creation art plays at least as great a role as science. Yet I am also a museum curator, and from that perspective I am keenly aware that nothing brings the past alive in the public's eye like a well crafted reconstruction. For the average person, fossil bones are static things; beautiful or majestic, perhaps, but hard to imbue with the attributes of a living, breathing form.

When I was given the responsibility of curating the American Museum of Natural History's new hall of Human Biology and Evolution, it was therefore evident to me and to Willard Whitson, the designer of the hall, that we needed to include some reconstructions of early humans in the exhibition. Furthermore, we wanted to portray these figures dynamically in the context of situations that our ancestors might have faced long ago. Only thus, we thought, could we truly bring these long-departed relatives back to some semblance of life. We hoped that

clever sculpting and modern casting materials could provide us with a level of realism rivaling that of the spectacular dioramas of modern animals in the adjacent galleries.

What I had failed to consider, however, was the extraordinary number of awkward decisions that would become necessary as work progressed. We scientists customarily deal with objective matters, and we are happiest when reaching judgments based on testable reality. Untestable speculation makes most of us acutely uncomfortable. As our experience with preparing the exhibits for the hall illustrates such speculations are inescapable in any attempt to depict extinct human species.

We had decided to recreate three distinct species of human ancestor reaching back several million years in evolution. the most recent of these is Homo neanderthalensis, shown at a site in France where Neanderthals lived about 50,000 years ago. (Although I and many other paleontologists are confident that Neanderthals were a species apart from modern Homo Sapiens, some workers still dispute the point.) Despite their geologic

youth and general similarity to us, the Neanderthals presented a typical set of difficult decisions.

The first choice to be made — what the three individuals in the scene should be doing — was the easiest because archaeologists have learned a fair amount about how Neanderthals lived. For example, the characteristic wear on their stone tools tell us that Neanderthals used flints to cut wood and scrape hides. We settled on showing a male sharpening a wooden spear while a young female scraped a hide and an older female offered advice. Because neanderthal front teeth were usually very heavily worn, we felt safe showing the young woman with one end of the hide held in her teeth.

For realism in the bodies of our Neanderthals, we chose to case the bodies of living human models. Neanderthals were more robust than modern humans, but their body proportions were generally similar to ours, and sculptural modifications could accommodate any slight differences between them and us — their longer thumbs and the shape of their shoulder blades, for instance — that might have been detectable in the flesh.

Paleontologists have found a reasonably large number of fairly complete Neanderthal skulls that well convey the general dimensions of the head. Neanderthals had a braincase as big as ours, but it was long, low and lacked the modern high forehead. Their face protruded along its vertical midline; the nasal orifice was large, and the cheekbones receded. Heavy bony ridges overhung the eye sockets.

To determine the appearance of the features overlying the skull, my museum colleagues Gary Sawyer and Steve Brois and I faced some unfamiliar problems. Modeling an extinct human depends on skills much like those of the forensic artists who reconstruct the faces of unidentified murder victims. The key is that muscles leave telltale impressions on the bones to which they attach. From the size and depth of these impressions, one can deduce the dimensions of the muscles. On the cast of the dry skull, an artist or paleontologist can then reconstitute the overlying soft tissues layer by layer, starting with the deep tissues and proceeding outward.

Forensic artists, however, have a great advantage; they can study the soft anatomy of living people. Paleontologists have no living Neanderthal standards for comparison. As a result, the shapes of the nose, lips, ears and other features that are so vital for establishing the character of a Neanderthal face boil down to educated guesswork. In our models we tried to avoid unreasonably exaggerating these soft tissue features.

Our arbitrary decisions were still only beginning. Once I would have laughed if anyone had predicted that I would spend weeks agonizing over Neanderthal eyebrows. Did they even have eyebrows? (Our closest living nonhuman relatives the chimpanzees do

not. If Neanderthals did have eyebrows, where wer they on those bulbous browridges? Similarly, how long would untended Neanderthal beards have grown? How much body hair did men and women have? What was its color and texture? What was the skin color? All these details offered endless scope for quibbling.

We based our final decisions on the best information available. The high-latitude habitat of these people suggested that Cathy Leone and our other preparators should give them a light skin color like that of modern Europeans because the European climate was cold long before the peak of the last Ice Age about 10,000 years ago. We also felt she should give them body hair well toward the heavy end of the modern range.

Neanderthal humans knew how to prepare hides, and it seems virtually certain that they wore hide clothing, but of what kind? These people became extinct almost 10,000 years before the earliest known bone needles were in use, so neatly tailored clothing was clearly out the question for them. That exclusion still left many possibilities. Basic issues, such as whether fur was worn on the inside or the outside of the clothing or whether it was scraped off altogether were unanswerable.

In the absence of scientific solutions, aesthetics was allowed to take over. During the months of work, all of us involved in the project grew accustomed to the evolving appearance of each reconstructed figure. The first major shock always came with the addition of hair to a formerly glabrous head — an old friend was suddenly transformed into a stranger. Draping the naked figures with animal hides that obscured the results of hours of laborious body detailing was equally jarring for us. Predictably, perhaps, our Neanderthals ended up with

rather minimalist clothing.

Clothing was not an issue in the second diorama which is set in the Turkana basin of northern Kenya about 1.7 million years ago. That tableau features a close relative of Homo erectus that should probably be called Homo ergaster, "working man". It was given this name because it used tools although it was not the first human to do so. Deciding what activities to depict in this scene was more problematic than for the Neanderthals because so much less is known about the behavior of earlier humans.

We had room in the diorama for only two figures, and we chose to use one male and one female, We did not want to suggest that they represented a nuclear family, however, because there is no evidence that these early people lived in such groupings. Compounding our difficulties was archaeologists uncertainty about whether H. erectus and its relatives hunted anything bigger than the smallest game. We wanted to show a carcass being butchered with crude stone implements — we know these people did that much — but we did not want to imply that our protagonists had necessarily made the kill.

An exhibit label might list as many ceveats as we liked but a concrete representation of any behavior would inevitably seem to make a rather definitive statement. In the end, we chose to show the female warding off a jackal with a piece of bone while the male, attacking an impala carcass with a stone flake, is startled by a vulture overhead.

The figures themselves were a little less problematic from the neck down these early people were quite similar to us. We gave them a dark skin for protection against the tropic sun and sparse body hair, which would have facilitated cooling through the evaporation of sweat. Reconstructing the faces of these small-brained people was trickier than doing so for the Neanderthals because their evolutionary separation from modern humans is so much greater. Nevertheless, the challenges were similar.

The third diorama takes museum visitors even further back in time. It depicts the making of the famous footprint trails at the Tanzanian site of Laetoli about 3.5 million years ago. A nearby volcano had puffed out a cloud of ash that settled on the landscape and was dampened by rain. Across this muddy surface walked two early humans and possibly a third (although that idea is disputed). The variety of hominoid that made the prints is also disputed, but the only known candidate is Australopithecus afarensis, the earliest member of the human lineage yet described.

The prints themselves, hardened in the mud and miraculously preserved, show an astonishingly modern upright bipedal gait. The two parallel tracks are those of a large and a small individual walking in step and so close to one another that they must have been in physical contact. Were they an adult and an adolescent, or perhaps a male and a female? Fossils show that A. afarensis males were much larger than the females.

From the tracks alone, we cannot know. For visual interest we opted once more for an adult male and female, the male with one arm draped over the female's shoulder. Perhaps that pose is too anthropomorphic for some tastes but the gifted English sculptor John Holmes produced such a vivacious result for

us in his finished Laetoli figures that we frankly didn't care.

No complete fossil skull of A. afarensis has been found. Composite reconstructions based on skull fragments reveal that it had a rather chimpanzeelike head with a small braincase and a large face. The soft tissue features, of course, posed the usual problems in even more acute form; we had to decide, for example, whether to give the figures a chimpanzeelike nose or something more human.

Moreover, uncertainty about the body structure of these early humans was also a factor. From the known fossils we could infer that they had shorter legs and longer feet than our own. That fact could imply that their gait differed from that of modern people. Yet such an idea conflicts with the evidence of the laetoli tracks. Caught between the fossils and the humanlike prints, we opted to favor the latter — but that is not the kind of decision that most scientists are comfortable making.

The skin color and body hair of the figures were also highly conjectural. A. afarensis lived in the tropics of East Africa well after the ancient forest began to give way to open grassland. This bipedal hominoid probably exploited the relatively rich zone where the forest and savanna intersected. Its skin might have been deeply pigmented for protection against the intense solar radiation of the open country, but if A. afarensis spend much of its life in the shade of trees such an adaptation might have been less necessary. It might have had the dense coat of hair typical of forest

primates. Alternatively like our own savanna adapted species, it might have swapped its coat for a mechanism that shed body heat through evaporation of sweat.

There is simply no way to be certain. The A. afarensis figures we created have a darkish skin and a fairly sparse covering of hair, but I am sure that many of my colleagues will find good reasons to disagree with our choices.

We had a similarly hard time deciding how closely our male and female A. afarensis should resemble each other. Chimpanzees do not show major sex differences in their facial hair, and neither do our figures — partly because we felt that the moustache with which the male was initially endowed with made him look a little too much like a Lothario from a 1920s movie. Once again, the decision is a hard one to justify on strictly scientific grounds.

You might conclude that the cumulative result of these unscientific decisions would be purely fantastic figures. Not so. Although I would not stake my life on many of the details I have mentioned, through careful sculpture and respect for the measured skull and body proportions we have produced evocations of these vanished humans that bring them to life without sacrificing reasonable scientific accuracy. Moreover, while visitors to the American Museum of Natural History are looking at the dioramas, they will also have before them replicas of the actual fossils on which these recreations are based — the best of both worlds. □

Answers on page 204

Questions:

1. Why did primates leave very few fossils? Relate to examples of primate structure and function.

2. Why did you think there will always be some major gaps in the knowledge of ancestors leading to human evolution?

3. Do you think that the production of models is a purely scientific solution to answering the questions of human evolution?

36

Anthropologists have long held a view that Homo erectus is the direct evolutionary link between our earliest ancestor, Homo habilis, and modern-day Homo sapiens. Recently that long-held view has been challenged. Investigators now propose as many as three contrasting theories to explain human evolution. In addition to support for the theory of human origin from Homo erectus, another viewpoint proposes the split of this species into two new species. Only one of these evolved into modern humans. A third explanation proposes that Homo erectus evolved into a diverse group apart from humans, therefore leaving the questions about human ancestry unanswered. Disagreements over the course of human evolution are due mainly to the incomplete, sparse succession of human fossils that have been discovered to date. There is also disagreement over how to differentiate among species through the study of the fossil record. The species has long been recognized as the unit of evolution. However, the process of recognizing current animal species is also being called into question by scientists. This confusion further confounds the challenge of applying the species concept to all populations, either modern or ancient.

Erectus Unhinged

by Bruce Bower

For more than 40 years, anthropologists have generally agreed that Homo erectus served as an evolutionary link between our earliest direct ancestor, Homo habilis, and modern Homo sapiens. This view holds that a hardy breed of H. erectus spread from Africa to Asia and Europe and lived from approximately 1.8 million to 400,000 years ago.

But in the last few years, H. erectus has suffered an identity crisis. Leading investigators now propose three contrasting theories of human evolution that would give any ancient ancestor cause for concern. One proposal advocates sticking with a single, widespread H. erectus; another calls for splitting H. erectus into at least two species, only one of which evolved into modern humans; and a third seeks to abolish H. erectus altogether, placing its fossil remains within an anatomically diverse group of H. sapiens that split off from habilis about 2 million years ago.

Disagreements of this sort stem from a fundamental parting of the ways about how to discern a species in the fossil record. Most anthropologists accept the species as the basic unit of evolution, while acknowledging that defining a species, even among living animals, often presents problems. Thus, different theories about how best to sort out extinct species based on the features preserved in ancient bones fuel the dispute over H. erectus and other members of the human evolutionary family, known as hominids.

However, some researchers stand outside the fray, viewing any attempt to nail down fossil species as an unscientific, arbitrary exercise in cataloguing the ambiguous bits of anatomy surviving in fossil bones.

"There's a growing diversity as to how species are perceived in modern and ancient populations," asserts Erik Trinkaus of the University of New Mexico in Albuquerque. "[Researchers] often end up talking past each other."

In April, Trinkaus and others debated various approaches to understanding H. erectus and fellow hominid species at the annual meeting of the American Association of Physical Anthropologists in Las Vegas and in interviews with SCIENCE NEWS.

The roots of this sometimes confusing clash extend back 100 years, when the first H. erectus fragments turned up in Java. Initially classified as Pithecanthropus, or ape-man, these Asian specimens and most ensuing hominid finds received a unique species designation from their discoverers. In the early 1950s, anthropologists realized that human evolution made no sense if virtually all fossil discoveries represented different species. Taking the view that an ancestral species with a wide array of skeletal features gradually transforms into a descendant species, researchers proceed to group fossils into a much smaller number of species.

So-called "lumping" of specimens led to a picture of human evolution as a series of three progressive steps, with H. habilis begetting H. erectus begetting H. sapiens.

But by the early 1980s, these ancestral lumps had begun to stick in the throats of some anthropologists. At the same time, concern grew that the definition of a species used by biologists and often borrowed by anthropologists — namely, characterizing a species as a group of organisms that reproduce only among themselves — offered no help in evaluating fossils.

Another approach — called cladistic, or phylogenetic, analysis — rapidly gained popularity. This view holds that new species evolve relatively quickly rather than in a series of gradual adjustments within ancestral species. Specifically, cladistics assumes that although most members of a population of related organisms display the "primitive" skeletal features that arose early in their evolutionary history, some members of the population sport "derived," or advanced, anatomical features that appeared later. A consistent pattern of unique derived features on a group of fossils serves as a species marker.

Phylogenetic studies indicate that H. erectus fossils actually encompass two species, one in Asia that became extinct and another in Africa that evolved into modern humans. Peter Andrews of the Natural History Museum in London reported in 1984 that most skeletal features commonly accepted as unique derived traits of H. erectus are actually primitive retentions shared by earlier Homo species. Moreover, the seven derived characteristics exclusive to H. erectus appear predominantly among Asian fossils. These include an angling of the cranium that produced a bony ridge at the top of the head, thick cranial bones, a cleft in the bone just behind the ear and a plateau-like bony swelling at the back of the head.

Since these features appear in only one geographically restricted set of fossils and do not turn up later in modern humans, Andrews suggests that Asian H. erectus met extinction on a side branch of human evolution. A separate species of African hominids living at the same time evolved into H. sapiens, he posits. Andrews' analysis dovetails with the theory that modern humans originated in Africa around 200,000 years ago and then spread throughout the world.

Bernard Wood of the University of Liverpool, England, has elaborated on Andrews' phylogenetic thesis. In the February 27 NATURE, Wood presents a cladogram — a tree diagram organizing hominid species according to the number of derived features shared by groups of fossils — based on analysis of 90 cranial, jaw and tooth measurements. Wood concludes that sometime before 2 million years ago at least three Homo species emerged in Africa: the relatively small-brained H. habilis, a group with larger brains and teeth which he calls H. rudolfensis, and H. ergaster, represented by the fossils that Andrews separated from Asian H. erectus.

The three species apparently shared an unidentified common ancestor, with H. ergaster serving as the precursor of H. sapiens, Wood argues.

Wood splits up early Homo species in a reasonable way, notes Ian Tattersall of the American Museum of Natural History in New York City. But neither phylogenetic theory nor any other approach offers practical help to fossil species hunters, Tattersall maintains. Closely related living primate species often differ in only one or a few subtle anatomical features, which may not show up in a set of bones, he points out (SN: 4/13/91, p.230). Thus, cladistic analysis tends to lump together some hominid species that share derived anatomical characteristics, he holds.

In an article accepted for publication in the JOURNAL OF HUMAN EVOLUTION later this year, Tattersall advises investigators to use the phylogenetic approach to identify groups of fossils with derived features that signal either a distinct species or possibly a clutch of related species. Lumping inevitably occurs, but the general pattern of human evolution remains unobscured, he argues.

H. sapiens also requires splitting when viewed under this modified phylogenetic lens, Tattersall contends. He places several partial skulls found at European sites and usually assigned to early, or "archaic," H. sapiens (mostly dating to around 200,000 to 400,000 years ago) in a new species, H. heidelbergensis.

"It's a virtual certainty that speciations have been much more common in hominid biological history than many paleoanthropologists have been willing to admit," he asserts.

Tattersall has it exactly backwards according to adherents of the theory of "multiregional evolution." The phylogenetic approach fails to appreciate the anatomical diversity that arises within different populations belonging to the same species argues Milford H. Wolpoff of the University of Michigan in Ann Arbor. Wolpoff and his colleagues champion an evolutionary perspective in which each hominid species encompasses one or more populations that share the same common ancestor, follow the same evolutionary patterns over time and yield anatomical evidence of a historical beginning and end.

H. erectus clearly splits off from H. habilis, but it gives no sign of an evolutionary demise, according to a study conducted by Wolpoff and Alan G. Thorne of Australian National University in Canberra. Instead, H. erectus gradually merges into the range of skeletal characteristics observed in regional populations of early H. sapiens, Wolpoff and Thorne argue. Of the 23 derived anatomical traits that distinguish H. erectus from H. habilis, 17 consistently turn up on H. sapiens fossils, they assert.

In other words, H. erectus never existed and H. sapiens has evolved in several parts of the world for approximately 2 million years, Wolpoff and Thorne maintain.

Evolutionary patterns observed in four different regions — Africa, Europe, China and Australia-Indonesia — show continuous, gradual change from about 2 million years ago to the most recent human populations, with no evidence of Africans replacing the other groups, Wolpoff and Thorne contend. They also hold that H. sapiens encompasses most, perhaps all, specimens now classified as Neanderthal (SN: 6/8/91, p.360).

The few anatomical idiosyncrasies separating H. sapiens from fossil remains widely attributed to H. erectus — such as greater cranial volume, smaller teeth and lighter limb bones — reflect evolutionary trends in the former species toward larger brains and a greater reliance on tools and other technologies spawned by increasing cultural complexity, Wolpoff argues.

In Wolpoff's view, the merging of H. erectus into H. sapiens (first proposed in the 1940s by German anatomist Franz Weidenreich, who continues to inspire the multiregional approach) forces scientists to take a closer look at anatomical changes that have occurred over time within our species. It also exposes the need for a workable definition of "anatomically modern humans," he says.

Between those vying to split or to sink H. erectus stand some stalwart defenders of its status as a unified species. "I see Homo erectus as a single species that spread across the Old World," says G. Philip Rightmire of the State University of New York at Binghamton. H. erectus probably gave rise to modern humans in a restricted geographical area, for example Europe, where temperatures cooled dramatically around 400,000 years ago, or possibly in Africa, Rightmire suggests. H. erectus populations apparently survived for a while in Asia, whereas H. sapiens thrived elsewhere, he says.

To buttress his theory, Rightmire offers a reassessment of a group of fossil skulls and skull fragments found at the Ngandong site in central Java. Multiregional theorists such as Wolpoff view the anatomy of these skulls as intermediate between H. erectus and H. sapiens, indicating a long, gradual evolution toward modern humans in that part of the world.

However, the Ngandong fossils — poorly dated, but generally placed between 100,000 and 250,000 years old — clearly fall within the range of anatomy observed in older H. erectus skulls from Java and elsewhere, Rightmire contends. This holds for the size and shape of Ngandong braincases, the thickness of the cranial bones and other features, he points out.

In contrast, the earliest H. sapiens specimens display marked increased in brain size, changes in cranial bones that signify shifts in brain organization, and a more flexed cranial base, indicating a vocal tract capable of producing a greater variety of speech sounds — all signs of substantial genetic changes that produced a new species in a relatively short time, Rightmire holds.

Another study, conducted by Steven R. Leigh of Northwestern University in Evanston, Ill., lends some support to Rightmire's contention that a measurable split occurs between H. erectus and H. sapiens. Leigh examined 20 H. erectus skulls from Africa, China and Indonesia that span a broad time range, as well as 10 early H. sapiens skulls. Significant expansion of brain size from the oldest to the most recent specimens occurs in the latter group, whereas the three regional samples of H. erectus show no such increases, Leigh reports in the January AMERICAN JOURNAL OF PHYSICAL ANTHROPOLOGY.

However, analysis of the Chinese and Indonesian skulls reveals substantial brain-size increases that do not necessarily coincide with Rightmire's view of an anatomically stable H. erectus inhabiting the entire Old World, Leigh points out.

The single-species view get further ammunition from another study of 70 hominid craniums, mainly H. erectus and H. sapiens specimens. The seven derived features considered unique to Asian H. erectus by Peter Andrews also appear on many African fossils attributed to H. erectus, as well as on a significant number of H. habilis and early H. sapiens specimens, according to Gunter Brauer of the University of Hamburg, Germany, and Emma Mbua of the National Museums of Kenya and Nairobi.

Although additional anatomical features need study, cladistic procedures mistakenly assume that unique derived traits are either present or absent in all members of a species, Brauer and Mbua contend in the February JOURNAL OF HUMAN EVOLUTION. They emphasize Tatersall's point that the same derived features may occur to a greater or lesser extent in different hominid species. Investigators need better data on variations in the skeletal anatomy of living primates and fossil hominids, they conclude.

Some anthropologists take a dim view of the entire controversy surrounding hominid species. "These fights over species classification are somewhat of a waste of time," says Alan Mann of the University of Pennsylvania in Philadelphia. "Most researchers see Homo erectus as a single species that evolved into Homo sapiens."

Others argue that fossil bones provide too little evidence for teasing out hominid species.

"Fossil species are mental constructs," contends Glenn C. Conroy of Washington University in St. Louis, who directed an expedition that recently found an approximately 13-million-year-old primate jaw in southern Africa (SN:6/29/91, p.405). "Clasdistic approaches try to separate species out of a vast array of biological variability over a vast time range, and I don't think they're capable of doing that."

Conroy prefers to group hominid fossils into "grades," or related groups tied together by general signs of anatomical unity with no evidence of sharp breaks between species. Thus, an Australopithecus grade (which includes the more than 3 million year-old "Lucy" and her kin) merges into a grade composed of H. erectus fossils and then shades into a H. sapiens grade, in Conroy's view.

"I'd put our limited funding into looking for new fossil primates or studying living primates, rather than pushing cladograms or arguing about the number of Homo species," he asserts.

But anthropologists wrangling over H. erectus and other hominid species find room for optimism amid their discord.

"The really interesting question isn't whether H. erectus existed," remarks William H. Kimbel of the Institute of Human Origins in Berkeley, California, a proponent of phylogenetic analysis. "For the first time in years, we're taking a step back and asking about the theories that underlie our work and the units we use to establish evolutionary relationships. It's a healthy sign that we're debating these questions vigorously." □

Questions: *Answers on page 204*

1. Why do you think that the species concept is difficult to apply to ancient populations?

2. Why do you think that the species concept is difficult to modern populations?

3. Why is cladistic, or phylogenetic, analysis gaining popularity to explain the evolution of humans and other animals species?

37

Paranthropus boisei, an ancestral hominid, maintained a distinctive anatomical appearance for nearly one million years. Distinctive features included: huge jaws, small front teeth, immense back teeth, flattened teeth, flared cheekbones, and a visor-like crest over the eyes. P. boisei belonged to a group of robust australopithecines appearing about 2.6 million years ago. After little evolutionary change of any type for one million years, it became extinct. This hominid was therefore not an ancestor to humans. Study of this species, however, has shed light on the evolution of modern humans. Many biologists think that hominids directly ancestral to modern humans also experienced few anatomical changes before undergoing abrupt changes to succeeding species leading to modern humans.

Some Hominids Show Fidelity to the Tooth

by B. Bower

Among the ancient members of the human evolutionary family, called fossil hominids by anthropologists, Paranthropus boisei cuts a striking profile. Its skull revolves around huge jaws that encase small front and immense, peg-shaped back teeth. A flattened face and flared cheekbones slope back to a visor-crest over the eyes. A bony ridge runs over the top of the head, where it meets a small, triangular brain-case.

A new study now indicates that P. boisei also exhibited a remarkably stubbon devotion to its distinctive look for more than 1 million years, until the Paranthropus lineage hit an evolutionary dead end. The basic features of P. boisei jaws and teeth remained unchanged during a time of marked brain growth and tooth-size reduction in direct human ancestors, contends anthropologist Bernard Wood of the University of Liverpool in England.

"I suspect P. boisei underwent little evolutionary change of any kind," Wood asserts.

The finding coincides with Wood's view that hominid species directly ancestral to modern humans also experienced few anatomical changes before their relatively abrupt evolution to succeeding species (SN:6/20/92,p.408).

P. boisei belonged to a group of African hominids, referred to as robust australopithecines by some investigators, which first appeared about 2.6 million years ago. P. boisei lived in east Africa from around 2.2 million to 1 million years ago, in Wood's view. Some anthropologists argue that the discovery of the so-called black skull extends the antiquity of P. boisei to 2.5 million years ago, a claim that continues to spark controversy (SN:1/24/87,p.58).

Wood studied 144 fossil jaws and teeth that belonged to P. boisei at various points in its evolutionary history. He could not conduct a similar anatomical survey of lower-body bones, because fossil hunters have found a scant collection of such specimens for P. boisei.

Only a few, marginally important changes took place over time in the extinct hominid's jaws and teeth. Wood reported last week in San Francisco at the annual meeting of the Americn Anthropological Association. Nine out of 47 anatomical features measured by Wood displayed significant change in a comparison of early and late P. boisei specimens.

Both the thickness and the height of the hominid's powerful lower jaw stayed constant over time, Wood notes. The overall size of the lower jaw increased slightly in later specimens, but the British scientist calls this trend "weak."

In contrast, a few teeth features underwent significant change without altering the crucial aspects of P. boisei dentition, Wood asserts. For example, canines enlarged from early to later specimens, but these teeth played a minor role in chewing and grinding, which was handled largely by the heavily enameled molars at the back of the mouth, he maintains.

Although premolar teeth just behind the canines also became larger, no general trends in dental evolution accompany this change, Wood holds.

Anatomical disparities among the teeth and jaws of some P. boisei fossils probably reflect differences between the sexes, he says. □

Answers on page 204

Questions:

1. What other lines of evidence do you think can lend support to the fossil record when studying human evolution?

2. What drawbacks exist for studying evolution when relying on the fossil record alone?

3. Do you think that the study of braincase sizes reveals information on the intelligence of ancestral species?

38

Normally DNA, and the organic tissue containing it, decomposes rapidly after the death of an organism. However, it often remains well-preserved when encased in amber. Amber-encased fossils have served as a source for ancient, preserved insects and their DNA repositories. Insects preserved in amber are one good source. Scientists have learned how to amplify small fragments of DNA from this source, increasing the number of copies for study. The DNA of current insect species can then be compared to the genetic makeup of the ancestors for similarities and differences. In addition to amber fossils, well-preserved bones often provide another source for the study of preserved DNA. Bones from the La Brea tar pits have provided preserved DNA of ancestors to the modern great cats. DNA from the scales of fossil fish is yet another new-found source.

Brushing the Dust Off Ancient DNA

by Kathryn Hoppe

The oldest reported DNA comes from some bugs that stepped in the wrong place about 30 million years ago.

This dramatic evidence of DNA's durability emerged last month in two papers announcing the successful extraction of DNA from fossil insects. Descriptions of DNA extracted from a fossil bee by California researchers appeared in the September MEDICAL SCIENTIFIC RESEARCH. A similar report by researchers at the American Museum of Natural History in New York City — focusing on an extinct termite — followed in the Sept. 25 SCIENCE.

In each case, scientists managed to amplify small fragments of DNA with a molecular copying process known as polymerase chain reaction (SN:4/23/88,p.262). And both teams examined insects preserved in pieces of amber from the Dominican Republic, one of the world's most significant sources of this gem.

Amber-encased fossils, long valued for their excellent three-dimensional detail, are particularly suited for molecular studies. Some "specimens are so well preserved that you can identify cellular structures" under the microscope, says Raul J. Cano, a microbiologist at California Polytechnic State University in San Luis Obispo, who took part in the bee study.

People have long recognized that amber, a form of fossilized tree resin, preserves organic tissue extremely well. The ancient Egyptians used crushed amber to preserve mummies, notes Ward Wheeler, who coauthored the termite report.

Normally, organic tissue — and the DNA it contains — degrades rapidly after an animal dies. When sealed in amber, however, tissues remain isolated from the decay-promoting effects of external air and water. Amber not only acts as a natural antibiotic that prevents the growth of microbes, but it also dries out the creatures it entombs to form natural mummies. An animal trapped in a glob of sap "is there for good," Wheeler says.

Such prisoners include "pretty much anything you can imagine that would be on the side of a tree — small frogs, small lizards, bird feathers, land snails, and a tremendous variety of insects," he says. The animals thus trapped are typically small, since the largest pieces of amber reach only about 6 inches across.

Wheeler's group studied an extinct termite called Mastotermes electrodominicus. Considered by some "a missing link between cockroaches and termites," this particular bug had the potential to solve "an interesting evolutionary question," explains entomologist David Grimaldi, who participated in the investigation. In the past, researchers had debated whether termites evolved from cockroaches or in parallel with them. Using sequenced fragments of DNA from M. electrodominicus, the New York scientists determined that their ancient bug was more termite than roach, suggesting separate origins for the two groups.

The California researchers extracted DNA from an extinct species of stingless bee known as Proplebeia dominicana. They hope their sample will reveal details about the evolution of this bee's modern relatives and provide a reference point for measuring evolutionary changes over time, says Cano.

Such investigations can provide detailed information that may not be available from studies of modern groups or from the anatomical details of fossils. In the future, "amber is going to be looked at for a wealth of information" about ancient lineages, predicts study coauthor George O. Poinar Jr., a paleontologist at the University of California, Berkeley.

Amber fossils represent only a small fraction of the many potential sources of prehistoric DNA.

Other researchers have examined a wide range of preserved bones for information about animals much larger than those trapped in amber. As with the amber-entombed insects, however, only tissues protected from weathering and microbial decay yield remnants of their original DNA.

These remains must be naturally mummified — as in the case of 3,300-year-old bird bones discovered in a dry cave (SN:9/19/92,p.183) — or preserved in rare deposits such as the tar pits of Rancho La Brea in Los Angeles.

Although not nearly as ancient as some of the amber reserves, the La Brea tar pits represent one of the world's richest fossil deposits. They have yielded approximately 2 million specimens representing more than 460 animal species, some of which date back almost 40,000 years.

The first report of sequenced DNA from bones preserved in these tar pits appeared in the Oct. 15 PROCEEDINGS OF THE NATIONAL ACADEMY OF SCIENCES. The study focused on the 14,000-year-old bones of an extinct saber-toothed cat, known as Smilodon fatalis, in the collection of the George C. Page Museum of La Brea Discovering in Los Angeles.

These animals, which brandished long, knife-like canine teeth, have been placed in several different groups of carnivores over the years. The DNA sequencing results now indicate that they belonged to the same family as modern cats and where closely related to the great cats, such as lions, leopards, and tigers, according to a research group led by Stephen J. O'Brien of the National Cancer Institute's Laboratory of Viral Carcinogenesis in Frederick, Md.

Additional studies of fossil DNA may clarify the evolutionary histories of other extinct animals from the La Brea deposits, among them mammoths, mastodons, giant ground sloths, and dire wolves. But such investigations may also have "implications beyond paleontology," asserts O'Brien.

DNA studies "offer the prospect of relating evolutionary adaptation to gene sequences and open the door for the search for ancient pathogens that may have contributed to species extinction," he says. For example, if a virus drove these animals to extinction, then traces of viral DNA might appear in their remains. O'Brien hopes such pathogenic studies will also allow researchers to test hypotheses about the relationship between modern epidemics and modern species.

Regardless of the exact information DNA sequencing may provide the future, it seem certain that the current chronological record holders will not reign for long. Specimens in amber date back approximately 100 million years, providing the potential for DNA studies of animals that lived during the time of the dinosaurs.

Moreover, researchers have achieved "positive results" in their preliminary attempts to obtain DNA from the scales of fossil fish preserved in lake sediments approximately 200 million years old, says Amy R. McCune, a paleontologist at Cornell University. McCune hopes that DNA from these ancient species will not only show their connections to modern species, but also clarify their relationships to one another.

On the other hand, the new excitement over ancient DNA won't put conventional paleontologists out of work, predicts Wheeler. DNA extraction "is a different tool, but not necessarily one that will supersede morphologic or anatomical analyses," he says. Not only are the majority of fossils preserved in conditions that do not protect DNA, but anatomical comparisons will continue to provide overall details that an isolated fragment of DNA may not reveal.

Scientists might not go through the elaborate effort needed to sequence ancient DNA when their questions can be answered through more traditional comparisons. But in some cases the new tool may provide answers to formerly unanswerable questions.

Although "there are a lot of problems and difficulties associated with working with fossil material," says O'Brien, "the prospect of being able to look at the DNA sequences of species that are no longer alive makes this kind of exercise worth it." ☐

39

Did the European Neandertals of 130,000 to 35,000 years ago actually talk? A study of their anatomy reveals that, at least, they were capable of voice patterns. Through an analysis of their preserved skulls, as well as the analysis of modern human skulls, scientists have concluded that a full and modern capacity for speech existed in these human ancestors. Although this ability existed, the evidence that actual speech occurred is lacking.

Neanderthals to Investigators: Can We Talk?

by B. Bower

European Neandertals, who lived from about 130,000 to 35,000 years ago, possessed all the anatomical tools needed for speaking as modern humans do, according to a report presented at the annual meeting of the American Association of Physical Anthropologists in Las Vegas last week.

The new analysis of Neandertal and modern human skulls, conducted by David W. Frayer of the University of Kansas in Lawrence, enters a debate over Neandertal vocal capacities that began in the 1970s. Arguments intensified recently with the discovery of a small neck bone said by its discoverers to demonstrate a fully modern facility for speech among Neandertals (SN:7/8/89,p.24).

"Neandertal speech and language ability was equivalent to ours," Frayer maintains. "Whether they indeed did speak is another issue."

Frayer studied the degree of bend in the base, or basicranium, of Neandertal and modern human skulls. A flat basicranium — ubiquitous in non human animals — indicates that the larynx, or voice box, sits high in the neck. An arched cranial base signified a lower larynx and a vocal tract capable of producing the sounds of modern human speech.

Often, important features of the basicranium are poorly preserved on ancient fossils. in his study, Frayer relied on a measurement of the angle from a relatively easily determined point near the center of the basicranium to a point at the front of the upper jaw.

The extent of basicranial flattening in four European Neandertal specimens falls within the range observed in a sample of modern human skulls dating from 25,000 years ago to medieval times, Frayer contends. In fact, some of the older modern skulls display flatter skull bases than the Neandertals, he says. The evidence supports theories of a close evolutionary link between Neandertals and modern humans, he adds.

One of the Neandertal skulls studied by Frayer was reconstructed in 1989 by a French anthropologist who also argued that the angle of its basicranium falls within the range of modern humans.

Other researchers, led by anatomist Jeffrey T. Laitman of Mount Sinai School of Medicine in New York City and linguist Philip Lieberman of Brown University in Providence, R.I., discern a flatter cranial base and more restricted speech ability in European Neandertals than in modern humans. Laitman's group estimates the position of several anatomical markers on fossils to determine four basicranial angles from the back of the head to the jaw, Lieberman devised a computer model of the Neandertal vocal tract based on the skull that was later reconfigured by the French investigator.

Although Neandertals had the ability to vocalize, their speech quality feel short of that exhibited by modern humans, Laitman asserted at the Las Vegas meeting. "I'd advise caution in measuring only one angle on the basicranium, as Frayer did," he says.

Frayer cites the poor preservation of basicranial features as the prime reason for using his study method. "I'm uncomfortable with how much of the cranial base is missing on Neandertal specimens," he remarks. □

Answers on page 205

Questions:

1. Do you think that an analysis of skull anatomy is sufficient to conclude that speech in the Neandertals was possible?

2. What other evidence is needed to conclude that speech was possible?

3. If they could speak, how would your compare Neandertal language with ours?

150

Part Seven

Interactions -
Behavior, Ecology, and
Environment

40

The bacterium <u>Bacillus thuringiensis</u> has been an effective weapon of biological pest control. It produces at least nine proteins that kill moth and butterfly caterpillars. As an alternative to chemical insecticides, toxins of this microbe break down quickly in the soil. By lasting only a short period of time, insect populations cannot evolve a resistance to the microbial toxins. Molecular biologists are trying to insert genes that make Batt-toxins directly into the genes of plants, converting the plants into pest-proof organisms. Recent breeding experiments, however, have produced a caterpillar that can tolerate all toxins of the microorganism. More studies are needed to how Batt toxins work and how insects can counter it before the full potential of thus mode of biological pest control is known.

Lab Insect Thwarts Potent Natural Toxins

by E. Pennisi

Often touted as one of the most effective weapons in biological pest control, bacteria called Bacillus thuringiensis (Batt) may have met their match.

Batt strains produce at least nine proteins that can kill moth and butterfly caterpillars. For years, organic farmers have sprayed Batt formulations on their crops. Unlike most chemical insecticides, Batt toxins quickly break down in soil, so insects do not develop resistance to them as fast. And when they do, that resistance tends to thwart only one protein in Batt's chemical arsenal, says Fred Gould, an entomologist at North Carolina State University in Raleigh.

But Gould and his colleagues have now bred a caterpillar that seems to totally disarm these mighty microbes. Their findings may complicate efforts by genetic engineers to make plants pest-proof with Batt toxins, Gould says.

Molecular biologists have focused on inserting Batt-toxin genes in plants for several reasons, says Gould. Because genes code directly for these proteins, the genetic alteration is more straight-forward than for other natural insecticides, which require genes encoding several enzymes that then produce the toxins. Also, unlike most proteins, Batt toxins do not disintegrate in the gut and so maintain their potency. And these proteins affect only the pest insects.

"It's a very special toxin; there are not going to be lots of others like it," Gould says. "We should not use it unwisely."

In 1988, as part of their efforts to determine the best way to use Batt, Gould and his colleagues collected eggs produced by Heliothis Virescens, a cotton, soybean, tobacco, and tomato pest sometimes called the tobacco budworm. As the researchers raised successive generations, they fed the insects enough of one Batt toxin to kill 75 percent of that generation. By the 17th generation, the researchers needed to use at least 50 percent more toxin, they report in the September 1, PROCEEDINGS OF THE NATIONAL ACADEMY OF SCIENCES.

Previous research done elsewhere on different insects had led Gould to expect these caterpillars to develop resistance against that toxin.

The toxin works by binding to receptors in the insect's gut and disrupting the gut lining. Typically, resistant insects change the receptor so that the toxin cannot attach.

But these insects were also resistant to other Batt toxins, including one very different one. "Not only was [this toxin] unrelated in terms of its amino acid sequence, but recent work shows that its mode of toxicity was different," says Gould. "And that level of resistance was as high as or higher than [the resistance to] the original [toxic] compound."

Some researchers had thought that by switching to different Batt toxins or engineering two toxins into a crop, farmers could stay one step ahead of resistant strains and maintain the effectiveness of this insecticide. "Now they will have to combine that strategy with other measures in the field to combat resistance," says Pamela G. Marrone, an entomologist with Novo Nordisk Entotech, Inc., in Davis California. For example, she says, if farmers were to grow patches of unaltered varieties in with the genetically engineered crops, some pests would stay vulnerable and by mating with resistant individuals, would slow overall resistance.

At the Monsanto facility in Chesterfield, Mo., researchers are working on varieties of corn, potato, and cotton that make their own Batt toxins. While the company says it depends on scientists such as Gould for guidance, "[this report] will probably not make too much impact in the way we view resistance management," says Monsanto entomologist Steven R. Sims. "This is only one laboratory experiment, which may not necessarily correlate well to what happens in the field, "Both he and Marrone have conducted experiments like Gould's, and their results highlight the need to know more about how Batt works and how insects counter it. "What we're seeing is that there is a lot more variation [in types of resistance] than we expected when we started," says Sims. □

Questions: *Answers on page 205*

1. How is the battle between microbial toxins and insects a possible example of co-evolution?

2. Why is biological pest control preferable to chemical pest control?

3. What advantage does the genetic engineering of the Batt gene have in producing pest-proof plants? Use specific examples of gene structure and function.

41

Up until recently wilderness corridors were hailed as an efficient solution for preserving wild populations. A wilderness corridor is a track of natural habitat that is allowed to remain between developed areas such as farms and commercial malls. These greenways in theory would provide a undisturbed habitat where animals can survive and travel, coexisting with adjacent human population centers. Recent studies with salamanders for example have revealed that some of these amphibians in a population planted in the corridors did not remain in the greenways, wandering into the developed areas instead. Many died in these developed areas. These findings indicate that corridors may not be the solution when located next to poor undesirable areas of a landscape. Although corridors can be made available, there is a lack of evidence that animals will use them.

Wilderness Corridors May Not Benefit All

by C. Ezzell

In the 1970's, a new idea began catching on among conservation ecologists: Leave strips of natural habitat between developed areas, such as farms and shopping malls, so that animals can travel undisturbed among otherwise isolated clumps of surrounding wilderness.

Researchers reasoned that such corridors, often called greenways, would allow wildlife and human populations to live intertwined without cloistering wild creatures into islands where they would eventually run out of resources and become dangerously inbred.

Greenways have become a cornerstone of conservation management in the past two decades. Now, however, a new study questions the efficacy of the corridor concept in preserving wild populations.

In one of the first controlled studies of wild animals' propensity to travel through corridors, researchers led by Daniel K. Rosenberg have found that 30 percent of an experimental group of salamanders failed to use corridors while journeying between clumps of moist, woodland habitat. Instead, these salamanders wandered into bare, inhospitable tracts where many died, the researchers found.

Rosenberg — a wildlife ecologist with the U.S. Forest Service's Redwood Sciences Laboratory in Arcata, California and a doctoral student at Oregon State University in Corvallis — presented the findings in Honolulu this month at the joint annual meeting of the American Institute of Biological Sciences and the Ecological Society of America.

His group evaluated corridor use by placing 50 salamanders into each of two test plots shaped like plus signs. The first plot, which served as a control, consisted of five small clumps of shady, damp forest connected by narrow corridors of bare, scraped earth. The researchers left two corridors in the second, experimental plot filled with natural vegetation. They enclosed both plots with aluminum sheets pounded 10 inches into the ground to keep the salamanders from escaping.

After two weeks, they found that nearly twice as many salamanders in the experimental plot used natural rather than bare corridors to travel to neighboring forest clumps. However, roughly one-third of the animals in the experimental plot strayed instead into the bare, dry tracts, where they either died of desiccation or fell into pitfall traps consisting of buried coffee cans. In contrast, in the control plot roughly equal numbers of salamanders traveled down each of the bare corridors, ruling out the possibility that the animals simply prefer to migrate in a given direction.

Rosenberg says the study, though preliminary, demonstrates the importance of the environment surrounding corridors, because many animals won't find their way into greenways. "Corridors may be a very good thing in some landscapes, but our study cautions against using corridors as a panacea in a poor landscape," he concludes.

Thomas C. Edwards Jr. of the Utah Cooperative Fish and Wildlife Research Unit at Utah State University in Logan calls the new study "very intriguing." "There are tons of anecdotes about animals moving through corridors," he says, "but there have been few experimental studies to prove that animals consistently use them."

Larry D. Harris, a conservation biologist at the University of Florida in Gainesville, argues that "it doesn't matter" what proportion of wildlife uses a corridor. "Even if only 1 percent makes it through," he says, "that's 1 percent that wouldn't have made it otherwise." □

Questions: *Answers on page 205*

1. Do you think that corridors are still valuable even if their use is not 100 per cent by a planted population?

2. What would happen to wild populations if wilderness corridors were not created?

3. Why is it a mistake to use human reasoning to predict what animal populations may do for selecting a habitat?

42

Computer forecasts paint a bleak picture for the recovery in the size of the loggerhead sea turtle population off the coast of the southeastern United States. Biologists think that a tenfold population increase in the current population will be needed to save this threatened species. One factor that has been diminishing their numbers has been their entrapment is shrimp-trawling nets. The smaller turtles of reproductive age are most vulnerable, thus further damaging the ability of the population to sustain its numbers. Use of a trap door excluder device (TED) in shrimp nets can remove one threat to the turtles. The harvesting of sargassum, a seedweed used to make pharmaceuticals, is also a threat to the loggerhead sea turtles. This seedweed is important to the ecological balance of their habitat.

Turtle Recovery Could Take Many Decades
by C. Ezzell

Under current protective laws, it would take at least 70 years to achieve a 10-fold increase in populations of the loggerhead sea turtle off the coast of the southeastern United States, according to a new computer forecast. Wider use of turtle excluder devices (TEDs) — trapdoor-like mechanisms that allow sea turtles to escape shrimp-trawling nets — would reduce this recovery time by only 30 to 40 years, the forecast indicates.

Biologists consider a 10-fold population increase crucial to saving this threatened species.

The new computer model is the first to take into account the differing effects of TEDs on the survival of loggerhead turtles of various ages and sizes, says Selina Heppell, a biologist at North Carolina State University in Raleigh. Heppell described the model last week at the joint annual meeting of the American Institute of Biological Sciences and the Ecological Society of America, held in Honolulu.

Because TEDs are less effective in saving smaller, younger turtles, she says, the model suggests that fewer than anticipated numbers of loggerheads will survive to reproductive age.

Most TEDs consist of a panel of metal bars inserted into a trawling net at an angle leading up to a hole in the top of the net. The bars are designed to allow shrimp to pass through and accumulate in the sack-like end of the net, while diverting larger marine animals — such as sea turtles — up and out of the net through the hole.

The U.S. National Marine Fisheries Service (NMFS) estimates that TEDs have slashed by 67 percent the annual mortality rate of sea turtles caught in trawling nets in U.S. costal waters since the 1988 enactment of a federal law requiring shrimp trawlers to use the devices during selected times of the year in most offshore areas. Despite the use of TEDs, however, the NMFS projects that at least 4,360 sea turtles inhabiting U.S. coastal waters will drown in trawling nets this year.

Heppell's forecast rests on the assumption that TEDs reduce the mortality rate of net-trapped juvenile loggerheads by only 34 percent each year, because these smaller turtles sometimes get swept between the bars of the devices and become caught in the nets. This potentially fatal generation gap skews the demographics of the loggerhead population toward older turtles with few remaining reproductive years, Heppell says. This, in turn, slows the recovery rate of the species as a whole, she asserts.

"TEDs have had a very positive effect on increasing the population [of loggerhead turtles]," Heppell says, "but 70 years is a longer time to see a 10-fold population recovery than we'd expected." She adds that although the number of nesting loggerheads has increased within the past three years, indicating the overall benefits of TEDs, "we have to be a little cautious in saying we've found the answer to saving the [loggerhead] turtle population.... It's going to take a long time."

Last April, the NMFS proposed new regulations that would expand the requirements for TED use by mandating trawlers to use TEDs year-round at all U.S. inshore and offshore locations. This would achieve a 97 percent reduction in trawler-related turtle mortality, according to agency estimates.

However, the proposal does not call for new TED designs less likely to trap juvenile loggerheads. Nor does it address what Heppell describes as a potential threat to all sea turtles: the harvesting of sargassum, a type of seaweed that grows in floating mats.

Commercial harvesting of these vast mats for use in pharmaceuticals or livestock feed could destroy an important habitat for "small juvenile" sea turtles, Heppell says. According to her model, this category of juveniles constitutes the second most important age group for the recovery of the loggerhead population. Increasing the mortality rate of small juveniles could delay a 10-fold increase in loggerheads to 140 years, Heppell projects. □

Questions: *Answers on page 205*

1. How does the association between loggerhead turtles and sargassum offer a good lesson in ecology?

2. Why is it particularly important to preserve the juvenile part of the turtle population?

3. How does the TED work?

43

Members of the honeybee colony exhibit an incredible division of labor. Specific groups perform specific tasks, combining efforts into a smoothly functioning unit or colony. What is the mechanism for coordinating and integrating these tasks? The key for coordination seems to emanate from the signals sent by the queen of the colony. This so-called "queenright" is induced by a powerful blend of chemical signals, called pheromones, that are discharged from the queen bee and spread through the members of the colony by physical contact. The influence of these chemical signals can be either stimulatory or inhibitory. The chemical language among honeybees consists of at least 36 pheromones. Each elicits a specific behavior from a colony member. Chemical signals from the queen bee are called primary pheromones because of their broad effects on the behavior of colony workers and colony reproduction. The chemical identity of these primary pheromones is just beginning to be recognized.

The Essence of Royalty: Honey Bee Queen Pheromone

Beekeeping, like most skills, has developed its own shorthand language, by which a word or simple phrase can evoke complex concepts that quickly are understood by those who practice the craft. One such word is "queenright," meaning that a colony's queen is present and thousands of her worker offspring are collaborating diligently on communal tasks. Hidden in this beekeeper jargon, however, lies the most fundamental mystery of social insects: the mechanisms of coordination and integration that mediate the tasks of thousands of individuals into the smoothly functioning unit of the colony. In this article, we shall describe some new findings concerning one aspect of these integrative mechanisms, the honey bee queen's pheromonal influence over worker bees. The discovery of the nature and function of this "essence of royalty" has proved profound insights into the functioning of social-insect colonies, and also has led to some significant commercial applications for crop pollination and beekeeping.

The study of social-insect pheromones, or sociochemistry, is not a new discipline. Many pheromones are known throughout the social insects that are used in myriad tasks such as alarm behavior, orientation, trail marking, nest recognition, mate attraction and others (Bradshaw and Howse 1984; Duffield, Wheeler and Eickwort 1984; Free 1987; Hölldobler and Wilson 1990; Howse 1984; Wilson 1971; Winston 1987). The honey bees, for example, are estimated to produce at least 36 pheromones, which together constitute a chemical language of some intricacy. However, virtually all of the known chemicals are secretions that "release," or elicit, a specific behavior. The queen sociochemicals belong to another class, called primer pheromones, that exercise a more fundamental level of control by mediating worker and colony reproduction and influencing broad aspects of foraging and other behaviors. Further, they can be both stimulatory and inhibitory, depending on the specific function. the existence of primer sociochemicals is well known, but their identification and synthesis have proved difficult.

Indeed, the only primer pheromone that has been identified comes from the queen honey bee (Apis mellifera L.); it is secreted by the mandibular glands, a paired set of glands on either side of the queen's head. Here we discuss the research that led to the pheromone's identification, new findings concerning its effects on worker bees and colony functions, transmission mechanisms within colonies, and some applications for crop pollination and beekeeping.

Early Research

For more than 30 years it has been recognized that queen pheromones mediate many colony activities. The pheromones were known to inhibit the rearing of new queens and the development of ovaries in worker bees; they were also observed to attract workers to swarm clusters, stimulate foraging, attract workers to a queen and allow

Reprinted by permission from AMERICAN SCIENTIST, Journal of Sigmi Xi, The Scientific Research Society.

workers to recognize their own queen.

Surprisingly, only two active substances had been identified. These two most abundant compounds in queen mandibular pheromone are the organic acids (E)-9-keto-2-decenoic acid, or 9ODA, which was identified in 1960 (Callow and Johnston 1960, Barbier and Lederer 1960) and (E)-9-hydroxy-2-decenoic acid, or (HDA, which was identified in 1964 (Callow, Chapman and Paton 1964; Butler and Fairey 1964).

Experiments with the synthetic acids, however, showed that they did not fully duplicate the effects of mandibular extracts or of the queen's presence (reviewed by Free 1987; Winston 1987). Although analysis of crushed bee heads produced laundry lists of chemicals, none was shown on closer examination to be active, and progress stalled.

One reason it took so long to identify queen mandibular pheromone is that bioassays for primer pheromones are difficult and time-consuming. The potency of candidate releaser pheromones can be evaluated by relatively simple means. Orientation pheromones, for example, can be evaluated by counting the number of bees entering a hive after pheromone is deposited at the hive opening. Demonstrating that a synthetic blend or a natural extract is fully equivalent to a primer pheromone, however, may require a battery of bioassays, some of which may take months to complete. To further complicate matters, there is often disagreement over a primer's function.

The chemical composition of queen mandibular pheromone and its biological effects were not the only controversial topics. Even the means by which the pheromone reaches worker bees was unknown. Is the pheromone transmitted when the workers exchange food? Does it volatilize and spread through the air? Or is it spread by body contact? Here the main obstacle was the minute amount involved: A queen produces less than four ten-thousandths of a gram per day, and a worker bee is able to sense a ten-millionth of the queen's daily production.

Discovery

One of the stereotyped behaviors thought to be elicited by queen pheromones is the retinue response. Worker bees turn toward the queen, forming a dynamic retinue around her. About 10 workers at a time contact her with their antennae, forelegs or mouthparts for periods of minutes. Then the workers leave the queen, groom themselves and move through the nest, making frequent reciprocated contacts with other workers for the next half-hour (Allen 1960; Seeley 1979).

bioassays involving retinue behavior had given particularly confusing results. Mandibular extracts seemed to be important to worker recognition of the queen and attraction to her. However, the most abundant component of the glands, 9ODA, did not by itself elicit retinue behavior. To add to the confusion, bees had been shown to exhibit the retinue response to queens whose mandibular glands had been surgically removed.

This is how matters stood in 1985 when, quite by accident, we put a glass lure coated with queen mandibular extract next to some stray workers on a laboratory bench. To our surprise, the workers formed a retinue around the lure. Clearly the extract had properties that its major known constituent did not. this was exciting because the retinue response is easily recognizable and takes place immediately. For both reasons it might serve as the basis for a practical bioassay for mandibular pheromone.

Indeed we were able to develop a bioassay based on the retinue response that was a particularly sensitive indicator of pheromonal activity. In a series of experiments we used this bioassay to evaluate complete or fractionated mandibular extracts. interestingly, fractions that contained a single substance were relatively inactive. We began to see retinue behavior around the lure only when we combined certain fractions. Eventually, we identified the components of the active fractions and devised a synthetic blend of five components that duplicated the activity of the natural mandibular secretion (Kaminski et al. 1990, Slessor et al. 1988).

The ingredients include two forms of 9HDA. This substance is chiral; It has a left-handed and a right-handed form. Earlier, we had demonstrated that one form, or enantiomer, is significantly more effective in evoking some behavioral responses than the other (Winston et al. 1982). We had also determined that one enantiomer predominates in the natural pheromone (Slessor et al. 1985). But it turned out that both enantiomers also have to be present to evoke the full retinue response.

The remaining missing ingredients were two small, aromatic compounds. One of the aromatic compounds, methyl p-hydroxybenzoate, or HOB, had been identified earlier (Pain, Hugel and Barbier 1960) but had not been considered to be active as a pheromone. The other aromatic molecule, 4-hydroxy-3-methoxyphenylethanol, or HVA, was first identified in the course of our experiments.

The pheromonal activity of the aromatic molecules was quite unexpected. Most insect pheromones are aliphatic compounds, built up out of straight or branched chains of carbon atoms; they are derived from fatty acids or terpenes. The decenoic acids in mandibular gland pheromone fit this description, for example. The aromatics, which feature closed rings of carbon atoms, are a different class of chemical entirely. Indeed we probably found them only because we reversed the customary experimental procedure. Instead of identifying substances in gland extracts and then screening them for pheromonal activity, we screened for activity and then identified the compounds we had isolated.

Our research into the mandibular gland's composition and activity cleared up several points of confusion. First, the complete ensemble of molecules is necessary for full efficacy. Removing any of the components reduced the pheromone's effect by as much as 50 percent, and individual components are inactive when they ar tested alone. This is why the pheromone's major component by itself had failed to elicit retinue behavior.

Second, although we managed to duplicate the effects of mandibular secretions, we should emphasize that we did not duplicate the queen's complete arsenal of pheromonal effects. The mandibular glands are not the only site of production for queen pheromones; they are also probably secreted by the tergite glands, which are located in some of the membranes between the queen's abdominal segments (Renner and Baumann 1964, Velthuis 1970). Moreover, the mandibular and other pheromones may have closely linked or overlapping effects, which may explain why demandibulated queens still evoked the retinue response.

Third, the pheromonal blend a queen produces varies with her age. A newly emerged virgin queen has virtually no pheromone in her glands, but when she begins to make some six days later, she is secreting the decenoic acids, and her glands contain about half the amount of

acid a mated queen's contain. The full blend, including the aromatic compounds, is secreted only after a queen has finished matting and begins to lay eggs (Slessor et al. 1990). This variation in the composition of the pheromone with the queen's age also may help to explain differing reports of effectiveness.

Finally, the mandibular pheromone is effective over a wide range of dosages. We defined the average amount of pheromone found in the glands of a mated queen to be one queen equivalent (Qeq). Worker bees respond to dosages as low as 10-7 Qeq, or a ten-millionth of her daily production (Kaminski et al. 1990, Slessor et al. 1988). This finding was particularly interesting in relation to hypotheses about pheromone transmission.

Functions

Once we were satisfied that we had completely identified queen mandibular pheromone, we began to test the effects of the pheromone on colony functions. One of the first pheromone-mediated behaviors we examined was the inhibition of queen rearing. This function can be easily demonstrated by removing a queen from her colony; the workers become agitated after half an hour and begin rearing new queens within 24 hours. Beekeepers call this "emergency queen rearing," because failure to promptly rear a new queen results in the death of the colony.

To rear queens, workers elongate the cells around a few newly hatched larvae. These larvae are fed a highly enriched food called royal jelly that workers produce in their brood-food glands. The specialized diet directs development away from the worker pathway and on to the queen pathway. Brood older than six days (three days as eggs and three as larvae) lose their ability to develop into queens. Since no new eggs are laid after the queen is lost, workers only have six days to

begin queen rearing before the colony loses its ability to produce a new, functional queen.

We examined the role of mandibular pheromone in emergency queen rearing by comparing queen rearing in queenright colonies, queenless colonies and queenless colonies receiving daily doses of synthetic mandibular pheromone (Winston et al 1989, 1990). The results were dramatic. When queenless colonies received one Qeq or more of pheromone per day, they made almost no attempt to rear new queens for four days after the queen was removed. Even six days after queen loss, there was no significant difference between the number of queen cells in queenless colonies treated with pheromone and the number of queenright control colonies. We concluded that the queen's mandibular pheromone is largely responsible for the inhibition of queen rearing in queenright colonies.

Colonies also rear new queens in preparation of reproductive swarming, the process of colony division in which a majority of the workers and the old queen leave the colony and search for a new nest. Left behind in the old colony are developing queens, some of which may issue with additional swarms once they emerge, and one of which will reign over the old nest once swarming is completed.

Swarming poses something of a puzzle because it does not occur unless new queens are being reared, but the old queen is present during the initial stages of queen rearing and is presumably still secreting her inhibitory chemicals (Butler 1959, 1969, 1961; Seeley and Fell 1981; Simpson 1958). One hypothesis is that the transmission of queen pheromones is slowed as colonies grow and become more congested prior to swarming. As the amounts of pheromone reaching workers diminish, the workers begin to escape the queen's control.

We tested the role of queen pheromone in reproductive swarming by supplementing the queen's normal secretion with additional, synthetic pheromone. Again, the pheromone had a significant effect on queen rearing. Colonies receiving a 10 Qeq per day swarmed an average of 25 days later than control colonies. The pheromone-treated colonies became extraordinarily crowded before they would swarm, indicating just how powerful a suppressant the queen pheromone can be.

For these experiments, the pheromone was presented on stationary lures, but queen bees move about in the nest. In another set of experiments, we tried to better mimic natural conditions by administering the pheromone as a spray. We found that a dose of one Qeq per day of supplemental pheromone was sufficient to suppress swarming. The fact that the better-dispersed spray was active at a much lower dose than the stationary lure supports the hypothesis that the slowing of the pheromonal dispersal in congested colonies triggers swarming (Winston et al. 1991).

The queen mandibular pheromone plays two other roles in swarming as well: It attracts flying workers to the swarm cluster, and it stabilizes the cluster, preventing workers from becoming restless and leaving before the swarm reaches a new next site. We found that the synthetic mandibular pheromone is almost as effective as the queen in attracting workers to the cluster and in keeping them together, at least for the first few hours following cluster formation.

Next we examined the effect of mandibular pheromone on worker ovary development and egg laying. Both are know to be suppressed by queen pheromones. indeed pheromonal suppression of worker

reproduction is one of the hallmarks of advanced insect societies. Primitive social insects suppress nest-mate reproduction primarily through dominance interaction. Many wasp and bumblebee queens, for example, seem to maintain dominance by biting and harassing the other females in the colony.

Honey bee workers have the potential to lay unfertilized eggs that can develop into males, or drones. However, some combination of pheromones secreted by the queen and the brood (the developing larvae or pupae) inhibits the maturation of the worker's ovaries, which usually remain small and non functional (Jay, 1972). Nevertheless, even in queenright colonies a few workers manage to lay eggs (Ratnieks and Visscher 1989, Visscher 1989) and many workers begin egg-laying two or three weeks after the queen and brood are removed (Winston 1987).

The ovary-suppressing substance produced by the queen was thought to be mandibular pheromone, particularly 9ODA (see reviews by Free 1987; Willis, Winston and Slessor 1990; Winston 1987). To test this hypothesis, we removed queens from groups of colonies and applied various doses of mandibular pheromone to the colonies for the next 43 days. Some colonies received daily doses as high as 10 Qeq. to our surprise, workers in queenless, pheromone-treated colonies developed ovaries at the same rate as workers in queenless, pheromone-free colonies (Willis, Winston and Slessor 1990). Apparently queen mandibular pheromone is not involved in the suppression of worker ovary development. Another primer pheromone must be responsible for this effect, possibly the unidentified queen tergite pheromone in combination with brood pheromone.

A final set of experiments demonstrated that queen mandibular pheromone is not just an inhibitor.

In addition to suppressing reproductive activity, it can stimulate foraging and brood rearing. To establish this, we applied pheromone to newly founded, queenright colonies and counted the number of foragers, the size of their nectar and pollen loads and the extent of brood rearing. The pheromone did not increase overall foraging activity, but colonies that received one Qeq of supplemental pheromone daily had more pollen foragers, and those foragers carried heavier pollen loads. Pheromone applications increased the amount of pollen entering the nest by 80 percent, and pheromone-treated colonies reared 18 percent more brood, possibly due to the increased amounts of protein-rich pollen being brought into those colonies (Higo et al. 1992).

Transmission

The set of experiments we have just described demonstrated that queen mandibular pheromone has a wide range of functions, but we still did not know how much is produced and secreted by the queen daily or how these important compounds are transmitted to workers. The chemical components of the pheromone are not particularly volatile, and so they probably cannot spread throughout the colony by diffusion alone. The number of workers in an established colony and the rapidity with which the pheromone is lost make it unlikely that the queen is the sole disseminator of the pheromone. Some experiments had suggested that retinue workers act as messengers, transmitting the queen's pheromones to other workers when they touch antennae or mouthparts (Juška, Seeley and Velthuis 1981; Seeley 1979). Pheromone could not be found on worker bees, however, possibly because it was present in concentrations below the limit of detection of the instruments then in use.

We were able to follow pheromone secretion and transmission using the more sensitive chromatographic and spectroscopic equipment now available, as well as radiolabelled pheromone provided by Glenn Prestwich and Francis Webster of the State University of New York at Stony Brook and Syracuse (Webster and Prestwich 1988). Quantitative measurements of rates of production, transfer and loss allowed us to construct a mathematical model of pheromone dispersal. Our results essentially confirm the messenger-bee hypothesis, but they raise many interesting questions as well (Naumann et al. 1991).

We first determined that a queen typically has five micrograms, or 0.001 Qeq. of pheromone on her body at any one time. To determine how much a queen secretes daily, we removed queens from colonies and measured the pheromone that built up on their cuticles. It turned out that a queen secretes between 0.2 and 2 Qeq of pheromone per day. These results fit well with our within-colony function studies, where pheromonal effects typically appeared at a dosage of about 1 Qeq.

We then quantified the processes by which pheromone is transferred or lost. For example, we allowed workers to contact the queen for different periods and then measured the amount of pheromone they had picked up. We found that each process could be approximated by a first-order rate equation. In other words, the amount of pheromone transferred during my short interval is proportional to the quantity present at the source during that interval; the constant of proportionality is called the rate constant. The first-order equation yields an exponential loss of material at the source; the "half-life" is inversely proportional to the rate constant. We calculated the rate constants for all of the transmission pathways from our experimental

data and then used the value we had obtained for the queen's daily secretion and the rate equations to determine the amount of pheromone that would reach worker bees each day.

Our confidence in the model was bolstered when it predicted rates of pheromone transfer that are consistent with bee behavior. For example, the model predicts a flux of about one Qeq of pheromone through the nest per day; in our function experiments, a similar dose was typically the most active. The model also predicts that the amount of pheromone passed between workers will drop below the detectable level (10-7 Qeq) about 30 minutes after it is picked up from the queen, which is about how long it takes for workers to become restless after the queen is removed. finally, the model predicts that pheromone deposited by the queen in the wax comb will become undetectable in a few hours, which is about how longs it takes the bees to begin rearing new queens.

The retinue workers do indeed remove the greatest fraction of the queen's pheromonal production. Not all bees in the retinue pick up and transfer the same amount of pheromone, however. There are two types of retinue bees, which we call licking and antennating messengers. The lickers touch the queen with their tongues, forelegs and mouthparts. The antennators brush the tips of the antennae lightly and quickly over her body. Only about 10 percent of the bees are lickers, but they pick up over half the queen's pheromonal secretion. Each antennating worker pickups up much less pheromone and passes on much less in subsequent contacts. Why some bees lick and others antennate is not known, nor do we understand the implications of this dual transmission system for the colony's social structure.

The wax comb also plays a role

in the transfer of pheromone, although the queen deposits only 1 percent of her production there. Workers pick up a little of the pheromone the queen deposits. The remaining pheromone is probably released slowly over the next few hours. The slow release may explain why empty comb that has had brood in it is attractive to adult workers. The queen spends most of her time in the brood area, and her pheromonal footprints would be concentrated there.

One of our more surprising findings is that pheromone is lost primarily through internalization. The queen herself internalizes some 36 percent of the pheromone she produces. She swallows some of it, some is absorbed on or bound to her cuticle, and some moves through the cuticle into the blood (hemolymph) system. What purpose internalization serves is unclear. Even if the internalized pheromone retains the same chemical identity, it is no longer available to the colony. It seems odd that such inefficient use is made of the queen's pheromone secretion. It is possible that the model somewhat overestimates the queen's internalization of pheromone, since this was the only process that did not seem to conform to a first-order rate equation.

On the other hand, the messenger bees also internalize pheromone at a surprising rate. The licker bees swallow 40 percent of the pheromone they pick up. Most of the rest is quickly transferred from the worker's head to her abdomen by a combination of passive transport and active grooming (Naumann 1991). There the pheromone passes into and through the cuticle. Some of it ends up in the blood system within 30 to 60 minutes, and some remains bound to or adsorbed on the cuticle. In short, between the queen and the workers, nearly all of the pheromone is eventually internalized.

The mechanisms by which

worker bees perceive the pheromone are not well understood. Sensory structures on a bee's antennae allow it to detect some orders at a concentration equal to a tenth or a hundredth of the concentration people can detect. The arrival of odorant at receptor cells inside the antennae causes associated nerve cells to fire, and the firing can initiate further behavioral and physiological changes. In most insects pheromones act by similar mechanism.

But what about the large amounts of pheromone that are internalized? One hypothesis is that once the pheromone is translocated across the cuticle, its constituents or their breakdown products become active as hormones rather than as pheromones. We are only beginning to explore this intriguing hypothesis.

It also is unclear why the queen produces such a large amount of mandibular pheromone. The answer may be rooted in queen-worker conflict and the evolution of queen dominance in highly social insects. As we have mentioned, there are almost no dominance interactions in a normally functioning honey bee colony. Instead the queen controls the workers entirely through her sophisticated arsenal of pheromones. Perhaps over evolutionary history the workers and queen engaged in a kind of chemical arms race, where the workers began to break down the queen's compounds more rapidly and the queen responded by secreting more pheromone. Thus the swallowing and absorbing of pheromone by workers that we find so puzzling may be an attempt to catabolize the queen's pheromonal dominance and escape her control. Although this idea is speculative, it is certainly plausible. Indeed, this concept is reminiscent of most families and societies, which exhibit a complex blend of cooperation and conflict, with some objectives in common but with each individual also having its own goals.

Commercial Applications

We have been investigating commercial applications of queen pheromone for beekeeping and crop pollination. Several beekeeping applications for queen mandibular pheromone are already close to commercial realization. For example, we have shown that the pheromone allows packages of worker bees to be shipped without queens (Naumann et al. 1990). Beekeepers often establish a colony by buying a kilogram of workers and a queen in a wire package, but queens are typically produced by beekeepers who specialize in this art, so they may come from a different source. The pheromone's ability to delay swarming also might be useful in bee management, since frequent swarming reduces honey production and can threaten colony survival. The pheromone's ability to stimulate pollen foraging and brood rearing might have commercial importance as well, since substantial increases in colony growth rates can be achieved by these means. Finally, we are investigating the use of mandibular pheromone as an attractant for swarms. Such an attractant would be useful for the monitoring and control of honey bee diseases and the Africanized bee (the so-called "killer bee").

The most significant application of queen mandibular pheromone, however, may be in assisting crop pollination. Managed crop pollination is economically the most important function of bees. Colonies of bees are moved to blooming crops in order to provide enough bees to ensure good pollination and seed set. Beekeepers receive up to $45 per colony for this service which is essential for the production of many fruits and vegetables. Not only does pollination affect yield, it also is often closely associated wit the quality of fruits, vegetables and seeds. Each year the honey bee pollinates crops worth at least $1.4

billion in Canada and $9.3 billion in the United States (Robinson, Nowogrodzki and Morse 1989; Scott and Winston 1984). Although hundreds of thousands of bee colonies are moved to crops each year, many crops are still inadequately pollinated. Reduced yields, occasional crop failures and lowered crop quality can all be attributed to this problem (Free 1970; Jay 1986; McGregor 1976; Robinson, Nowogrodzki and Morse 1989).

Since queen pheromone is highly attractive to flying workers, we thought it possible that pollination would be improved by spraying crops with a dilute blend of pheromone. To test this hypothesis we sprayed blocks of trees or berries with various concentrations of pheromone and monitored the number of workers attracted to the blocks and the corp yields. In almost all of the experiments, up to twice the number of bees visited treated blocks compared to untreated ones.

The effect of pheromone spraying on yield was more variable. In preliminary trials with apple trees we found no increases in yield or improvements in fruit quality. However, pear trees sprayed with pheromone produced fruit that was heavier by an average of 6 percent, which translated to an increase of 30 percent in profits or about $1,055 per hectare once spraying costs were deducted. Cranberries treated with pheromone yielded about 15 percent more berries by weight which increased net returns by $4,465 per hectare averaged over two years (Currie, Winston and Slessor 1992; Currie, Winston, Slessor and Mayer 1992). In some cases there was no improvement, whereas in others the yield increase was much greater than these average values. Clearly, if we ar able to exercise in the farmer's field the kind of sociochemical control the queen exercises in the nest, we can dramatically expand the economic value of this already beneficial social insect. □

Acknowledgements

We are grateful to all of our collaborators in the research this article describes, including: J.H. Bordon, S.J. Colley, R.W. Currie, H.A. Higo, L.-A. Kaminski, G.G.S. King, K. Naumann, T. Pankiw, E. Plettner, G.D. Prestwich, F.X. Webster, L.G. Willis and M.H. Wyborn. We would also like to acknowledge the technical assistance of P. LaFlamme. Funding for the research came from the Natural Sciences and Engineering Research Council of Canada, the Science Council of British Columbia, The Wright Institute and Simon Fraser University.

Bibliography

Allen, M.D. 1960. The honeybee queen and her attendants. Animal Behavior 8:201-08.

Barbier, J., and E. Lederer. 1960. Structure chimique de la "substance royale" de la reine d'abeille (Apis mellifica). Comptes Rendus des Séances de L'Académie de Science (Paris) 251:1131-35.

Bradshaw, J.W.S., and P.E. Howse, 1984. Sociochemicals of ants. In: Chemical Ecology of Insects, W.J. Bell and R.T. Carde (eds). Sunderland, Mass.:Sinauer.

Butler, C.G. 1959. Queen substance. Bee World 40:269-75.

Butler, C.G. 1960. The significance of queen substance in swarming and supersedure in honey-bee (Apis mellifera) colonies. Proceedings of the Royal Entomological Society (London) (A) 35:129-32.

Butler, C.G. 1961. The scent of queen honey beens (Apis mellifera) that causes partial inhibition of queen rearing. Journal of Insect Physiology 7:258-64.

Butler, C.G., and E.M. Fairey. 1964. Pheromones of the honey bee: biological studies of the mandibular gland secretion of the queen. Journal of Apicultural Research 3:65-76.

Callow, R.K., J.R. Chapman and P.N. Paton. 1964. Pheromones of the honey bee: Chemical studies of the mandibular gland secretion of the queen. Journal of Apicultural Research 3:77-89.

Callow, R.K., and N.C. Johnston. 1960. The chemical constitution and synthesis of queen substances of honey bees (Apis mellifera L.). Bee World 41:152-3.

Currie, R.W., M.T. Winston and K.N. Slessor. 1992. IMpact of synthetic queen mandiublar pheronome sprays on honey bee pollination of berry corps. Journal of Economic Entomology (in press).

Currie, R.W., M.L. Winston, K.N. Slessor and D.F. Mayer. 1992. (Effect of synthetic queen mandiublar pheromone sprays on pollination of fruit crops by honey bees. Journal of Economic Entomology (in press).

Duffield, R.M., J.W. Wheeler and G.C. Eickwort. 1984. Sociochemicals of bees. In: Chemical Ecology of Insects, W.J. Bell and R. T. Carde (eds). Sunderland, Mass.:Sinauer.

Free, J.B. 1970. Insect Pollination of Crops. London:Academic Press.

Free, J.B. 1987. Pheromones of Social Bees. Ithaca:Cornell Univeristy Press.

Higo, H.A., S.J. Colley, M.L. Winston and K.N. Slessor. 1992. Effects of honey bee queen mandibular gland pheromone on foraging and brood rearing. Canadian Entomologist 124:409-418.

Höllidobler, B., and E.O. Wilson. 1990. The Ants. Cambridge, Mass.:Harvard University Press.

Howse, P.E. 1984. Sociochemicals of termites. In: Chemical Ecology of Insects, W.J. Bell and R.T. Carde (eds). Sunderland, Mass.:Sinauer.

Jay. S.C. 1970. The effect of various combinations of immature queen and worker bees on the ovary development of worker honey bees in colonies with and without queens. Canadian Journal of Zoology 48:168-73.

Jay, S.C. 1972. Ovary development of worker honeybees when separated from worker brood by various methods. Canadian Journal of Zoology 50:661-4.

Jay, S.C. 1986. Spatial management of honey bees on corps. Annual Review of Entomology 31:49-66.

Juska, A., T.D. Seeley and H.H.W. Velthuis. 1981. How honeybee queen attendants become ordinary workers. Journal of Insect physiology 27:515-19.

Kaminski, L.-A., K.N. Slessor, M.L. Winston, N.W. Hay and J.H. Borden. 1990. Honey bee response to queen mandibular pheromone in a laboratory bioassay. Journal of Chemical Ecology 16:841-49.

McGregor. S.E. 1976. Insect Pollination of Cultivated Crop Plants. Agricultural Handbook No. 496. Washington, D.C.: U.S. Department of Agriculture.

Naumann, K. 1991. Grooming behaviors and the translocation of queen mandibular gland pheromone on worker honey bees. Apidologie 22:523-531.

Naumann, K., M.L. Winston, M.H. Wyborn and K.N. Slessor. 1990. Effects of synthetic honey bee (Hymenoptera: Apidae) queen mandibular gland pheromone on workers in packages. Journal of Economic Entomology 83:1271-75.

Naumann, K., M.L. Winston, K.N. Slessor, G.D. Prestwich and F.X. Webster. 1991. The production and transmission of honey bee queen (Apis mellitera L.) mandibular gland pheromone. Behavioral Ecology and Sociobiology. 29:321-32.

Pain, J., M.-F. Hugei and M.C. Barbier. 1960. Comptes Rendus des Seances de L'Academie de Science (Paris) 251:1046-8.

Ratnieks, F.L.W., and P.K. Visscher. 1989. Worker policing in the honey bee. Nature 342:796-97.

Renner, M., and M. Baumann. 1964. Uber komplexe von subepidermalen drusenzalien (Duftdrusen?) der bienenkonigin. Naturwissenschaften; 51:68-9.

Robinson, W.S., R. Nowogrodzki and R.A. Morse. 1989. The value of honey bees as pollinators of U.S. crops. American Bee Journal. 129:411-23, 477-87.

Scott, C.D., and M.L. Winston. 1984. The value of bee pollination to Canadian apiculture. Canadian Beekeeping 11:134.

Seeley, T.D. 1979. Queen substance dispersal by messenger workers in honey bee colonies. Behaviorial Ecology and Sociobiology 5:391-415.

Seeley, T.D., and R.D. Fell. 1981. Queen substance production in honey bee (Apil mellifera) colonies preparing to swarm. Journal of the Kansas Entomological Society 54:192-96.

Simpson, J. 1958. The factors which cause colonies of Apis mellifera to swarm. Insectes Sociaux 5:77-95.

Slessor, K.N., G.G.S. King, D.R. Miller, J.L. Winston and T.L. Cutforth. 1985. Determination of chirality of alcohol or latent alcohol semiochemicals in individual insects. Journal of Chemical Ecology 11:1659-67.

Slessor, K.N., L.-A. Kaminski, G.G.S. King, J. H. Borden and M.L. Winston. 1988. Semiochemical basis of the retinue response to queen honey bees. Nature 332:354-56.

Slessor, K.N., L.-A. Kaminski, G.G.S. King and M.L. Winston. 1990. Semiochemicals of the honey bee queen mandibular glands. Journal of Chemical Ecology 16:851-60.

Velthuis, H.H.W. 1970. Queen substances from the abdomen of the honey bee queen. Deitschrift —— vergierchende Physiology 70:210-22.

Visscher, P.K. 1989. A quantitative study of worker reproduction in honey bee colonies. Behavioral Ecology and Sociobiology 25:247-54.

Webster, F.X., and G.D. Prestwich. 1988. Synthesis of carrier-free tritium-labelled queen bee pheromone. Journal of Chemical Ecology 14:957-62.

Willis. L.G., M.L. Winston and K.N. Slessor 1990. Queen honey bee mandiular pheromone does not affect worker ovary development. Canadian Entomologist 122:1093-99.

Wilson. E.O. 1971. The Insect Socities. Cambridge. Mass.:Harvard University Press.

Winston, M.L., K.N. Slessor, M.J. Smirle and A.A. Kandil. 1982. The influence of a queen-produced substance. 9HDA, on swarm clustering behavior in the honey bee Apis mellifera L. Journal of Chemical Ecology 8:1283-88.

Winston, M.L. 1987. the Biology of the Honey Bee. Cambridge, Mass.:Harvard University Press.

Winston, M.L., K.N. Slessor, L.G. Willis, K. Naumann, H.A. Higo, M.H. Wyborn and L.-A. Kaminski. 1989. The influence of queen mandibular pheromones on worker attraction to swarm clusters and inhibition of queen rearing in the heony bee (Apis mellitera L.). Insectes Sociaux 36:15-27.

Winston, M.L., H.A. Higo and K.N. Slessor. 1990. Effect of various dosages of queen mandibular gland pheromone on the inhibition of queen rearing in the honey bee (Hymenoptera:Apidae). Annals of the Entomological Society of America 83:234-38.

Winston, M.L., H.A. Higo, S.J. Colley, T. Pankiw and K.N. Slessor. 1991. The role of queen mandibular pheromone and colony congestion in honey bee (Apis mellitera L.) reproductive swarming. Journal of Insect Behavior 4:649-659.

Questions: *Answers on page 206*

1. How does a knowledge of chemistry help to understand the division of labor in a honeybee colony?

2. Why can a pheromone be called an environmental hormone?

3. Do you think that the language of pheromones is confined to insect societies?

44

A microscopic unicellular organism lurks in the oceans, spelling dome for the survival of many fish species. Upon the approach of a school of fish, this dinoflagellate releases a potent neurotoxin into the water which often causes the massive death of fish populations. After its attack, the motile stage of this microorganism encysts, disappearing from the open waters by settling into bottom sediments where it awaits an opportunity for future attacks. During the attack the dinoflagellate feeds on the sloughed-off flesh of the dying fish. The chemical identity of the neurotoxin has not yet been determined. Scientists believe that it is responsible for many of the recent, mysterious fish kills throughout the world. The role of a neurotoxin for nutritional gain is a new concept, as its function in other organisms is for defense and not for active predation.

A Phantom of the Ocean

by Theodore J. Smayda

A charming explanation of so-called 'red tides' is to be found in the rich folklore of the seas, the characteristic phosphorescence (bioluminescence) and associated surface discolorations are, it is said, produced by an enormous fish. the Devil of the Waters, which inhabits the seabed, spitting fire to destroy its prey, and thus reddening the waters. Well, this fictitious monster would have met its match in the marine David described by Burkholder and her colleagues on page 407 of this issue[2]. They report the discovery of a microscopic, unicellular organism which, at one stage in its remarkable life cycle, secretes a powerful toxin which kills fish; most unusually, the organism uses dying fish to its nutritional advantage.

The organism concerned is a photosynthetic dinoflagellate, and it lurks dormant on the seabed until live fish approach. Then, sometimes within minutes, it excysts, releasing a motile, vegetative stage which swims, grows and swarms within the water column. There it secretes a potent, water-soluble neurotoxin which causes fish death, sometimes on a massive scale — in one episode

about a million menhaden (Brevoortia sp.) were killed. Several hours later, the dinoflagelate cells encyst, sink to the bottom sediments and await a renewed ichthyo-stimulated foray.

The authors appropriately liken this stunningly rapid and ephemeral sequence of lethal appearance and disappearance to that of a phantom. The chemical nature of the neurotoxin remains to be elucidated, as does the broader distribution of the dinoflagellate. But although it is known only from two estuaries in the United States, the authors predict that it is widespread and has been responsible for many of the enigmatic fish kills which occur in coastal regions.

Red tides (which may in fact be red, green, yellow or brown) are of course one manifestation of blooms of phytoplankton, population explosions undergone by dinoflagellates, diatoms and several other groups of photosynthetic microalgae. Such blooms have underpinned marine foodwebs for at least three billion years, but the red-tide blooms of nutritionally inadequate or toxic species have less

benign effects. That has been known for a long time. What is new is an apparent increase in bloom frequency and spread of noxious species; the occurrence of toxicity in species that had been thought to be harmless; and the growing number of reports of catastrophic mass deaths of marine mammals, including whales[3], and of fish and invertebrates.

Nor can the public at large stand aloof from these phenomena. In the best known of the toxic bloom cases paralytic shellfish poisons (PSP) toxicity, shellfish feeding upon certain dinoflagellates accumulate the neurotoxic saxitoxin and its analogues without harming themselves. But when the shellfish are eaten by humans, serious illness or death can occur: a recent PSP outbreak in Guatemala resulted in 26 deaths and 187[5] cases of illness. Likewise, reports of fish deaths caused by red-tide outbreaks are not new as such[4,6,7]. But we are seeing the emergence of hitherto unsuspected toxins and toxic syndromes, such as the 1987 discovery of the amnesic shellfish poisoning in Cardigan Bay, Nova Scotia, caused by the neuroexcitatory amino acid, domoic acid[8]. Before then, toxin production by diatoms, a major component of the phytoplankton, was unknown. This toxin also suddenly appeared in Monterey Bay, California, in 1991, where it caused the death of pelicans feeding on anchovy which had in turn accumulated domoic acid from its diatom prey[9].

Another recent development, underlining how easily the phytoplankton can spring surprises on us, was the report of a previously undescribed mode of feeding (termed dasmotrophy) by a toxic photosynthetic phytoplankter on other phytoplankton[10]. This toxic flagellate, also lethal to fish, invertebrates and macroalgea, is thought to produce membrane-puncturing compounds which allow it to extract nutrients from its prey.

It is against this background that the paper of Burkholder et al. must be viewed. With the possible exception of dasmotrophy, phytoplankton toxins have not been reported to have a nutritional role, rather, it is thought that they may serve as chemical deterrents against predation, even protecting shellfish which have sequestered them during feeding[11]. By contrast, the dinoflagellate described by Burkholder et al. does derive

nutritional gain during its kill — it digests flecks of sloughed-off fish tissue to which it attached by means of a peduncle. Another difference between the biology of this dinoflagellate and that of other toxin producers is in the surprising influence of phosphorus. The addition of phosphate, unlike nitrate or ammonia, stimulated the growth of gametes, whereas it is limitation of this nutrient that has been implicated in toxigenisis in other species[12].

Burkholder et al. point out that the apparent correlation between increasing nutrient enrichment of global coastal waters and increased incidences of harmful algal blooms is probably not a factor in blooms of the phantom dinoflagellate. But a sobering consequence of their discovery is that it further complicates resolution of the matter of whether harmful algal blooms in the sea are actually increasing, or whether the increase is due to improved monitoring of such events. However that may be, the public are subject to periodic alarms over the safety of seafood, and fisheries and aquaculture enterprises suffer considerable economic loss.

Several big questions need to be resolved. Are toxic blooms in general being triggered largely by increased nutrient loadings in

coastal waters, or by fisheries and aquacultural activities, or are they manifestations of long-term cyclical trends? Are they perhaps a worrying consequence of global deterioration of the marine environment? The next chance for all concerned to put their heads together and discuss these issues will be in October 1993, when the sixth international conference on toxic blooms takes place. We cannot hope to have the answers by then, but we can hope for more data to go on. Theodore J. Smadya is at the Graduate School of Oceanography, University of Rhode Island, Kingston, Rhode Island, 02881, USA. ☐

1. Morvan, F. Legends of the Sea (Crown, New York, 1980).
2. Burkholder, J.M., Noga, E.J., Hobbs, C.H. & Glasgow, H.B. Jr. Nature 358, 407-410 (1992).
3. Geraci, J.R. et al. Can. J. Fish. Aquat. Sci. 46, 1895-1898 (1089).
4. Smayda, T.J. in Food Chains: Yields, Models and Management of Large Marine Ecosystems (eds Sherman, K., Alexander, L.M. & Gold, B.D.) 275-307 (Westview, Boulder, 1989).
5. Rosales-Loessener, F., De Porras, E. & Dix, D.W. in Red Tides: Biology, Environmental Science and Toxicology (eds Okaichi, T., Anderson, D.M. & Nemoto, T.) 113-116 (Elsevier, New York, 1989).
6. Roberts, B.S. in Toxic Dinoflagellate Blooms (eds Taylor, D.L. & Seliger, H.H.) 199-202 (Elsevier/North Holland, New York, 1979).
7. Toxic Marine Phytoplankton (eds Graneli, E., Sundstrom, B., Edler, L. & Anderson, D.M.) 1-554 (Elsevier, New York, 1991).
8. Bates, S.S. et al. Can. J. Fish. aquat. Sci. 46, 1203-1215 (1989).
9. Work, T.M. et al. in Toxic Phytoplankton Blooms in the Sea (eds Smayda, T.J. & Shimizu, Y.) (Elsevier, Amsterdam, in the press).
10. Estep, K. & MacIntyre, F. Mar. Ecol. Prog. Ser. 57, 11-21 (1989).
11. Kvitek, R.G., De Grange, A.R. & Beitier, M.K. Limnol. Ocean. 36, 393-404 (1991).
12. Edvardsen, B., Moy, F. & Paasche, E. in Toxic Marine Phytoplankton (eds Graneli, E., Sundstrom, B., Edler, L. & Anderson, D.M.) 284-289 (Elsevier, New York, 1991).

Questions: *Answers on page 206*

1. How do you think a neurotoxin acts on a prey species?

2. Why do you think that this dinoflagellate has been difficult to discover previously? Refer to examples of its behavior and ecosystem.

3. Why do algal blooms represent another threat to fish populations in the ocean?

45

An epidemic of diabetes mellitus is occurring on Nauru, a Pacific islet. Its rate is over 60% among older adults. The type of diabetes mellitus developing in these Pacific inhabitants is type II, known as non-insulin-dependent diabetes. Type I diabetes, a less common type which is also known as juvenile diabetes, results from a deficiency of the hormone insulin. Type II develops due to the inability of target organs in the body to respond to the signal of this hormone. Therefore, insulin injections will not remedy the problem of Type II, also called maturity-onset diabetes. Type II diabetes has strong genetic and environmental factors. The environmental factors include lack of sufficient exercise, a high-caloric intake of supermarket food, and obesity. Its genetic component is not yet completely clear. The environmental factors, however, are indicative of the modern western lifestyle and represent behavior patterns of the citizens of many countries as they develop. Therefore, the study on Nauru may be a forecast of changes yet to appear in the carbohydrate metabolism of citizens of other developing countries as they engage in the modern, western lifestyle.

Diabetes Running Wild

by Jared M. Diamond

Studies of a diabetes epidemic on a Pacific islet, carried out by Paul Zimmet and colleagues[1-6], will make grim reading for public health professionals, geneticists and students of human evolution. Among older adults, the prevalence of diabetes on Nauru Island is over 60%, a horrifying figure which may be but a foretaste of a sharp rise in the incidence of diabetes in other developing countries. Events on Nauru may also illustrate natural selection in action, shifting human gene frequencies, while exemplifying the paradox of an epidemic of a non-infectious genetic disease.

At the outset, let us be clear about the differences between type II diabetes (NIDDM, non-insulin-dependent diabetes mellitus, the subject of this article) and the much less common type I diabetes (IDDM). The first differs in its genetic basis, age of onset (mainly in adults instead of juveniles), cellular manifestations (peripheral insulin resistance rather than autoimmune destruction of pancreatic islet B-cells), and treatment (often by diet, not by insulin injections)[7]. It has strong genetic and environmental components, the environmental risk factors including the hallmarks of a modern Western lifestyle: low physical activity, high calorie intake of a supermarket diet, and obesity[1]. It is the switch to such a lifestyle ('coca-colonization')[3] accompanying rising affluence that is now precipitating NIDDM epidemics in many parts of the developing world[1-6].

One of the most serious of these epidemics is on Nauru, a remote atoll occupied by 5,000 Micronesians whose formerly energetic lifestyle depended on fishing and subsistence farming. Colonization by Britain, Australia and New Zealand, and income from phosphate mining, transformed Nauruans into one of the world's wealthiest, most sedentary peoples. Virtually all food is now imported and energy-dense; calorie intake is more than double Australian recommended norms; and obesity is rampant. NIDDM used to be nonexistent but a severe form striking many young adults reached epidemic frequencies after 1950 and now affects almost two-thirds of adults by age 55-64. The disease now contributes to more non-accidental deaths on Nauru, with the paradoxical result that wealthy Nauru has one of the world's shortest human lifespans[5].

Analyses[2] of health surveys conducted on Nauru in 1975-76, 1982 and 1987 trace the 'course of' the epidemic. Although the age-standardized prevalence of NIDDM had increased from practically 0% to 28% in the preceding decades, the corresponding prevalence of impaired glucose tolerance (first stage in development of NIDDM) actually declined from 21% in 1975-76 to 9% in 1987, while the incidence of progression from normal health to impaired glucose tolerance of full-blown NIDDM more than halved in those same years. As a result, the age-standardized prevalence of NIDDM in 1987 had dropped below 1975-76 levels.

Evidently, the NIDDM epidemic on Nauru has passed it peak. But this outcome cannot be attributed to the most obvious explanation, a decline in environmental risk factors, because they remain at least as severe as before. Rather, it seems that NIDDM associated with the new lifestyle has already struck most of the genetically susceptible. Nauruans, leaving the unaffected population to consist mostly of genetically resistant individuals. Nauru's epidemic of a genetic/ environmental disease thus resembles an infectious disease epidemic running its course, but the decline in disease frequency stems from an entirely different mechanism; inter-individual variation in genetic susceptibility, rather than acquired immunity. It illustrates natural selection operating on humans under our very eyes.

This conclusion may surprise those who are accustomed to thinking of NIDDM as a disease without selective consequences because it supposedly strikes 'post-reproductive' adults. In traditional human societies, however, no adult ever becomes post-reproductive in a broad sense: adults over 50 remain crucial to the survival of their children and grandchildren. Even in a narrower sense, NIDDM is not a post-reproductive disease on Nauru, where it often strikes at peak reproductive age. Diabetic women on Nauru have more stillbirths and less than half as many live births as healthy controls[2]. Through natural selection, those narrow and broad effects would cause a diabetogenic genotype to decline in frequency in a diabetes-eliciting environment, as actually observed in successive generations of laboratory rats.

The diabetes epidemic on Nauru is not just a freak event, but a harbinger of events to progress throughout much of the developing world. Epidemics of NIDDM accompanying coca-colonization have already been noted among groups of Polynesians. American Indians and Aboriginal Australians[3,5]. Ominously, they have also

been reported among coca-colonized Chinese and Asian Indian emigrants to Mauritius, where the current NIDDM prevalence of over 20% in the Chinese contrasts with that of 1% in mainland China[4]. A sharp rise in the frequency of NIDDM also emerges from studies of Asian Indian emigrants to Fiji. South Africa and Britain, and of Chinese in Singapore, Taiwan and Hong Kong. As life-styles change in China and India themselves, by far the most populous countries in the world, NIDDM may create an enormous public health problem; Zimmet[6], estimates that there will be 50 million diabetics in these two nations by the year 2000.

Because the now-lethal NIDDM genotype somehow became widespread in many ethnic groups under their formerly spartan living conditions, what advantage did the genotype provide then? The most plausible interpretation is Neel's 'thrifty genotype' hypothesis[8]. According to this view, under conditions of fluctuating and usually meager food supply that prevailed throughout most of human history, those individuals with 'thrifty' adaptations such as hair triggered insulin release would convert more of their ingested calories into fat during occasional food gluts. They would thus be better able to survive bouts of starvation, and would develop diabetes only under modern conditions of constantly available high-calorie food and little exercise. This view is supported by the frequent development of diabetes in 'affluent' zoo-housed monkeys, and by the better ability of diabetic rats to survive periods of starvation.

Although most human populations would have been selected in this way for a thrifty genotype, selection would have been espe-

cially intense on Nauruans for three reasons: their ancestors reached the island by long canoe voyages on which initially fatter people would have been more likely to escape death by starvation; droughts and crop failures were common on Nauru in times past; and most of the population was forcibly deported (and over one-quarter died of starvation) during the Japanese occupation of 1942-45[2]. Because Nauruans used to be more starvation-prone than most people, and have now become wealthier faster, the frequency with which they suffer from NIDDM is now especially high. But they differ from the rest of humanity only by degree.

One anomaly remains: why, when NIDDM is striking so many peoples, is its incidence so low in whites of the industrialized nations (for example only 8% in the Unites States)? As an explanation, the Nauru studies point to natural selection operating in the past[2]. We term our modern, sedentary, high-calorie, overweight lifestyle a Western lifestyle precisely because it first arose in Western industrialized nations. Before modern medicine made NIDDM more manageable, genetically susceptible Europeans would have been gradually eliminated, bringing NIDDM to its present low frequency. If so, the Nauru epidemic gains additional interest, by telescoping into a few coca-colonized generations the operation of natural selection for so may centuries in Europe. □

1. Zimmet, P. Diabetes Care 15, 232-252 (1992).
2. Dowse, G.K. et al. Am. J. Epidemiol. 133, 1093-1104 (1991).
3. Zimmet, P. Diabetes Annual 6, 1-19 (1991).
4. Dowse, G.K. et al. Diabetes 39, 390-396 (1990).
5. Zimmet, P. et al. Diabetes/Metabolism Reviews 6, 91-124 (1990).
6. Zimmet, P. Int. Diabetes Fdn. Bull. 36, 30-32 (1991).
7. Vadheim, C. et al. in Principles and Practice of Medical Genetics, Vol. 2 (eds Emery, A. & Rimoin, D.) 1521-1558 (Churchill Livingstone. Edinburgh, 1990).
8. Neel, J.V. Am. J. Hum. Genet. 14, 353-362 (1962).

Questions: *Answers on page 206*

1. Why is the study on Nauru an example of natural selection in action?

2. Why is it a paradox to apply the term "epidemic" to a disease such as diabetes?

3. How will the gene frequencies for type II diabetes change on Nauru?

46

Ecology is a branch of biology which deals with the interactions experienced by organisms. These interactions are with the living, or biotic, environment as well as the nonliving, or abiotic, environment. Countless examples of each type of interaction are found in a unique type of ecosystem, the salt marsh. This ecosystem offers a harsh environment to the plants and animals that inhabit it. For example, the animals of the littoral zone spend part of their day underwater and the balance of the day exposed to the air as the shoreline level recedes with the tides. The members of a species in a salt marsh vary in their ability to tolerate conditions at different tide levels. The competition for the habitat of the marsh is won by the species members that are best adapted for survival to changing environmental conditions. Tide levels and other physical constraints of the environment regulate the distribution of organisms in the salt marsh. Whereas the interactions are sometimes competitive, at other times they are cooperative. Both types of interaction determine the distribution of organisms across the salt marsh.

The Ecology of a New England Salt Marsh

by Mark D. Bertness

The salt marshes of North America's coastal regions are home to some of the most productive biological communities in the world. In these broad expanses of mud and sand, covered by a lush carpet of grasses and rushes, life seems to thrive on the soothing rhythm of the tides. The thick growth hints that here is found fertile soil, washed by the sea — an environment where plants should thrive.

Yet a salt marsh is in fact a harsh environment, where survival is difficult for plants and animals alike. The receding tides leave soil soaked with salt; waves break violently over the plants during storms. At latitudes where winter wraps the marsh in ice, the plant carpet is easily uprooted by the movement of chunks of the ice sheet, pried loose by winter storms. The animals of the littoral zone — the shore between the high-tide and the low-tide lines — likewise endure environmental extremes, spending part of the day underwater and the rest of the day exposed to the air.

These physical constraints have a visible impact on the marsh. Plant life in a salt marsh is organized into zones, the grasses forming distinct strips between the tide marks. The most obvious explanation for the patterns that form across the intertidal landscape is a physical one: Organisms very in their ability to tolerate conditions at different tide levels, and the competition for space in the marsh is won by species that do best at a given tide level. The clear patterning of life in salt marshes makes them a good field laboratory for the study of the forces shaping vascular-plant communities, but the simplicity of these patterns is deceptive. To survive under extreme conditions, plants and animals often cooperate. The ecology of a salt marsh is shaped not simply by adaptation and competition, but by a combination of physical forces and biological interactions, including cooperation. I have spent more than a decade studying the thriving salt marsh at Rumstick Cove, near my home in Barrington, Rhode Island. Along with others who have taken a close look at salt-marsh communities in New England and elsewhere, I have come to appreciate their underlying complexity.

Walking the Marsh

A walk from land to sea at Rumstick Cove begins at the edge of the high marsh, an area flooded only by monthly extreme tides. Black rush (Juncus gerardi) dominates this zone. This deep-green grass is a dense turf, and is tall enough to scrape your calves. Partway through the high marsh, however, there is a line: The black rush ends, and salt-meadow hay (Spartina patens) begins. This tidal line marks the mean highest tides of each month. Salt-meadow hay is a light-green grass, about as tall as your lawn would be without mowing. A few other grasses, such as spikegrass (Distichlis spicata), also grow here. And there are distinct grass-free patches, often filled with a succulent annual — slender glasswort (Salicornia europaca) — that adds color to the high marsh. Slender glasswort is green until fall; then it turns reddish. But in any season, glasswort is a very salty plant that can be pickled or taken as is and

Reprinted by permission from AMERICAN SCIENTIST, Journal of Sigmi Xi, The Scientific Research Society.

placed in a salad.

Just beyond the salt-meadow hay there is another line. This is the boundary between the high marsh and the low marsh — the mean high-water line of the tide. Below this line, there is no salt-meadow hay. It is replaced by the cordgrass (Spartina alterniflora) that dominates the low marsh, the area covered by each day's high tides.

Cordgrass, the tall, wispy grass that rolls back and forth in waves under a gentle breeze, is the plant that dominates the common image of a marsh. At the edge of the border marked by the high-tide line, the cordgrass is short, less than 30 centimeters tall. But as you move closer to the sea the cordgrass becomes taller, sometimes reaching nearly two meters in height. As you bend down to look at the muddy soil around the roots of the tall cordgrass, small holes are evident in the ground, perhaps as many as 175 in a square meter of soil. These are openings to the burrows of fiddler crabs. The fiddler crab is small, less than three centimeters across, and the male has one very large claw that he holds the way a violinist carries his instrument — hence the name.

If you push through the last stems of cordgrass during low tide, you come to a muddy beach. Densely packed beds of ribbed mussels covered the transition from the cordgrass to the sea. Sometimes as many as 1,500 mussels can be found in a single square meter, attached near the roots of the cordgrass.

By the end of your journey, it is clear that the zonation of a salt marsh is precise, a distribution dictated by the limits of the tides. There is no cordgrass on the terrestrial side of the high marsh. There is no black rush near the boundary between the high marsh and the low marsh. Specific constraints must impose this arrangement.

Border Disputes

Early investigators believed that environmental variation caused the distribution of plants across marshes. The distinct borders between patches of salt-marsh vegetation were seen as clear examples of the varying ability of species to adapt to tidal conditions. Experiments on rocky beaches, however, suggested that other biological factors — specifically predation and competition among species — significantly affect the zonation of plants and animals in littoral habitats. Could the same forces create zones of plants in a salt marsh? Because herbivores had been shown to have little influence on the survival of marsh plants, I suggested that marsh-plant zonation might arise from a combination of environmental variation and competitive interactions.

I tested my ideas at Rumstick Cove. After describing the plant-zonation pattern in detail, my students and I experimentally examined the distribution of plants across the marsh; we transplanted salt-marsh plants from one zone to another, both with and without neighboring competitors. Through this work, we began to understand what controls the salt-marsh borders.

In one respect, cordgrass controls itself. Beneath tall cordgrass, near the water, the soil has little peat. But the growing cordgrass continuously produces more below-ground debris, and the burrowing crabs hasten its decomposition. As this material compacts, it becomes peat. Peat decreases substrate drainage, and the consequently waterlogged soil is low in oxygen. These factors combine to limit the productivity of cordgrass. As a stand of cordgrass matures, it effectively destroys its own habitat. This process creates the variation in the productivity of cordgrass that is

seen as you move down the shore. Higher on shore, cordgrass is short; its soil is rich in peat, and thus its growth is stunted. Cordgrass grows better near the water — uninhibited by deposits of peat — and pushes even farther into the sea.

Other interactions in a salt marsh involve competition between species. In fact, cordgrass is limited to the low marsh more by competition than by its production of peat. If cordgrass is transplanted to an area free of neighboring plants, it grows rather well in the high marsh. But if cordgrass is transplanted near salt-meadow hay or black rush, these plants quickly eliminate the cordgrass. Likewise, salt-meadow hay and spikegrass both grow best in the terrestrial portion of the high marsh, the black-rush zone; but black rush eliminates its competitors.

Aaron Ellison of Mount Hollyoke College and I showed that competitive relationships among salt-marsh plants are primarily determined by two factors: morphology and the timing of spring emergence. Both black rush and salt-meadow hay have dense mats of roots, rhizomes and aboveground tillers (shoots). Cordgrass and spikegrass tillers, however, are separated by relatively long lengths of below-ground runners. In most competitive interactions, turf morphology — the mats of roots, rhizomes and tillers — defeats runner morphology. As a consequence, salt-meadow hay and black rush exclude cordgrass from the high marsh, and spikegrass is limited to disturbed areas within the high marsh. On the other hand, runner morphologies are more mobile than turf morphologies; therefore, cordgrass and spikegrass can rapidly colonize an area. That increases the success of cordgrass in the ever changing low-marsh. These bits of information explain most zones, but leave one question: How does black rush exclude salt-meadow hay from the terrestrial portion of the high marsh? Black rush wins this battle through timing. It emerges in March, nearly two months ahead of any other perennial grass in the marsh, and thereby defeats salt-meadow hay, even though the two plants have similar morphologies.

Environmental variation alone does explain some patterns in the marsh. In some cases, a plant simply cannot tolerate a specific environment. For example, black rush, salt-meadow hay and spikegrass all die within a single season if transplanted to the low marsh, even if they are planted in an area with no competition with cordgrass.

Cooperation and Cordgrass

The low marsh is the most difficult marsh environment for plants. The soil is extremely salty and lacks oxygen. Irv Mendelssohn of Louisiana State University showed that anoxic soil can prevent plants from using the nutrients in the soil. Moreover, the low marsh is continually battered by physical disturbances, especially erosion and ice damage. The lapping of waves throughout the year eats away the shoreline. In Rhode Island I have found these sheets of ice to be 10 to 30 centimeters thick and frozen to the underlying cordgrass. During severe high tides, large chunks of ice, up to 10 meters across, and the incorporated substrate of the marsh can be torn away and rafted offshore. For the marsh to survive, the growth of cordgrass must exceed the destructive effects of erosion and ice damage.

Cordgrass survives in the low-oxygen soil of the low marsh largely because of a morphological specialization. Cordgrass contains aerenchymal tissue — an internal pathway that allows air to move from the tips of the leaves to the ends of the roots, oxygenating the

soil And dense stands of cordgrass move disproportionately more oxygen into the soil. In other words, cordgrass plants thrive by cooperation; dense stands do better than sparse plantings because the increased oxygen makes the soil more hospitable for the plants.

Nevertheless, cordgrass has other help in dealing with the low-oxygen soil. As I said earlier, fiddler crabs are prolific in the low marsh, and dig many burrows. A burrow is usually 10 to 30 centimeters deep and provides shelter from predators during high tides. These burrows, however, are in a constant state of flux. Some are abandoned, others collapse. Old burrows are often modified or enlarged. Through this process, the crabs work over much of the top 10 centimeters of soil during each season. This increases the drainage of the soil, the decomposition of below-ground debris and the oxygen content of the soil. The fiddler crab is the earthworm of the low marsh, and can be largely responsible for the high productivity of the tall cordgrass.

Likewise, ribbed mussels contribute to the survival of cordgrass. Tom Jordan of the Smithsonian Environmental Research Center and Ivan Valiela of Boston University calculated that a mussel can filter as much as five liters of seawater per hour in search of plankton. This results in the deposition of nitrogen-rich feces that can increase the growth of cordgrass by 50 percent in a single season. Moreover, mussels buffer cordgrass against physical disturbance. A special gland in mussels secretes strong, proteinaceous filaments called byssal threads. Mussels use a series of these threads to attach to the roots of cordgrass, binding marsh soil from erosion. The cordgrass, in response, grows more below-ground roots. This, indeed, is a team effort.

Dealing with Disturbance

When you are standing on the seaward edge, the low marsh seems a more difficult environment than the high marsh because its soil is waterlogged and constantly eroded by the tide. Nevertheless, walking through the high marsh you find bare patches in the midst of the stout green grasses. And particularly in the spring, large mats of entangled, dead cordgrass stems are scattered throughout the high marsh, where they can kill the underlying vegetation. Some of these mats are bigger than a football field and more than 10 centimeters thick. The high marsh, too, is a demanding habitat.

Floating plant debris creates chronic physical disturbances in the high marsh. During the winter, sheets of ice driven back and forth by the tides clip cordgrass stems. By early spring, floating mats of tangled cordgrass skeletons covered the intertidal low marsh. Extreme high tides in the spring raft these mats of dead cordgrass onto the high marsh and leave them stranded when the tide ebbs. The mats may clutter the high marsh for months before decaying or being washed away. Plants trapped under the stranded mat often die, leaving bare soil.

Although at first a bare patch is a competition-free environment, it rapidly becomes a challenging habitat. The sun heats the exposed surface. As the water in the soil evaporates, salinity increases, particularly near the surface. How high the salinity goes depends on the size and location of the bare patch. Larger bare patches have more surface exposed; more water evaporates, and the soil becomes more saline. The highest salinity is found in large bare patches near the border between the salt-meadow hay and the black rush. Here the salinity of the soil can be 30 times the soil salinity under the dense perennial vegetation in the high marsh. Closer to the sea, frequent flooding limits the accumulation of surface salt. And farther into the black rush, rainwater dilutes any salty soil.

The hypersaline soil in bare spots creates water-balance problems for most plants. Scott Shumway of Wheaton College demonstrated that a hypersaline environment prevents germination of the seeds of most marsh plants. Even a healthy seedling often dies if transplanted to hypersaline soil because the salty soil pulls water from the plant.

Glasswort, however, thrives in hypersaline bare patches. This plant readily germinates in extremely salty soil. Ellison and I found that seeds of seed-bearing skeletons of slender glasswort are often conveniently carried to the bare spots by the cordgrass mats. Because of these two factors, slender glasswort typically dominates bare spots during the first year of regrowth.

Glasswort's reign, however, is temporary. Spikegrass moves in shortly, invading bare spots through asexual means. This salt-tolerant New England marsh perennial produces long rhizomes, just a few centimeters below the surface, that invade the bare spots. Shumway showed that these invading rhizomes survive the saline soil and produce young shoots because, through the rhizomal connection, they receive water from surrounding plants growing in less saline soil. The spikegrass prospers and shades the soil, reducing evaporation and thereby reducing soil salinity. As the salinity of the soil decreases, the habitat becomes more hospitable to other high-marsh perennial, particularly those with turf morphologies. Within two to four years, either salt-meadow hay or black rush displaces both slender glasswort and spikegrass.

Much as the crabs, mussels and cordgrass cooperate to occupy the seaward edge of the low marsh, high-marsh perennial work together to reclaim hypersaline bare spots. Salt tolerance and competitive ability are inversely related in plants of the high marsh in New England. The less competitive but more salt-tolerant plants first invade bare spots. This facilitates succession, making the habitat livable for the more competitive but less salt-tolerant perennial. Here again, a cooperative effort among the plants emerges only if the bare spot is hypersaline, and thereby stressful. If a bare spot has a low level of salt. The spot is filled through strictly competitive interactions among the high-marsh plants. So local conditions determine whether a patch is invaded cooperatively or competitively.

Cooperation Is Important

Ecologists have overestimated the universal role of competition in nature. Phrases such as "survival of the fittest" certainly evoke and perpetuate this misconception. Over the past 20 years, many ecologists have examined natural communities in search of competition, and have found it to be particularly pervasive in physically mild habitats. Early in this century, ecologists accepted, uncritically, that species can cooperate to reclaim disturbed habitats. Because of this untested acceptance, contemporary ecologists largely reject this idea. Nonetheless, many recent experiments show that positive interactions among organisms often play a large role in natural communities.

Physically harsh environments generate, as a matter of course, cooperation among organisms. David Wood of California University at Chico and Roger Del Moral of the University of Washington found facilitation among plants during early succession in subalpine habitats on Mount St. Helens. Other reports show that interactions among the same organisms can be competitive in benign environments and cooperative in harsh environments. For example, Mark Hay of the University of North Carolina's Marine Science Center showed this in turf-forming seaweed. Examples such as these reveal that a New England salt marsh is one habitat among many in which biotic interactions are competitive under mild physical conditions and cooperative under harsh physical conditions. Both competitive and cooperative forces likely play major roles in the organization of most natural communities.

Bibliography

Bertness, Mark D. 1984. Ribbed mussels and the productivity of Spartina alterniflora in a New England salt marsh. Ecology 65:1794-1807.

Bertness, Mark D., and T. Miller. 1984. The distribution and dynamics of Uca Pugnax burrows in a New England salt marsh. Journal of Experimental and Marine Biology and Ecology 83:211-237.

Bertness, mark D. 1985. Fiddler crab regulation of Spartina alterniflora production on a New England salt marsh. Ecology 66:1042-1055.

Bertness, Mark D., and Aaron M. Ellison. 1987. Determinants of pattern in a New England salt marsh community. Ecological Monographs 57(2):129-147.

Bertness, Mark D. 1988. Peat accumulation and the success of marsh plants. Ecology 69:703-713.

Bertness, Mark D., L. Gaugh and Scott Shumway. In press. Salt tolerances and the distribution patterns of New England marsh plants. Ecology.

Ellison, Aaron M. 1987. Effects of competition, disturbance, and herbivory in Salicornia europaea. Ecology 68:576-586.

Harper, J.L. 1977. Population Biology of Plants. San Diego: Academic Press.

Hay, Mark. 1981. The functional morphology of turf-forming seaweeds: persistence in stressful marine habitat. Ecology 62:739-750.

Jordan, Tom E., and Ivan Valiela. 1982. A nitrogen budget of the ribbed mussel, Geukensia demissa, and its significance in nitrogen flow in a New England salt marsh. Limnology and Oceanography 27:75-90.

Mendelssohn, Irv A. 1979. Nitrogen metabolism of the height forms of Spartina alterniflora in North Carolina. Ecology 60:574-584.

Mendelssohn, Irv. A., K.L. McKee and W.H. Patrick. 1981. Oxygen deficiency in Spartina alterniflora roots: metabolic adaptation to anoxia. Science 214:439-441.

Shumway, Scott. 1991. Secondary Succession Patterns in a New England Salt Marsh. Dissertation. Brown University, Providence, Rhode Island.

Wood, David M., and Roger Del Moral. 1987. Mechanisms of early primary succession in subalpine habitats on Mount St. Helens. Ecology 68:780-790.

47

Human reactions to different wild animal species often determine whether they are conserved or destroyed. For example, the long-nosed bat, an endangered species, is being ignored for the most part. On the other hand, the slaughtering of the harp seal evoked considerable human concern. Although threatened, this species was never endangered.

Species receiving the most attention by conservation efforts are often the ones that are most attractive to humans. These species are considered "cute" and babylike. These features include: a large head, upright posture, flat face, round features, feet that resemble hands, large eyes, and soft fur. Consider the large eyes of the panda or soft fur of many animals and the positive human reactions they evoke.

In some cases, conserving an appealing species has led to positive results for ecosystems. In other cases, neglecting to care for species lacking this "star quality" has led to unfortunate consequences for endangered and now-extinct species.

Who's Cute, Cuddly, and Charismatic?

By Fiona Sunquist

We go "ahhhh" over animals that look like human babies, but is this any way to direct conservation efforts?

Look into the huge, dark eyes of a soft furred baby harp seal. Then consider the Mexican long-nosed bat, with a sharp, narrow snout ending in a growth that looks like a big wart. These wildly different animals give us a glimpse into one of the greatest puzzles confronting modern conservationists: the human heart.

All well-loved harp seals, marine mammals of the North Atlantic, have been hunted since the 1700s. In the 1960s several human organizations launched a campaign to ban sealing, and at least one group distributed close-up portraits of baby seals, their big eyes staring soulfully from a poster that urged people to "Stop the Killing." The image evoked such an overwhelming response — five million letters and postcards of protest to members of the European Parliament — that European countries banned the import of harp seal furs. The market for harp seal fur collapsed, and the population of the animals now stands at 2.5 million.

For all its problems, the harp seal never came close to being endangered. The Mexican long-nosed bat, on the other hand, is listed as endangered by the U.S. Fish and Wildlife Service. Yet legislators have not received five letters, much less five million, expressing concern for the bat. Folklore says that these bats such blood. People became frightened and destroyed most of the big colonies in Mexico.

The bats don't suck blood at all but eat flower nectar, cactus fruit and insects. Long-nosed bats are essential pollinators for some agaves and other important desert plants, and bat declines make biologists fear for the health of the ecosystem. Despite the rarity and importance of the bat, "people prefer animals with big, soft eyes like seal pups," says Stephen Humphrey, bat specialist and curator of mammals at the Florida Museum of National History. "It's a log easier to get support for a giant panda than a bat."

Thus human instincts about animals — even subtle, subconscious reactions — influence which conservation efforts succeed and which ones fail. This raises fundamental questions about what it is that we Homo sapiens find appealing about other animals. Why does our species so often like "cute, cuddly and charismatic" creatures, as biologist Tom Lovejoy calls them? Why do so many of us dismiss other species as "ugly" without a backward glance? And how do this whims affect conservation?

Nobel-prize-winning animal behaviorist Konrad Lorenz first explored these questions 50 years ago. He suggested that animals with large eyes, flat faces, founded bodies, stubby limbs and soft fur often trigger the human urge to protect and nurture. In other words, we respond positively to animals that remind us of our own babies.

Human behaviorists believe that this response helps preserve our species. "Babies cannot feed, move or protect themselves: they need to be attractive to their care givers to keep them nearby," says Shena MacLeod, clinical coordinator of the Nurturing Program at Shands

Hospital in Florida.

Humans don't change this frame of reference when looking an animals. "We are fooled by an evolved response to our own babies, and we transfer our reaction to the same set of features in other animals," says evolutionary biologist Stephen J. Gould of Harvard University.

But what does "babylike" mean? Lorenz noted major features that tend to evoke the "cute" reaction in humans. Since the, other scientists such as behaviorists Desmond Morris and Irene_us Eibl-Eibesfeldt have added details, but the list remains essentially the same:

_ **Big Head.** A head that seems too large for its body, a protruding forehead and puffy cheeks are all typical of human babies. Creators of Cabbage Patch dolls, Mickey Mouse, Garfield, and Chip and Dale have emphasized these infant-like characteristics with great success, and baby chimps, hamsters and puppies owe some of their appeal to these features. Even fish, which do not as a rule score high in personal appeal, can be elevated to semi-cute if they have round, full cheeks and big eyes like a puffer fish.

_ **Upright Posture.** The ability to sit upright or to walk on two limbs appeals strongly to people, so penguins, koalas, giant pandas, apes and monkeys score high points. Though mice, cats, chipmunks, turtles, pigs and panthers naturally walk on all fours, their cartoon images often stand upright. The posture is so compelling that we train our dogs to sit up and beg, and even teach circus elephants and tigers to stand on two legs.

_ **Flat Face.** Humans find flat faces more appealing than long pointed snouts, probably because our own faces are flattened. Witness the common preference for shorter-nosed hamsters and guinea pigs over

rats as pets, and the appeal of Persian cats and flat-faced dog breeds such as pugs. Pekinese, bulldogs, Lhasa apsos, Shih Tzus and boxers.

_ **Round Profile.** Among body shapes, humans often prefer animals with rounded outlines and short, thick limbs. Koalas fit the pattern almost perfectly, as do seals, sea otters, manatees, bears and elephants. Thin elongated animals like weasels have a tough PR problem. Not only are they the opposite of rounded, but they also have pointed noses and, in human terms, small eyes.

_ **Feet Like Hands.** Handling objects is a major human activity, and people are fascinated by animals that use their feet like human hands. Pandas, raccoons, primates and sea otters all use their forefeet to feed, and, among birds, parrots easily stand on one foot while holding food in the other. The elephant's ability to pick up and manipulate objects with its trunk undoubtedly helps its popularity rating.

_ **Large Eyes.** Big eyes are one of the most evocative features of a "cute" animal, creating the impression of childlike innocence. The advertising industry has exploited this effect for a long time, using pictures of tearful, big-eyed children and wide-eyed puppies and kittens to sell everything from greeting cards and candy to toilet paper and car tires. In the animal world, bush babies, lorises, owls and clouded leopards all have large eyes, not, of course, because they are innocent but because they are nocturnal. Other cute animals, like the giant panda, only appear to have large eyes. Its black "eyes" are mostly dark fur circles like make-up. The actual eyes of the giant panda are small and beady, a characteristic that most people find unappealing.

_ **Soft Fur.** Softness attracts people. Many animals just look soft and fluffy but feel prickly and rough. Bears' coats actually feel harsh, but they appear soft from a distance. Hedgehogs also trigger the "cute" response, probably because their spiny bodies look like little balls of fluff. Armadillos, rhinos and other animals with scales, shells or hard-looking skins are generally much less popular than furry creatures.

Few animals have all seven characteristics. Yet creatures like the giant panda and the koala, which possess many of the features, rank high among popular animals.

Other characteristics besides cuteness can endear an animal to humanity. Being large helps. Whales and the big sea turtles cannot really be classified as cute, but both are such big, dramatic animals that they have staunch support groups. We love and revere tigers, eagles, cranes and other so-called charismatic animals of symbols of human characteristics. According to surveys by Stephen Kellert of Yale University, who studies public attitudes toward wildlife, "People feel affection for these species because they are believed to be intelligent, mysterious or noble," Snakes and alligators have their own followings, but here the fascination may be based on the thrill of being close to a potentially dangerous animal. In all these cases, people adore certain species for arbitrary, human reasons, which have nothing to do with the animal's place in nature.

These human-centered preferences skew conservation efforts, according to Wayne King, curator of herpetology at the Florida Museum of Natural History. "Appealing animals definitely get more conservation attention," he says. "Animal charisma translates to dollars and cents."

Geoffrey Mellor from Wildlife

Conservation International (WCI), an arm of the New York Zoological Society, ticks off examples. WCI's Rhino Rescue Fund has attracted about $400,000, and the Forest Elephant Fund almost $1 million.

However "you'd have a hard time raising money for the babirusa," says Mellor. Even though WCI supports ground-breaking work on these Indonesian relatives of pigs, "Ugliness is this species' most obvious train, so fund-raising would be an uphill battle," Likewise Mellor hasn't started a Save-the-Cassowary Fund. He touts the importance of the work WCI underwrites on these critical seed-dispersers in the rain forests of New Guinea, but for attracting public money, the cassowary, which looks like a giant turkey with a plucked, deformed head, is "a pretty tough sell."

In the United States, this bias toward species that appeal to people extends event into the wording of the Endangered Species Act, laments Nature Conservancy scientist Larry Morse. Under the act, each distinct population of a vertebrate can be protected. Hence grizzlies in Yellowstone or bald eagles in the Lower 48 states get special treatment, even though both species thrive in Alaska. However, the act does not offer the same umbrella to distinct populations of insects, spiders, shellfish or plants. "I know of no scientific basis for that," Morse says.

The impact of people's biases on global conservation efforts is similar, according to Whitney Tilt of the National Fish and Wildlife Foundation. "Almost our entire endangered species effort has been driven by charismatic megafauna. The health of the ecosystem and the loss of habitat is almost an aside."

In some cases, concentrating the appealing species has been an effective way to conserve ecosystems. When the animal attracting the attention and money is a top predator, like the tiger or jaguar, then hundreds of smaller species benefit because huge areas of habitat fall under the umbrella of concern. The public's affection for gorillas and giant pandas has resulted in parks and reserves — Rwanda's Volcanoes National Park for example, or China's Wolong Natural Reserve.

But unfortunately some of the most endangered ecosystems do not have animals with the right star quality. Huge areas of temperate rain forest in Chile and Argentina are home to nothing larger than the poodle-sized pudu deer and a small wild cat called the kodkod.

In other cases, the focal animal may prove to have not quite enough charisma. Whitney Tilt points to the spotted owl/logging debate in the northwestern United States as a good example of how the charismatic-species approach can backfire. "In this case the public believes that it is a question of owls or jobs for loggers, whereas the real point is if we continue to cut these old-growth forests as we have been, then these guys will be out of a job in five years instead of tomorrow. The entire system is about to be wiped out, and this will take out the logger and the owl."

No one can say for sure which species are most important in terms of keeping the Earth's ecosystem's healthy and functioning, but cuteness and eye-appeal to humans are probably not the measure. According to E.O. Wilson of Harvard University, a strong advocate of protecting biodiversity, "It is the little things that run the world. The truth is we need invertebrates but they don't need us. If human beings were to disappear tomorrow, the world would go on with little change. But if invertebrates were to disappear, I doubt that the human species could last more than a few months." □

Questions: *Answers on page 206*

1. Evaluate the measures humans use to decide which species are worth keeping in the ecosystems of the Earth.

2. What does the following comment from Wilson of Harvard mean: "It is the little things that run the world."

3. How has human preferences for conservation affect some animals in the tropical rain forest?

48

A team of "rodent busters" in India is successfully battling populations of rodent pests without the use of poisons or chemicals. This Rat and Terminal Squad (RATS) is saving millions of tons of grain from the crop fields of India through its successful efforts.

Rats are as abundant in India as cattle. Each rat, for example, can eat an ounce of rice each day. Members of the Irula tribe have eradicated much of this crop threat by becoming master rat catchers. Once caught, the rats become a source of food for the people of the tribe.

Winning the Race in India
By Zai Whitaker

Without using pesticides, Irula tribespeople rid their fields of rodents and reap a bonanza in rice.

Grinning and bearing it, an Irula tribesman named Raman holds up his catch: Indian mole rats seized from a rice paddy. Skills like his go a long way in a country where rodents eat nearly a quarter of all stored and standing grain.

Dust Hangs in the air as Raman, his back gleaming with sweat, immerses himself in his work. The wiry Irula tribesman is plodding through a rice field near Madras in southern India, scrutinizing a vast network of tunnels carved out by hundreds of burrowing rats.

He creeps toward a burrow, eyes down, then lunges toward a darkened opening. Dirt flies everywhere during the scramble as he snatches a fat female rat by the back of the neck, expertly avoiding the animal's long, curved, brown-stained incisors. To keep the rat from biting or chewing its way free, Raman hooks it tail behind its lower teeth and, with a flick of this finger, snaps them off.

He plops the animal into a bag his wife, Lakshmi, is holding, then thrusts his arm back in the tunnel up to his shoulder. In seconds, face beaming, he holds up another prize, a half-grown juvenile. "Should be eight of these, at least," he says, then, sure enough, proceeds to pull out seven more of the furry pests, one by one.

If the world wants a better mousetrap, it should beat a path to Raman's door. A rat catcher all his life, he knows his rodents. And that kind of knowledge goes a long way in a land where, every year, swarms of Indian mole rats, rice rats, gerbils and field mice steal almost a quarter of all stored and standing grain — enough to feed the country's 900 million people for three months.

Elsewhere, farmers battle rodent pests with an arsenal of deadly chemicals that also poison people and the environment. Raman's methods — which, as it turns out, also produce meat for his family — are environmentally benign and even more efficient, suggesting there may be a bright side to an otherwise grim and desperate struggle.

Chingleput District, home of the Irula tribe, was once a vast forest dominated by thorny acacias and teeming with such species as black buck, axis deer, wild pigs and leopards. But in recent decades, growing numbers of people moving through the area have drastically altered the habitat, hacking down the forests and decimating the wildlife. Today, what isn't irrigated farmland is largely desert, with islands of scrub overrun by mesquite and other weeds.

Such conditions are ideal for rats. Not so for a tribe of 28,000 nomadic hunters once known for their knowledge of the forest. Most Irulas still survive by collecting edible roots, berries and nuts. They fish where they can and continue to go after small prey like mongooses, monitor lizards and the occasional hare. Fortunately, they have turned their hunting prowess to advantage in recent years. Like Raman, many of them have become India's master rat catchers.

Years ago, the Irulas earned a reputation (and a living) as snake catchers. They caught cobras, rat snakes and other large serpents and sold them to the snakeskin industry. In the mid-1970s, the government banned the trade, so a group of

Irulas started the Snake Catchers Cooperative Society and began extracting snake venom for antivenin production. A few years ago, about 75 organized themselves into a team of rodent busters called the Rat and Termite Squad (RATS). Their mission: to search fields and destroy crop-eating rodents.

In terms of biomass (the combined weight of a group of animals in an area), rats are as abundant as cattle in India, and they multiply as a dizzying rate. A female has perhaps ten young at a time and can produce as many as seven litters in one year. People inadvertently help the creatures by converting forest and grassland into rice fields, which supply all the food, water and shelter a rat could want.

In return for this hospitality, the rodents have brought destruction of staggering proportions. Every year, tea estates and fruit plantations in India lose crops worth millions of rupees to rats and the erosion that follows devastation of farm fields. A recent (and expensive) short circuit at the nuclear plant near Madras was sparked by a rat with a taste for wire insulation.

The average rat can gobble almost an ounce of rice per day, and that adds up quickly in areas teeming with the animals. During a recent infestation in the state of Rajasthan, for example, a single acre swarmed with as many as 400 rats, which consumed nearly a ton of rice in three months.

Enter the Irulas, whose generations-old knowledge of rodent natural history makes them the most efficient rat catchers ever. For them, the importance of the hunt is twofold: Where else can you rid your fields of a nuisance animal and shop for dinner at the same time?

It takes a lot of rats to feed a hungry family. Some Irula chefs curry the animals and eat them with rice, while others simply roast them on an open fire. A hunting party might fire up a spontaneous barbecue if hunger hits on the way home. To the Irula way of thinking, rats are just another game animal. Nothing repugnant about them. (But you'll never see an Irula eat frog legs or, for religious reasons, beef.)

The Irulas' knack for nabbing rodents is something to behold. One February morning, a tribesman named Anamalai and his family got together with others from the RATS team for a day of ratting on a 5-acre spread of rice fields. Work began at eight o'clock. By noon, their bags were bulging with 242 jumping, squeaking gerbils, mole rats and field mice.

Because the Irulas manage to catch such an amazing number of rats — and without a speck of chemicals — the Oxfam Trust (an international aid organization) decided to fund a pilot project to pay them to do what they do best. In 50 hunts, the catchers nabbed several thousand rats at a cost of about 1.50 rupees, or 5 cents (U.S.), per animal. In parallel trials elsewhere in the country, the per-rat cost using pesticides was 14 rupees, or about 50 cents.

Officials with India's Department of Science and Technology were so impressed with the feat that they granted the RATS team 750,000 rupees ($27,000 U.S.) for a three-year rat-extermination program. During the first year, Irula ratters combed fields and farmlands of Chingleput District, catching the animals alive by hand or whacking them with sticks as they ran out of their burrows. The bounty: more than 98,000 rats.

One overcast morning finds Raman and Lakshmi and two other Irula couples walking along the muddy bunds, or mounds, in a 10-acre rice field. The ripening stalks are aglow with vivid green, the morning air filled with the calls of bee-eaters, meadowlarks and a pair of noisy rollers flashing their electric-blue wings. On the far edge of the flooded paddy, a white-breasted kingfisher lands with a splat, skewering a fat paddy frog. The shrill cries of two spotted owlets burst from the crown of an elegant black palmyra, while large male garden lizards lazily make their way to sunny basking spots. It's harvest season, the time when the rodents really take over. Tension grips the air, a feeling that always accompanies the hunt — even for rats.

Clipboard and pen in hand, Gopal, the RATS team supervisor, confers with the catchers on strategies for the attack. The, while village women harvest rice from the edges of the field, the rat men and their wives begin work at the threshing platform.

Raman walks over to a half-dozen gerbil burrows, kneels and studies the brushlike marks in the sand made by furry tails. With a few artfully aimed blows of his crowbar, he exposes a section of one tunnel, then scrapes a bit of the earth onto his palm. He solemnly points to a few light-colored hairs and some lice, which always occupy rats and their homes.

The catchers string out nets in a big rung around the holes. While the men dig, the Irula women poke the ground with sticks a couple of yards away to find the animals' escape hatches. Gerbils typically dig a secret burrow within an inch or so of the surface. If a rat snake or other predator enters the burrow, the rodents break through to the surface and skip out the back. Once these exits are located, the wives lay small sock nets over them.

Work stops for a moment after a flurry at one hole, where an 8-inch black scorpion has bumbled into the stark sunlight. Workers stand back and allow the creature to proceed past bare toes into the shade of a thornbush.

Suddenly — more action, and

from every direction. Two adult gerbils dash out of their holes and straight into the nets. Raman has another in his hand, and two other Irulas have scooped up bunches of young ones. The women and busy with several rodents bouncing around in the sock nets. That makes a total of 14 gerbils, the rodents with the most highly esteemed meat. But the greatest challenge lies ahead: breaking up the mole-rate colony that has been multiplying apace as rice plants mature and making serious inroads into the growing grain. After a quick consultation in Tamil, the teams gets to work on the grass-and-earth bunds.

Village women in colorful saris continue harvesting, kicking up the rich scent of ripe grain and paddy straw as the rat busters descend on the main tunnel system. In a moment, a rat catcher named Murugan calls for a group of visitors to come have a look. Reflecting sunlight off the shiny blade of his crowbar, he illuminates the dark interior of a rat hole. Lying inside is the tightly coiled body of a medium-sized rat snake. Murugan nods meaningfully toward the women, then pokes at the snake, which shoots out of the hole and into the rice stands, drawing shrieks from the harvesters. Murugan smiles at his prank, then goes back to work.

In a burst of furry energy, a half-dozen or so rats dart for cover. Irula women dispatch some of the escapees with accurate blows from the thin, flexible sticks they carry. But in a maze of diabolically designed burrows and escape routes such as this, a rat hunt can quickly turn into a comic nightmare, with rodents popping out everywhere and fleeing into nearby bushes. Strong measures are in order.

Raman takes two clay pots and, with a sharp stick, punches a small hole in the bottom of each. He fills them with dry grass and leaves, then inserts a burning twig to start them smoldering. Using strong lung power, two of his comrades soon have the pots billowing smoke. They carefully align the pots over two of the main exits of the burrow system, pack dirt around them to make an airtight seal, then blow smoke into the rodent hideouts. In a moment, wisps of smoke circle up from cracks along the bund like a range of smoldering miniature volcanoes.

Then they wait. Raman and the other men light up strong, leaf-rolled cigarettes called "beedis" while the women huddle and talk softly. In a few minutes, the men again take up their crowbars and start digging. Limp carcasses of rats, asphyxiated by the smoke, pile up on the bund. After a while, the workers excavate and measure the burrow system, then tally their harvest: 27 mole rats from nearly 100 feet of tunnels.

Raman scoops handfuls of rice from three nest chambers, which double as granaries for the hording rodents. In all, he and his mates recover about 11 pounds of rice from the rat burrows. Not long ago, Irula rat man Chockalingam and his wife returned from a hunt in millet fields lugging a gunny sack stuffed with nearly 15 kilograms (33 pounds) of the food grain hidden underground in rat granaries.

Back at home, the rat patrol hands over its catch and its report on the hunt to program managers Dravidamani and Shymala. Most of the rodents will be sold to the Madras Crocodile Bank, where mugger and saltwater crocodiles, among other species, will feast on the Irulas' labors.

As the sun begins to sink over the Bay of Bengal, visitors who got over their squeamishness long ago join tribespeople as they gather round a fire in front of an Irula hut. A transistor radio blares from Vadanemmeli village not far away, but the sound is not too loud to drown out the chorus of yapping jackals looking for crabs and turtle eggs on a nearby beach. Darkness settles in as the aroma of sizzling meat wafts to your nostrils and you reach, with pleasure, for another roasted rat.

Zai Whitaker teaches at Kodaikanal International School in South India. She and her husband, Rom, founded the Madras Crocodile Bank, where they live with their two sons and 10,000 crocodiles. London-based Michael Freeman traveled to India for International Wildlife. □

Questions: *Answers on page 207.*

1.   What is the biological means of control used by the Indian tribe people?

2.   Why is a biological means of control preferable to a chemical means of control of populations?

3.   Why not simply feed rats with rice and grow them intentionally as a source of food?

49

Only a decade ago the air pollution from power plants in Germany was considerable. Cars added to this air pollution. Since then the citizens of the country have succeeded at cleaning up their environment. Not only has this been a sound step ecologically, it has given Germany a competitive edge over industrial rivals in the global market place. Cleaning up the environment has stimulated economic and industrial growth.

Advances include: efficient recycling of junk car parts, retrofitting all power plants, diminishing harmful atmospheric gases, and labeling environmental safe products. Results in Germany show that positive steps for ecology and industry can go hand in hand.

Down Germany's Road to a Clean Tomorrow
By Curtis A. Moore

The world's future begins in Germany, down a country lane and through a gate into tomorrow. You get to it on a two-lane road that winds across a field, then disappears over the crest of a hill.

There, nestled in a hollow near the tiny village of Neunburg just 20 miles from the border with Czechoslovakia, stands a glimmering complex of yellow and white, glass and steel, which converts solar radiation into electricity and, in turn, into hydrogen fuel. It is the experimental $38 million Solar Wasserstoff power plant.

By utility-company standards the futuristic plant, owned by a consortium of governments and industries, is minuscule; it generates only enough energy to supply the equivalent of 50 or 60 households. By every other standard, however, it is revolutionary. The fuel it produces to run cars and furnaces is zero-polluting and virtually limitless.

Even more important, Solar Wasserstoff symbolizes a strategy that may catapult Germany into a position of global economic dominance for years to come. Like

dozens of other environmental miracles currently being nurtured in that country, the Wasserstoff operation is lean, mean, competitive and poised to take advantage of an expanding global marketplace — precisely because it is clean.

More than anywhere else on Earth, Germany is demonstrating that the greening of industry, far from being an impediment to commerce, is in fact a stimulus. From Bonn to Berlin, Germany's citizens, businesses and government have concluded that a robust economy and a safe environment are like the chicken and the egg: One leads to the other. "What we are doing here is economic policy, not environmental policy, says Edda M_ller, chief aide to Germany's minister for the environment.

The impetus for that economic policy, however, began with an environmental shock wave: a phenomenon called Waldsterben, or "forest death." The term refers to a mysterious tree die-off, first noticed around 1979 or 1980, which has been linked to air pollution. In a nation with an almost mythic connection between its people and

the forests, the specter of losing woods ranging from the fabled Black Forest of Hansel and Gretel to the graceful lindens of Berlin generated an irrepressible demand for action.

The result was the burgeoning of the world's first environmentally based Green political party. At one point the Greens claimed roughly one of every 12 members of the parliament, or Bundestag, and their grass-roots pressure pushed politicians inexorably toward tighter and tighter environmental controls.

Then, just as this upwelling of environmental fervor seemed to be subsiding, the meltdown and explosion in 1986 of the Soviet Union's Chernobyl nuclear power plant revived and strengthened it. As mothers fearfully kept their children indoors to avoid the hazard of nuclear fallout, and, as rumors of two-headed calves being born to exposed cattle swept the nation, environmentalism became a deep-seated national value.

Quickly, the zeal to make environmental improvements spread from smokestacks to cars, then to ozone-destroying chemicals, hazardous wastes and global warming. As a result, says Alan Miller of the Center for Global Change, a policy-analysis institute at the University of Maryland, "There is virtually no field of environmental protection where Germany does not stand out. It has the most rigorous controls of any nation, bar none — that includes the United States."

In the space of ten years, Germany has vaulted into the vanguard of global environmental leadership, eclipsing the United States, Canada, Japan and even ecological hotbeds like Sweden. Although the Green Party has waned nationally, it policies have been usurped by the political mainstream and nurtured by a politician regarded as among the world's most conservative, Chancellor Helmut Kohl. As a result, Germany today is a nation which is:

_ **Revolutionizing the junk business.** In the car industry, a "take-back" program requires car companies to pick up and recycle junked cars of their make. New cars will soon roll off the assembly like with bar-coded parts and predesigned disassembly plants capable of dismantling an auto in 20 minutes. By 1994, similar requirements will be imposed on products ranging from yogurt containers to cameras.

_ **Retrofitting all power plants.** While politicians in North America were arguing about whether acid rain was fact of fantasy, Germany adopted rules requiring every power plant within its borders to slash by 90 percent the air pollutants that cause acid rain. By 1990, the German retrofit was complete. Today, while companies in the United States continue to bicker over the details of an acid rain program, Germans are selling Americans and the rest of the world antipollution technology and know-how.

_ **Phasing out harmful atmo-**spheric gases. In 1989, Germany mandated a ban by 1995 — five years before the rest of the world — on CFC gases, the primary culprits in the destruction of the ozone layer that protects Earth from the sun's ultraviolet radiation. The country had also committed to reducing emissions of carbon dioxide, principal cause of global warming, by 25 percent by the year 2005. These are the swiftest and toughest phasedowns in the world, and they required German industries to respond quickly.

_ **Labeling environmentally friendly products.** In 1977, the government's environment ministry began a labeling program to alert consumers to product brands that are less harmful than those of competing brands. The symbol, known as the Blue Angel, is a laurel wreath encircling a blue figure with outstretched arms. The government licenses use of the label for about 3,500 products selected by an independent nine-member Environmental Label Jury. The Blue Angel program has unleashed a torrent of innovation among manufacturers, spawning new environmentally friendly products ranging from low-polluting paints to mercury-free batteries.

With such economic efficiencies now in place, Germany faces the tantalizing prospect of stealing the competitive edge over its industrial rivals for global markets. "Any country that does not emulate Germany's strategy will be at a competitive disadvantage in 10 or 20 years," says Konrad von Moltke, a senior fellow at the World Wildlife Fund who has written extensively on the country's new policies.

This cross-fertilization between environmental protection, government regulation and economic development has already begun making an impact in the marketplace. For example:

_ At the ford auto plant in Cologne, managers complied with new requirements by modernizing the paint-spray line, cutting pollution by 70 percent and the cost of painting a car by about $60 — a savings that makes German-made cars marginally more saleable.

_ At the "4P" plastic-film manufacturing and printing plant in Forchheim, where plastic bags for frozen french fries and other foods are printed and stamped by the millions, officials were forced by strict new pollution laws to cut emissions by 70 percent. They installed a recycling system that reclaims up to 90 percent of the plant solvents, saving so much money that the 4P pollution controls will not only pay for themselves but will actually start saving the company money as the price of solvents rises. A sister plant with a similar system already makes money by recapturing solvents, once again lowering the price of its service and increasing profit.

_ At the Knauf gypsum manufacturing plant in Iphofen, "scrubber sludge" (acidic wastes neutralized in a slurry of limestone and water) from nearby power plants is manufactured into wallboard, concrete and a wide array of other building materials. Knauf's program has proven so successful that it recently opened a new plant — at Sittingbourne, England, on the banks of the Thames River. There, waste exported from German power plants is manufactured into wall-board for the English market, generating annual sales of $48 million a year for Knauf. In the United States, such byproducts from pollution controls are dumped as useless waste.

_ At Siemens in Munich, the German conglomerate has made numerous inroads into foreign markets. Siemens makes the world's cleanest and most efficient gas turbines for generating electricity. Two have already been installed in

Delaware at a Delmarva Power and Light facility, and the company is pushing hard to close sales in San Diego, California. Siemens also manufactures compact, high-efficiency fluorescent light bulbs, which were initially developed in the United States with funding from the Department of Energy. But it is the Germans who are making a profit selling the bulbs.

_ At Deutsche Voest-Alpine, based in Dusseldorf, engineers have developed and built the world's first cokeless steel mill. Long considered an essential step in the manufacture of steel, coking — the heating of scarce and expensive metallurgical coal in massive, airtight ovens — spews a noxious mixture of cancer-causing pollution. A second full-size facility has already been build in South Africa, and negotiations are underway to build another in the United States.

Such breakthroughs are likely to continue as Germany edges closer to a self-imposed target of reducing carbon dioxide emissions from the former West Germany by 25 percent and from the former East Germany by 30 percent, both by the year 2005. As the already efficient economy slims itself down further, this will mean:

_ **Harnessing wasted energy.** In most power plants and factories, only about one-third of the energy in coal, oil or gas is converted into electricity. The rest is vented as waste heat. Now, the German government is preparing regulations that require large and medium industries and utilities to market this waste energy. The energy can be used to heat or cool homes and factories, operate paper mills and chemical plants, and even generate additional kilowatts in power plants. Officials estimate that by using this waste heat, efficiency can be boosted to roughly 90 percent and air pollution — already at the world's lowest level — can be chopped in half or more.

_ **Cutting auto use.** Other regulations on the drawing board will reduce pollution by forcing drivers out of gas-guzzling cars and onto public transit. Inner cities are systematically being closed to auto traffic, while fees for highway and bridge uses are being raised. The government hopes that the number of bicycle riders in Germany, already one of every 20 commuters, will double.

_ **Recycling more than cars.** Because products made from virgin raw materials can require 10 to 100 times the energy of those made from recycled goods, the car company "take-back" requirements now applicable to new vehicles are being extended to virtually all other products. By 1995, 72 percent of glass and metals must be recycled, along with 64 percent of paperboard, plastics and laminates. incineration, even if used to generate power, is ruled out. Goods made from recycled materials require up to 95 percent less energy — savings that can make products less expensive and more competitive.

_ **"Taking back" used products.** Consumers will be required to begin paying deposits on bottles, paint cans and even soap boxes. Such products will bear a "green point" recycling symbol. "The ferocity of the new regulations is extraordinary," said the international business magazine The Economist.

To be sure, there are critics of the German strategy. Many question whether the country can achieve its environmental goals while maintaining a high standard of living and a social welfare system that is among the world's most comprehensive — all while undertaking the mammoth costs of reunifying the former East and West Germanies.

Gasoline prices have already been hiked by roughly 45 cents a gallon to help pay the estimated $128 billion cost of bringing the former East Germany up to West Germany's stringent environmental standards. New taxes on toxic waste, carbon dioxide and other pollutants are imminent. Now some industries are resisting.

When the government proposed yet another turn of the environmental screw in late 1991, the chairman of Hoechst, Germany's largest chemical company, complained bitterly, saying the government "had lost all sense of proportion." Wolfgang Hilger said stringent requirements had already forced Hoechst to halt production of some dyes and chemicals at cost $60 million, and another $48 million were threatened. Chemical firms, Hilger warned, "now have to study each new legislative proposal to see whether we can still afford to invest in Germany."

It is exactly these kinds of fears that have slowed cleanup in the United States, Canada and other developed countries. In the United States, for instance, the prevailing view among Bush Administration officials is that the inevitable consequence of environmental protection is a weakened economy and loss of jobs.

Increasingly, however, the marketplace results of the German experiment are discounting this view. Now many experts feel that if the United State and other industrial giants fail to take actions of their own to catch up, the Germans will be catapulted into the lead in many areas.

This makes sense to economists who have attempted to explain the dynamics underlying the debate. "Tough standards trigger innovation and upgrading," says Harvard Business School economist Michael Porter, whose 855-page multinational study of industrial economies, The Competitive Advantage of Nations, examined the impact of environmental regulations on competitiveness. "Nations with the most rigorous requirements often lead in exports of affected products," he adds, citing both Germany and Japan's air-pollution requirements.

"Although the U.S. once clearly led in setting standards," Porter continues, "that position has been slipping away. Today the U.S. remains the only industrialized country without a policy on carbon dioxide, and our leadership in setting environmental standards has been lost in many areas."

Roger Gale, a former senior official at both the U.S. Environmental Protection Agency, agrees. Investments in new technologies of the sort being deployed in Germany "accomplish the twin goals of improving environmental quality while improving competitiveness," says Gale, who now advises foreign and domestic utilities on the competitive implications of environmental regulations. "Gains in efficiency from investment in new technologies and services will provide a huge, long-term competitive advantage."

Meantime, there is no sign that the German public is growing weary of the pursuit of green policies. In fact, the government is now launching a series of innovative initiatives that capture the imagination. The most environmentally elegant may be an aggressive and focused program to commercialize what many energy experts consider to be the two perfect fuels: solar electricity and hydrogen.

Electricity can be generated from sunlight through the use of photovoltaic panels — devices commonly used on a small scale to power calculators and watches. Solar electricity is utterly silent, nonpolluting and more reliable than coal, oil or natural gas.

Trouble is, solar electricity costs up to five times that made from coal or oil. So, to bring down costs and gain hands-on experience with engineering details, Germany launched the Thousand Roofs program. Its aim is to install residential-scale solar-electric panels on roofs throughout the country by providing government purchase subsidies of up to 75 percent. The wildly successful program quickly doubled after the inclusion of roofs in the former East Germany.

For other uses, hydrogen is an equally perfect fuel: It can be produced by splitting water with electricity, a process called electrolysis. Then, in advanced engines, it produces only pure water and pure energy. Again, however, there's a hitch: While there's little doubt that hydrogen can be used for everything from home furnaces to cars, the infrastructure of pipelines, storage tanks and the like is lacking, as is extensive experience in using the highly explosive gas safely.

To gain this experience, the governments of Germany and Bavaria teamed with a handful of industrial partners, including BMW

and aerospace giant Messerschmitt-Boelkow-Blohm (MBB), to build the Solar Wasserstoff plant. Using its outputs of zero-polluting electricity and hydrogen. German engineers are experimenting with different types of furnaces, cars, storage systems and other equipment to eliminate the devilish kinks that can spell the difference between success and failure. During off hours, the excess electricity is sold to the local utility and used to power nearby homes.

In addition, Germany's government is feeding money to Mercedes-Benz and BMW to hasten development of hydrogen-powered cars and trucks. If the ultimate goal of using solar-derived electricity to decompose water into oxygen and hydrogen is realized, Germany will convert itself to utterly nonpolluting fuels: solar electricity to run homes, shops and factories, with solar-derived hydrogen fueling the nation's cars, trucks, locomotives, planes and even ships and submarines.

I may be two or three decades before the investment in the Solar Wasserstoff plant pays off. But if threats such as global warming and ozone depletion prove to be the peril to human survival that many scientists predict, the path to Germany's future may prove to have been a road of another sort: a yellow brick road, bathed in sunlight — and paved with gold. □

Questions: *Answers on page 207*

1. How does the U.S. rank worldwide in setting environmental standards?

2. What major problem must be overcome for using solar energy as the energy of the future?

3. What major problem must be overcome for using hydrogen as the energy of the future?

50

Numerous immigrant species are colonizing the Great Lakes. The best-known recent immigrant is the zebra mussel. Beyond the 94,560 square miles of the five Great Lakes, it has expanded its range as far south as Tennessee. Since th 1800s, 136 exotic plants and animals have invaded and settled in the Great Lakes. As exotics, the effects they create on the ecology of this major watershed have been detrimental. Not having coevolved with the other predators and prey present in the lakes for generations, often a natural check and balance system is missing to control the numbers of the new, exotic species. Without any regulation on their numbers, exotics can often spread rapidly and crowd out more established species. One of the more worrisome intruder to the Great Lakes is a five-inch fish, the ruffe. This aggressive competitor tends to dominate any waterway that it enters. Its appearance has correlated with the decline of 20 to 30 other species of fish. Ths spiny water flea represents another possible threat to the Great Lakes.

From Tough Ruffer to Quagga
by Janet Raloff

Waves of immigrants are colonizing the Great Lakes, threatening the harmony — and sometimes the very existence — of local residents. Arriving in trickles, their initial request may seem benign: to compete for a place to eat and sleep. But these newcomers aren't like the natives. They're exotic species.

"Exotic" usually connotes characters possessing traits at once unusual and desirable. In ecology, however, the terms takes on a different, generally malignant connotation.

As aliens unchecked by the usual predator-anmd-prey balance that tends to develop in nature, exotics may quickly overwhelm their new environment, much like a cancer. And like a cancer, pioneering members of an exotic invasion often evade detection until their population swells to ungovernable proportions.

At a minimum, exotics may constitute a muisance, like the alewives that washed up dead and stinking on the beaches of Lake Michigan in the 1970s. Other species have wreaked economic

havoc, such as the zebra mussels that clogged pipes to power plants and water-treatment facilities. But to indigenous species, the introduction of each new exotic threatens not only a dramatic change in their way of life, but also extinction.

Natural resource managers also recognize keenly the threat that exotics pose. That may explain Dennis M. Pratt's initial reaction to learning an aggressive perch-like fish had just invaded an estuary he oversees: "It was like finding out that your wife has AIDS," the Wisconsin state fisheries biologist recalls.

No freshwater reservoir covers a greater surface — 94,560 square miles — than the five interconnected Great Lakes of Superior, Michigan Huron, Erie, and Ontario. This waterway, which provides ocean-going vessels access to ports 1,300 miles inland, carries more shipping than any other freshwater system in the world.

While this traffic has brought commerce and prosperity to many North American ports, it has also served as a major conduit for

introducing a host of living pollutants.

Since the 1800s, 136 exotic plants and animals have settled in the Great Lakes. According to a recent report issued by the Great Lakes Fishery Commission in Ann Arbor, Mich., 32 percent have been released by ship traffic. Indeed, the report notes, more than one-third of the Great Lakes' alien plants and animals have arrived over the past three decades, "a surge coinciding with the opening of the St. Lawrence Seaway."

The most notorious recent immigrant remains the zebra mussel, Dreissena polymorpha (SN:5/4/91,p.282). Since entering the Great Lakes, probably in 1986, this proliferating mollusk has fanned into the Hudson, Susquehanna, and Mississippi rivers. To date, it's been spotted as far south as Tennessee, notes marine ecologist James T. Carlton of Williams College-Mystic (Conn.) Seaport.

Carlton cites U.S. Fish and Wildlife Service estimates indicating that a 10-year effort now underway to control the mussel's further spread — it has access to most remaining U.S. waterways — amy cost $5 billion. Speaking at a Smithsonian Institution conference in Washington, D.C., last November, Carlton described this exotic's successful settlement in U.S. waters "as one of the outstanding invasions of North America in the past 200 years."

Several less well-known ecological "weeds" took up residence about the same time. One that's got more than a few biologists worried is the ruffe (Gymnocephalus .cernuus), a fish whose common name rhymes with tough. They measure only about 5 inches long — significantly shorter than the minimum 7-inch perch that most anglers consider worth cleaning.

First spotted in western Lake Superior's St. Louis estuary in 1987, the ruffe, says Pratt, "is one invation that started as far upstream as you could go" — a mere hop, skip, and splash away from the Duluth-Superior harbor, the Great Lakes' second busiest port. Within four years, an estimated 1.8 million were spawning in the 11,500-acre estuary and this European cousin of perch and walleyes has become the most abundant fish in the adjoining St. Louis River.

Its North American debut so far inland suggest that the ruffe stowed away aboard some European freighter, Pratt says, probably in the ballast water used to balance and stabilize a ship.

As large freighters unload their cargo, they fill ballast tanks and floodable cargo holds with water from the harbor. One or more ports of call later, as they prepared to take on new cargo, the crew will release that ballast — and any freeloading aquatic life it may be carrying — into the locat waters. Indeed, Carlton's analyses indicate, of all ship-related introductions of exotic species to the Great Lakes, an estimated 56 percent entered in

ballast water, another 34 percent in solid ballast such as sand or rocks.

An aggressive competitor with a range that extends well north of the Arctic Circle, the ruffe tends to dominate whatever ecosystem it enters. How it conquers each new domain remains an open question, but its reproductive cycle certainly provides one major advantage.

Unlike most members of the perch family, the ruffe can mature early — by age 1. Females generally carry two batches of eggs, spawning them a week or two apart. And unlike many of their North American relatives, this fish "is very nonselective about where it spawns," Pratt says.

Since the ruffe appeared, he adds, "We've seen declines in the great majority of the other 20 to 30 species [of fish] common in the St. Louis estuary," including all age classes of walleyes, a particularly popular game fish.

Moreover, the ruffe is on the move. Some have already been caught about 180 miles north and east of Duluth-Superior, inside Thunder Bay, Ontario. And from there? "It's farily easy to predict that in the future they'll spread to the rest of North America" — including northern Canada and Alaska, Pratt says. "There are just too many connections between the Great Lakes and the rest of our freshwater ecosystems."

For now, Wisconsin and Minnesota have outlawed possession of the ruffe, a move aimed as limiting the fish's further spread. (If people were allowed to take the fish home, they might clean them on the dock — a move that could inadvertently seed their home waters with viable eggs.) In addition, both states have reduced by two-thirds the allowable catch of potential predators — such as walleyes, muskies, northern pike, and large-mouth bass — that anglers can take from ruffe-ionfested waters. The states have

also increased the rate at which they cooperatively stock predators in the St. Louis estuary — from 4.5 fish per acre in 1988 to 17 per acre last year.

"The good news is that the predators we have there do eat ruffe. The bad news is that they don't eat enough to have shown any impact yet," Pratt notes. In fact, his studies show, "Given a choice, these predators would eat anything else."

The spiny water flea, Bythotrephes cederstroemi (BC), represents another potential exotic threat to the Great Lakes. Zoologist W. Gary Sprules of the University of Toronto in Mississauga, Ontario, suspects the invader crossed the Atlantic in ballast water picked up in Leningrad by Soviet freighters dispatched to carry cargo home from North America. The flea made its North American debut in Lake Huron during December 1984. The next year, it showed up in lakes Erie and Ontario, entering lakes Michigan by 1986 and Superior by 1987.

Just 2 to 3 millimeters long (10 mm if you count its spine), BC preferentially dines on the even smaller Daphnia, a microscopic algae-feeding crustacean. Because small fish also consider Daphnia a popular dinner entree, aquatic biologists worry that BC's overgrazing could starve out some fish.

How likely is that? In laboratory studies, Sprules and his co-workers have shown that individual BC fleas consistently clean Daphnia out of 0.4 to 1.1 liters of water daily. But Sprules has found that Daphnia can double its population far quicker than BC — "in at most 10 days." This would seem to indicate, he says, that "the prey can grow much faster than BC can impose any kind of predation pressure on them."

So why worry about BC? Because field surveys by others in the Great Lakes have recorded sharp declines in the openwater populations of certain sizes of Daphnia and in their biodiversity following the spiny flea's arrival. For instance, Sprules observes, before BC entered Lake Michigan, surveys of deep, open waters recvorded three Daphnia species. Two years after the water flea's arrival, surveys turned up just one Daphnia species. Moreover, he notes, the smallest Daphnia "actually disappeared during 1987, the year that BC numbers increased substantially."

Explains Sprules, this change in Daphnia's size distribution is "consistent with predation by an invertebrate [such as the BC flea]." Fish, by contrast, would initially target the biggest Daphnia, leaving the smallest for last.

Clever biologists will keep a sharp eye out for still more exotic invations, Carlton maintains. And that's just what it took — a sharp eye — to discern what may be the Great Lakes' newest invader.

At a glance, the new bivalves resemble the zebra mussel. Unlike the typical D. polymorpha, however, the new critters' ventral shell is not flat, but round, notes Edward L. Mills of Cornell University's biological field station in Bridge port, N.Y. His colleague, Donna L. Dustin, spotted the new mussels in the fall of 1990 among specimens collected from trawls of deep waters in the southern basin of Lake Ontario. At the time, Mills recalls, "We didn't put too much credence on the [importance of those shells]." Small and found in deep water, he said, "They just appeared to be zebra mussels growing abnormally.,"

Last summer, his team discovered more in the Erie Canal and realized they probably signified a new animal — a suspicion that has just been confirmed by genetic studies.

Trawling runs by federal biologists along the southern shore of Lake Ontario last month have identified regions where this new "quagga" mussel — nicknamed after

an extincy relative of the zebra (SN:8/3/85,p70) — constitutes roughly 50 percent of the zebra-like bivalves present. Up to 3 centimeters across, they tend to be 20 to 50 percent larger than true zebra mussels. Mills notes that Cornell geneticists have also just confirmed the presence of quaggas among zebra-like mussels collected in the Black Sea by colleagues in the former Soviet Union.

No one has yet established whether or how the quagga's habits differ from those of the true zebra mussel. It's something Mills wants to investigate.

Next on the exotics horizon? At the American Association fo rthe Advancement of Science meeting in Chicago earlier this year, Carlton identified a crustacean 7 mm long and a snail 10 mm long as two of the most likely candidates to invade the Great Lakes — again through shipping. Initially hailing from the Caspian Sea and New Zealand, respectively, both creatures are now firmly established in western European ports frequented by Great Lakes haulers.

Biologists are laying out no welcome mats for either.

The tiney, shrimp-like crustaciean, Corophium curvispinum, dwells in little tubes of mud it cements together with mucus. This amphipod has already gained notoriety throughout western Europe as a biofouling organism. Producing up to three generations a year, the animals quickly set up communities that coat underwater surfaces — from boat hussl and docks to zebra mussels — sometimes to densities reaching 100,000 animals per square meter.

Potamopyrgus antipodarum, the snail, also enjoys crowds, sometimes congregating in choking densities of up to 800,000 per square meter. And it's not very sensitive to temperature, surviving environments of 36° to 80° F.

Beginning this November, transoceanic shippers much exchange freshwater ballast for seawater before entering the Great Lakes — a move aimed at preventing the transport of freshwater exotics from one continent to another. This should slow the immigration of new species, Carlton notes: "If we had ballast exchange in place in 1980, my guess is that we would not have had the zebra-mussel invasion."

However, he warns, this new program is no panacea. Stowaways that can survive brief periods in salty or brackish water — as both C. curvispinum and P. antipodarum can — may still move in. Even more likely is their entry via a "back door," Carlton predicts.

The Nonindigenous Aquatic Nuisance Prevention and Control Act of 1990 mandates ballast exchange only for shs entering the Great Lakes. Such vessels remain free to discharge millions of gallons of water — into other U.S. freshwater systems. An donce an exotic species "get a toehold in North America,"Carlton says, "it will eventually colonize the Great Lakes too."

Manye xotics have already used that back door route: The Eurasian goldfish (Carrassius auratus) may have entere dfrom residential ponds through seasonal streams feeding Great Lakes tributaries. The oriental weatherfish (Misgurnus anguillicaudatus) made its escape from an aquarium supply house through a river draining into Lake Huron. Anglers probably released the ghose shiner (Notropis buchanani), first observed in the Great Lakes 13 years ago, as discarded bait while fishing lake tributaries. And the purple loosestrife (Lythrum salicaria) — which has edged out cattails and other prime waterfowl habitat along much of the Great Lakes shoreline — may have arrived in the early 1800s with sheep or as a cultivated plant.

"As a result of the invasions that we've been seeing here at the end of the 20th century," Carlton says, "we're beginning to see clear, direct movement about trying to reduce the amount of ballast water — which acts as a major mediator of invasions." His lab, for example, has just begun the National Biological Invasions Shipping Study — to guage the amount and source of ballast water entering freshwater systems throughout the United States. "We will also examine in detail proposed control options for ballast water," he reports.

The Canadians and Australians "also are very interested in doing something about ballast water," Carlton adds, and the U.S. Coast Guard has formally called for the voluntary noational adoption of ballast-water exchange for ships entering all U.S. ports from foreign waters.

"I think we are where we were many years ago with [the problem of shipps discharging] oil into the ocean," Carlton observes: "There is a growing realization that things have to change." ☐

Questions: *Answers on page 207*

1.   Why has the ruffe not been controlled to date by predation?

2.   Why is the appearance of the spiny water flea a possible threat to the Great Lakes?

3.   What is the newest invader to the Great Lakes and the problem it presents?

51

The northern mockingbird is unrivaled as a species that can copy and vocalize the sounds of its envrionment. Scientists have learned that these vocalization patterns are necessary to attract a mate. Changes in the singing of the male bird coincide with hormonal changes that are necessary for mating and nesting. This vocal display is equivalent in function to the visual display of some brillantly-colored male birds. Thus it is important for the survival and perpetuation of the species. Analysis of sonograms of the mockingbird's repetoire show that the male can assemble up to 400 songs each mating season. Its ability to mimic the frequency and timing of imitated sound is highly accurate.

Listen to the Mockingbird
By Doug Harbrecht

Samuel Grimes can still remember the first time he heard the song of the mockingbird. "I was five years old, sitting on the porch of my family home in Kentucky," recalls the 80-year -old photographer. "And this bird was in a tree just a few feet away, singing so clear and so close. It amazed me."

Captivated by that memory, Grimes put down his camera two decades ago and picked up a tape recorder. he crisscrossed the country, collecting 45 hours of the sweet song of his favorite avian species. The result was "The Vocally Versatile Mockingbird," a sort of "greatest hits" record documenting the talents of Mimus polyglottos (Latin translation: "many-tongued mimic").

You name it — other bird calls, sirens, bells, frogs, crickets, squirrels, a home alarm, rusty gate, the whirring and squeaks of a washing machine — and this extrovert of lawns and hedges will imitate the sound with grace and skill. Grimes tells of one mockingbird near Miami that mimicked an alarm clock so perfectly that it awakened residents early every morning.

In truth, no other animal on Earth can touch the northern mockingbird, as the North American species is called, when it comes to copying the sounds of its environment. Parrots can be taught to mimic, but only in captivity. Thrashers and catbirds — mockingbird relatives — will get off a good imitation only once in awhile.

So why do mockingbirds mock, anyway? Just who, besides Grimes, is listening? Behavioral researchers believe they have the answer. Mockingbirds apparently sing for the same reason humans turn the lights low and put Frank Sinatra or Luther Vandross on the stereo.

But while getting in the mood may be mere romance for people, it is essential to the mockingbirds' survival. Scientists have found that fluctuations in the birds' singing coincide strongly with hormonal changes that are necessary for mating and nesting in the spring and summer.

Unrestrained singing by males not only tracks production of their own testosterone, but studies suggest it may also serve to "reset" the female's reproductive system. What's so sexy about imitating the ping of a bell, the scream of a red-tailed hawk or the chattering of a chipmunk? Kim Derrickson, an ornithologist at the Smithsonian Institution, has been studying mockingbirds almost exclusively for several years in Washington, D.C.'s Rock Creek Park. As he explains, "the mockingbird's song is the vocal equivalent of a peacock's tail."

For males, which do all the singing during the breeding season, an elaborate repertoire sends a scintillating message to a female. Life is tough in the wild, where predators, disease and starvation are constant threats. The variety in a male's songbook is a signal to females that he is wise to the world, has survived for awhile and established a territory with plenty of food.

That's important for the survival of not only the female but also of future fledglings. "Male mockingbirds are better fathers than the females are mothers," says Randall Breitwisch, a University of Dayton ornithologist. Males build the nests and provide most of the food for the nestlings. And it is mostly the males, with a little help from females, that drive away cats, jays, snakes and other predators from the nesting territory.

Mockingbirds are unusual among birds in their unrelenting focus on breeding. It's not uncommon for a pair of mockers to carry out the breeding process four or five times in a season, with the loss of a nest to a predator triggering a new outburst of song to start things all over again. While most other North American birds become silent or migrate in the fall, mockingbirds keep on singing in hopes of landing a mate for next spring.

Only in recent years have scientists discovered how adaptive the gray-and-white creatures are among North American birdlife, especially compared with other species threatened by disappearing habitat. The northern variety is one of several mockingbird species found around the world, particularly in the tropics and South America. While originally native to open scrub pine and oak forests of the West and Southeast (there is an eastern race and a slightly larger western race), mockingbirds have recently increased their numbers in the northern part of their range, as they push into previously uninhabited territory.

The key reason is simple: They like what suburbanites have done to huge swaths of the continental United States and southern Canada. "Mockingbirds love mowed lawns and ornamental shrubs," says Peter G. Merritt, a Miami ornithologist who did much of his doctoral work on the species. "These are great places for them to forage for insects and berries. Man has created a habitat to which they were pre-adapted."

The birds' propensity of suburban living has been a boon, not just for them but for people, too. It's the mockingbird's personality, perhaps more than its song, that makes the species so distinctive, endearing and, from time to time, obnoxious. Official state bird of Texas, Florida, Arkansas, Tennessee

and Mississippi, the mockingbird "has qualities we admire and talk about," as Texas naturalist Roy Bedichek puts it.

Long before the Europeans came to this continent, the mockingbird's song captured the attention of native Indians, who believed the creature was ridiculing other birds in the forest. One tribe of Algonquins called it Cencontlatolly, or "400 tongues." The Biloxi Indians believed it "mocked one's words," while the Choctaws called it the bird "that speaks a foreign tongue." In Hopi myth, the mockingbird gave the tribe the gift of language.

The mocker was quick to catch the fancy of new comers to North America. "He imitateth all the birds in the woods ...[and] singeth not only in the day but also at all hours of the night," observed colonist Thomas Glover in his "Account of Virginia" in 1676. President Thomas Jefferson kept one as a pet. Poet Walt Whitman wrote sadly of a mockingbird near his Long Island home that, Whitman theorized, sang all the time because it had lost its mate.

Waxing poetic about the mockingbird is fine — up to a point. Before the advent of air conditioning, it was a rare southerner who never felt the urge to throw a pot or shoe at a lovesick mockingbird at 3:30 on a steamy spring morning. Scientists who have studied the birds say the extended, eleventh-hour singing is exclusively the work of unmated males. Alas, these crooners often become piles of feathers ont he lawn the next morning, victims of predators.

Most of the singing stops during the cold months. Them, as winter recedes and the days grow longer, the symphony begins anew, and the medley becomes more frequent and more intricate as spring approaches. What begins softly in February, when the birds prefer to remain

hidden in shrubs and evergreens, is by April a sonata with elaborate themes. Males belt out their acquired mimicry as they fly around their territory, hoping to attract a female.

Derrickson, who has analyzed sonograms of mockingbird songs, says there appears to be no limit to the number of songs a mocker can pick up. Most will master at least 180 songs in a few months. But "if you followed a bird for an entire mating season, you would end up with more than 400 song types," he says. "There is no point at which their repertoire flattens out. They just keep adding. Some they will forget or not use; others they will remember into the next breeding season."

A mockingbird's song can match the frequency and timing of the imitated sound with astonishing accuracy. Derrickson has recorded a mockingbird that mastered not only the call of a male red-winged blackbird, but also the paired response of a female red-wing. In other words, the mockingbird was performing a duet with itself. A virtuoso bird can run through its repertoire for almost an hour and repeat itself only occasionally.

Preliminary research suggests that the regional dialects of mockingbirds in the South may be moving northward, passed from bird to bird. According to Derrickson, though, mockers in the North singing the songs of the South may merely be mimicking the calls of migratory species passing through the area.

At first, some students of mockingbird behavior thought mimicry might serve as a kind of defensive gesture, designed to drive away rival bird species. A flock of blue jays flying through mockingbird territory, for example, invariably provokes mockers to mimic and attack. But Derrickson and others now believe the jays simply "remind" the mockingbird that it has already mastered the jay's call.

As soon as a female enters a male's territory, the suitor tries to win her over with a full recital of his repertoire. Once a bond is established, the singing tapers off — if only temporarily. The male bursts once more into song during nesting and breeding, but never more robustly than during copulation itself. Once the eggs are laid, singing declines and another dramatic aspect of mockingbird behavior comes to the forefront — the bird's penchant for guarding its turf.

While the youngsters are under parental care, heaven help any intruder wandering near the fledglings or the nest, which is usually built only 3 to 8 feed off the ground. A few years ago, mail carriers in Houston refused to deliver letters into one neighborhood after they were assaulted by aggressive mockers.

Because mockingbirds don't migrate, establishing territory is essential and plays a major role in their survival. Researchers have discovered that mockingbirds establish two types of territory. One is the spring and summer breeding territory, which a monogamous pair generally holds and defends jointly. While most bird species try to breed once or twice in a season, mocking-

birds make as many as five attempts with a full burst of song by the male that appears to trigger the female reproductive cycle.

Mockingbird pairs, particularly in the norther part of their range, usually split up for the fall and winter. Each bird then may establish a separate, smaller territory around a food source, such as a group of holly bushes or winterberry.

Still, the males sing. Why? Cheryl Logan, a psychologist at the University of North Carolina at Greensboro who has studies mockingbird mating habits, theorizes that the birds "use the luxury of not migrating to get a head start on the mating effort int he spring. They are showing that they can hold a territory, even under difficult conditions."

Logan's research has shown that in the world of mockingbirds, males had better measure up to female expectations — or else. Females that have lost their mates during breeding season have been found sharing "married" males in adjoining territories, moving back and forth from one to the other. Females will also abandon a mate and find another if an attempt at nesting ends in failure.

The mockingbird's charm does not end with its singing ability. While studying mockingbirds at the

University of Miami, Peter Merritt was startled to discover that the birds learned to recognize him. They picked him out of a crowd as a familiar intruder and attacked him on sight. Some mockingbirds, in fact, remember intruders from season to season. Kim Derrickson says he has tried wearing a different baseball cap every time he enters the territory of a bird whose nest he has studied. But, he admits, even these clever disguises don't fool the sharp-eyed birds. "They still recognize me."

Cheryl Logan tells the story of one female mockingbird she was observing that spend enormous amounts of time lingering in a campus parking lot. "We couldn't figure it out," she says. "We assumed parking lots were just dead space for mockingbirds." Then she took a closer look: The bird was pulling freshly squashed bugs off car windshields.

In the 1975 Pulitzer Prize-winning book Pilgrim at Tinker Creek, Annie Dillard wrote, "The mockingbird's invention is limitless. He strews newness about as casually as a god." Maybe that's why we all stop, look and listen, as enthralled as young Samuel Grimes was 75 years ago when he first listened to the mockingbird on the front porch of his old Kentucky home. □

Questions: *Answers on page 207*

1. What has allowed the mockingbird to adjust so well to previously-uninhabited northern terrorities?

2. Could the song pattern of a mockingbird also be a defensive display at times?

3. How are the nervous and endocrine systems related to the behavioral patterns of the mockingbird?

52

Recently the use of a chemical has been banned for only one reason: it posed a threat to wildlife. The pesticide called granular carbofuran was banned in the state of Virginia. Although it did not pose a direct threat to humans it was poisoning bald eagles and other wild bird species. The EPA states that the chemical will be taken off the market in crop-by-crop stages, with very little being sold by 1994.

Taking Aim at a Deadly Chemical
By Diana West

When a potent farm pesticide began mowing down wild birds, Virginia conservationists fought for a ban — and made legal history

The old-timer plunged into the piney thicket of the Slash, a densely wooded tract on Curles Neck Farm in Henrico County, Virginia. Great armies once crisscrossed this quiet land: Union troops sweeping through en-route to Richmond; Confederates driving the foe to the banks of the nearby James River. But on a fine April day in 1985, the blood and powder of pitched battle are unimaginable. Only a metal detector, like the one wielded by the old man, can uncover the telltale Minie balls and brass buttons lodged in the forest floor.

As the relic hunter — whom locals knew only as "Mr. Phelps" — trekked deeper into the wood, he glimpsed the white head and dusky winds of a balk eagle, one of the endangered birds that nest in this remote stretch of Tidewater Virginia.

But this eagle was on its back, wings extended, legs outstretched. Dead. Staring down at the carcass,

the old man opened his penknife, cut off the raptor's leg and pocketed the bird's aluminum identification band. While not a Civil War souvenir, the band might be worth keeping.

Trudging out of the woods that day in 1985, old Mr. Phelps had no idea that his chance discovery would ignite a new war in Virginia, one that would pit agricultural interests against protectors of wildlife, big business against state government — all over the use of granular carbofuran, a potent poison developed in 1970 by the FMC Corporation of Philadelphia.

Last year, Virginia environmentalists declared victory in that war when state regulators, swayed by support from Governor L. Douglas Wilder, voted to ban the sale and use of granular carbofuran in the Old Dominion. For the first time anywhere, a chemical had been banned solely because it imperiled wildlife. Even DDT, notorious for ravaging the nation's eagle, falcon and pelican populations and outlawed in the country in 1972, remained on the market until evidence mounted that it threatened

humans as well as animals.

Had it not been for a handful of Virginia conservationists mobilized by the death of the Curles Neck eagle, the carbofuran story might have ended in political impasse. For while Virginians took aim at the poison, the feds dawdled. By 1989, the U.S. Environmental Protection Agency (EPA) had the facts: Carbofuran was killing two million birds every year in this country. Since 1972, carbofuran-poisoned birds have been found in 23 states. Supported by the U.S. Fish and Wildlife Service and groups such as the National Wildlife Federation, the EPA recommended canceling the pesticide's registration — that is, banning it. But nothing happened.

There was no action, says Robert Irvin, an NWF attorney, because "The process for canceling pesticides is weighted heavily in favor of the manufacturers — which is backwards. It should be weighted in favor of the environment."

"If a chemical is not going to cause a problem for humans," adds John Bascietto, a former EPA biologist, "it generally won't be regulated for wildlife concerns."

Ultimately, Virginia would get its ban and the federal government would be stirred into action, with EPA crafting an agreement with FMC designed to eliminate granular carbofuran's use throughout the United States. But in 1985, when the dead eagle turned up a Curles Neck, victory was a long way off.

"Our position on this thing right from the start was: Forget EPA. They're sitting on their hands," says Robert Duncan, chief of the wildlife division at the Virginia Department of Game and Inland Fisheries. One of the band that pushed to cancel carbofuran, Duncan, says, "This was a Virginia problem, one that was documented in Virginia, and Virginia was going to take care of it — i.e., get rid of it. But that was more easily said than done."

Carbofuran is the name of the poison in Furadan 15G, a pinky-purple insecticide that farmers throughout the United States sow by the sackful to protect corn, rice and other crops from root-eating nematodes and insects. It is deadly stuff. One of carbofuran's ingredients is methyl isocyanate, the compound that killed some 2,000 people in 1984 when a gas cloud escaped from a Union Carbide plant in Bhopal, India. Although agricultural carbofuran poses no threat to people, a single granule can kill a songbird.

U.S. Fish and Wildlife Service special agent Don Patterson had scarcely heard of the stuff when he was called in to investigate the Curles Neck eagle's death. As he followed old Mr. Phelps (who had told authorities of his find), he could smell the eagle before he saw it, rapidly decaying in the springtime warmth near the base of its nesting tree. Inside the nest, 60 feed above the ground, he found a dead eaglet. He also found the remains of pigeons and blackbirds, evidence the eagles had probably died from eating poisoned carrion. Patterson shipped the eagle to a pathology lab for a necropsy.

The verdict, "death by carbofuran," rang a bell. Patterson remembered that Virginia wildlife officials recently had picked up a sick female eagle on the south side of the James River. The raptor could barely stand, let alone fly.

Officials ferried the eagle in a large dog carrier to the Wildlife Center of Virginia, a private veterinary hospital and research center in the Shenandoah Valley. Founded by conservationist Ed Clark and veterinarian Stuart Porter, the Wildlife Center treats nearly 2,000 wild animals a year, providing researchers an opportunity to monitor the region's environmental problems at close range. Treated and released, the eagle had given Virginia its first warning of carbofuran poisoning.

After reading the necropsy report on the state's second-known carbofuran victim — the Curles Neck eagle — Don Patterson alerted farm owner Richard Watkins. "We stopped using Furadan right away," says Watkins, who hasn't used insecticides on his 400 acres of corn at Curles Neck since hearing the news. "We haven't noticed any change in our yield," he adds.

But there were hundreds of thousands of acres of Virginia farmland where Furadan remained in use. Patterson sent copies of the necropsy report to Robert Duncan and Ed Clark, two men he though could help. "In our wonderful naivet_," says Clark, a career conservationist soon to be appointed to the Virginia Council on the Environment, an advisory board, "we thought now that we knew something was poisoning eagles, somebody will do something about it." But stopping a killer chemical would prove much harder than expected.

Spring planting season arrived, and Patterson picked up four more poisoned eagles. By now, Furadan's maker acknowledged a problem with wildlife. But FMC faulted farmers, accusing them of misusing the poison to bait predators, or of mishandling it by spilling it. "We [sponsored] an extensive steward-ship program to make sure applicators and growers were sensitive to the risks," FMC spokesman Jeffrey Jacoby says of company efforts in Virginia. "We all certainly didn't want to kill any bald eagles."

Patterson once guided some FMC representatives through Virginia eagle country. "While there, we saw 12 [living] eagles," he says. "And one [FMC representative's] comment that came to me, was, 'Hell, we're always going to have a problem here; there's too many eagles!'" Patterson bursts into a wheezy guffaw at the memory.

He also tells of a Virginia farm family that awoke one morning several years ago to find their yard sprinkled with the vivid yellow carcasses of between 50 and 70 goldfinches. "Cats and dogs were already carrying them off," says Patterson, who arrived after carcasses had been collected. Surrounding corn fields, it turned out, had been laced with Furadan.

But it was not until 1990 that officials learned of the event that, in Virginia, would mark the beginning of the end for Furadan. That year, a massive bird kill occurred in eastern Virginia near the Rappahannock River. Agents combing corn fields picked up more than 200 dead sparrows, eastern bluebirds, red-winged blackbirds, blue jays, goldfinches, starlings and other species. "It was right quiet," recalls Bob Duncan, describing the macabre afternoon he visited the site.

The aftermath was anything but quiet. The bird die-off thrust the carbofuran issue before a new state Pesticide Control Board, a ten-member regulatory panel that had come into being a year earlier. The board scheduled public hearings throughout the summer and fall of 1990. Both sides of the pesticide war clashed, noisy and clear. When the last verbal shots had been fired, the board decided to give the manufacturer one more chance.

FMC insisted new safety measures could reduce the risk to birds. The pesticide board told the company to put those measure to work during the 1991 planting season in a 32 country region east of interstate 95 — the heart of Virginia eagle country. The board also asked Duncan's state wildlife division to monitor the measures' effectiveness.

Come spring, FMC launched a publicity blitz. A soaring eagle glided across billboards and baseball caps FMC distributed across Virginia, tools in a campaign urging customers to "Protect as you plant." The company drew up new instructions for Furadan's use, prescribing smaller doses than previously suggested and improving application techniques.

For all FMC's noisy drills and maneuvers, the real battle was a quiet one, bird against poison, waged in the freshly planted fields of Virginia. State government monitors would determine the outcome by counting casualties.

"We were a little nervous going into it," says Elizabeth R. Stinson, the state game department biologist who devised the monitoring program. Finding a bird as big as an eagle is hard enough. Spotting a small Savannah sparrow in a vast farm field can be next to impossible. Stinson helped train a team of ten monitors to search corn fields on eleven farms. Through tests, she established that monitors could find 30 to 50 percent of all dead birds.

Monitors combed fields before and after Furadan's application. As the toll mounted, carcasses of kestrels, robins, water pipits,

cowbirds, meadowlarks, cardinals and other birds were sent to labs to confirm the cause of death.

On 900 acres, Stinson's team gathered 62 bird carcasses, 10 ailing birds and 47 "feather spots" — piles of feathers that indicated a bird had been scavenged. Assuming that monitors were, at best, picking up half the carcasses to be found, and considering the hundreds of thousands of acres of Furadan-treated farmland across the state, the monitoring program indicated tens of thousands of birds were dying of pesticide poisoning every spring.

In May 1991, the Pesticide Control Board deliberated on the results of the monitoring program. Virginia Governor Wilder threw his support behind a Furadan ban. "We thought we had it," says Ed Clark, who went home feeling relieved. News headlines one day before the board would vote changed his feelings.

In what Clark calls a "public relations masterpiece," FMC officials announced their own eleventh-hour decision to stop selling Furadan in Virginia. Pesticide Control Board chairman George Gilliam accepted the offer, saying a voluntary withdrawal was as good as an official ban.

But what FMC could withdraw, FMC could reintroduce. Besides, even if Furadan 15G stayed off the shelf in Virginia, what was to prevent a farmer from hopping the border to buy pesticides?

"Those of us who knew what was going on behind the scenes went ballistic," says Clark. Impassioned calls went to the governor's office, the state agriculture department, the media. "My phone was melted by the end of the day," Clark says of the last-minute lobbying blitz.

Ultimately, the pesticide board accepted FMC's offer — but also imposed its own emergency ban on Furadan 15G, beginning June 1,

1991. "Sort of a belts-and-suspenders approach," quips Chairman Gilliam, who ended up favoring a ban. Gilliam notes that the state pesticide board banned Furadan after focusing on the issue only one year, while "The EPA has been on this issue since 1985 . . . . I feel if we hadn't had the backbone to act, the EPA may very well have taken another several years."

Instead, EPA took just days to announce it had reached a multiphase agreement with FMC. ("It took Virginia a year to do something and they only had to worry about one state," an EPA official said of the years of agency deliberation that preceded the agreement. "We had to think about the whole nation.")

The EPA agreement stipulates that FMC will take Furadan 15G off the U.S. market in crop-by-crop stages, beginning with corn. By September 1994, only 2,500 pounds may be sold annually, a thimbleful compared to an estimated 10 million pounds that crossed the counter each year.

As some people see it, the long-awaited agreement contains loopholes. For one thing, FMC retains the right to appeal to the EPA before each stage of the phaseout kicks in. Also, a U.S. phase-out by itself provides incomplete protection.

"Many of the same birds we were seeking to protect here are going to be exposed to this chemical in other regions as they migrate," Ed Clark says. In fact, Tarry Lalonde, manager of a West Virginia plant that makes Furadan 15G, told the Charleston Sunday Gazette-Mail that exports of the insecticide will actually increase to balance shrinking domestic sales.

And those exports will not be tainted by the stigma of an EPA ban. "If something is banned in the United States," explains former EPA biologist John Bascietto, "it becomes a serious problem for overseas regulators not to do anything about it. If it's not completely banned then foreign regulators are under less pressure to look at it more closely."

But thanks to Virginia conservationists, FMC cannot tell customers that Furadan 15G isn't banned anywhere. "Virginians were well served," Robert Duncan says of the battle to banish the bird-killing chemical from the Old Dominion. "And the Pesticide Control Board members deserve credit for that," he drawls, "even if they had to be persuaded to bring it about." □

Questions: *Answers on page 207*

1. Why is it often difficult to convince humans about the dangers to them from a chemical such as granular carbofuran?

2. Why is it important to save the wild bird populations from an ecological standpoint?

3. Are there other reasons to save wild bird populations aside from an ecological viewpoint?

Answers to Articles

1. Wet, Wild, and Weird

1. Some substances are hydrophobic, water-hating. They do not dissolve in water and actually separate from water if an attempt is made to mix them with the liquid state of this substance.

2. Some substances are hydrophilic, water-loving. They easily dissolve in water.

3. Hydrogen bonds to oxygen in the water molecule through polar covalent bonds. Through the unequal sharing of electrons, there is a negative end where oxygen is located and a positive end where hydrogen is located. Polar substances, possessing molecules with charged ends, dissolve well in water for this reason.

2. Protein Identified in Dinosaur Fossils

1. Proteins do not stay preserved for a long time and the molecule may have been a bacterial protein that inhabited the dinosaur instead.

2. On an Earth that is four to five billion years old, the last two million years is only a very small part of the total picture.

3. Genes make proteins in the organism. If two proteins are similar, the genes making them are similar. If the proteins are different, this reveals a genetic divergence.

3. Actin in Cell Attachment

1. They slide between each other in the sarcomeres of the muscle cells. This shortens the sarcomeres. As the sarcomeres shorten, the muscle cells shorten. As the cells shorten, the entire muscle shortens or contracts.

2. In either case some sort of organized movement is occurring.

3. It removes the normal protein in the host cell it attacks. Apparently this is a necessary step for its attachment and attack.

4. The Case of Ghost Molecules

1. It is a very simple molecule, consisting of only three atoms. Is it complex enough to encode information?

2. There probably is a biochemical basis but the molecule must be more complex that water to encode information. Nucleic acids and proteins have the necessary complexity.

3. The basis for memory is molecular in some way, a chemical phenomenon. The idea of signaling through the EM spectrum takes a page from physics.

5. Vitamin A-like Drug May Ward Off Cancers

1. A relationship between dietary dosage and cancer prevention has not yet been proven scientifically.

2. Liver damage and side effects on body systems can result.

3. They are not a cure but represent a possible source of prevention.

---

6. Ockham's Razor and Bayesian Analysis

1. Any area where a simple explanation might work is acceptable. Stating a simple relationship between the structure and function in a plant or animal part is one example.

2. A relationship involving multiple interactions might not be acceptable. Population dynamics and other branches of ecology offer examples.

3. Statistics allows a means to measure the probability that a hypothesis is simplest and most likely to be correct.

---

7. The Handmade Cell

1. The climate and environment of the primitive earth was ideal in providing the conditions for the biochemical reactions needed to produce cells. The conditions today are very different and are not ideal. The temperatures are not warm enough and the atmosphere contains large amounts of oxygen.

2. The cell is the fundamental unit of life, serving as the step for all higher levels of biological organization. For example, an organ such as the pancreas consists of several tissues. One of these tissue types consists of specialized cells that make insulin, which is a molecule. Understanding the function of the entire organ requires understanding its cellular and biochemical makeup.

3. RNA has the necessary complexity, consisting of many possible sequences of four building blocks (bases), to carry the instructions to direct life processes. It addition to functioning as a genetic blueprint, it can also build copies of itself. Reproduction is a basic property of life.

---

8. End of the Needle

1. It would lower the level of glucose in the blood.

2. Protenoids and cells are both membrane-bound bodies packing biologically-active molecules.

3. Most of these new, genetically-engineered products are proteins. Proteins, unprotected, are digested in the stomach.

---

9. Liposomes

1. It offers the ability to deliver a drug to a specific site rather than in a free form.

2. The arrangement of lipids in the liposome membrane establishes hydrophilic and hydrophobic ends, thus defining a biological layer or membrane.

198

Liposomes (con't)

3. Any requirement for administering a substance in a direct concentrated way is a possibility.

---

10. A Time to Live, a Time to Die

1. It is not an either-or question. A gene, or series of genes, is probably present to signal cell death. They are probably turned on, however, by different types of environmental prompts.

2. Unless normal cell growth occurs, usually the rate of cell division is counterbalanced by the rate of cell death for a given cell type, therefore maintaining an equilibrium. During cancer, the rate of new cell production greatly outpaces the rate of cell destruction.

3. Cells must die to produce a separation of parts, such is in the webbed foot of a human fetus to produce digits. Sometimes cells must be rearranged radically to produce new forms such as in metamorphosis.

---

11. Two Human Chromosomes Entirely Mapped

1. Scientists identify each gene on a chromosome and can describe their sequence and location along this structure.

2. Each gene will be identified and its effect will be known. This may help in the strategy to alter its effect, particularly if it causes a disease.

3. No, as their are segments between genes coding for proteins that serve as breakpoints.

---

12. Dark Bogs, DNA and the Mummy

1. Through mitosis all body cells are produced from the first, developing cell of an organism. Therefore, all body cells contain the same chromosome complement and array of genes. A bone cell has the same genetic picture as a skin cell or liver cell. Only a small number of these genes are expressed however.

2. It is a language with four letters (bases) assembled into three-letter words (codons). The base sequences in DNA exist in an almost infinite array of possibilities.

3. The lower the pH value, the more acidic the environmental conditions. Unprotected DNA is quickly destroyed is acidic conditions, removing it as a source for study.

---

13. Cancer Gene May be Relatively Common

1. Cancerous cells multiply too rapidly. The basic timing to produce cells at a rate for normal cell replacement is lost.

2. Usually many modifier genes are also involved. Most human characteristics are polygenic.

3. A mutated gene makes a different protein product. It this protein is a regulatory protein, it can change the timing mechanism of normal cell division.

---

14. The African Gene

1. It is hypertensive, as values above 140 (systolic) and 90 (diastolic) fall into the ranges of high blood pressure.

2. It puts extra force on the wall of blood vessels, damaging them. It can also lead to the formation of clots and heart damage.

3. More salt in the blood draws water from the tissue spaces by osmosis. As more water is added to the circulation this way, the fluid buildup elevates the pressure in the vessels.

---

15. Gene in a Bottle

1. As with most behavior patterns, there is an environmental component as well as a genetic component. A genetic component, however, may make a person more susceptible to environmental influences.

2. It the gene is identified in a susceptible person, their environment can be modified to prevent the tendency toward this genetic disease.

3. The statistics cited in this article offer arguments for both sides. As with most issues, scientists seldom agree entirely. More research is necessary.

---

16. Better Than the Real Thing

1. The range of possibilities include positive acceptance to larger, more desirable crops on the one hand, to concern over food purity and gene splicing on the other hand.

2. Possibly a hardy, mutated form growing wild could threaten the growth of normal plants or present a health hazard by the production of undesirable products.

3. Hormones such as the auxins and ethylene control the growth and development of plants. By gene splicing the effects of these signals can be enhanced, if beneficial, and blocked if undesirable.

---

17. The Mysterious Virus Called "Isn't"

1. It only shows properties of life when inhabiting its host. Outside of the host it crystallizes into a nucleic acid and protein.

2. Either there is a third AIDS virus or the cause it not viral in origin.

3. Further investigation is needed to isolate and identify an actual virus or the correct causative agent.

---

18. Body Temperature: Don't Look for 98.6°F

1. Various metabolic processes vary under the influence of some poorly-understood biological clock.

2. There are genetic variations between people that affect better-understood traits, ranging from height to intelligence.

3. The Centigrade or Celsius is a more appropriate, metrically-based scale.

---

19. Anti-inflammatory Drugs May Quell Asthma

1. It would close the air tract, blocking the air flow.

2. The data is preliminary and more studies are needed for verification.

3. They could affect the growth of the child, causing significant weight gain.

---

20. Excess Iron Linked to Heart Disease

1. No, as a small amount is necessary for red blood cell production. The absence of iron in the diet leads to anemia.

2. Differences in the levels of sex hormones is a possibility. Also, different environmental conditions ranging from exercise preferences to work-related conditions should be investigated.

3. Avoid iron supplements but eat a well-balanced diet to consume normal, needed levels of iron.

---

21. Clues to the Brain's Knowledge System

1. Different convolutions on the cerebral cortex have unique functional abilities.

2. Many of its functions are very abstract and are not well understood. To date the origin of a human thought remains unknown. Functions such as reasoning and memory remain only poorly understood.

3. They instead consist of the integrative or association areas that coordinate functions of the mapped sensory and motor areas.

---

22. Brain Gain: Drugs That Boost Intelligence

1. Any drug must pass through the digestive tract before being absorbed into the bloodstream. What will its makeup and effect be by the time the circulation carries it to the brain?

2. Brain cells signal one another synaptically through the discharge of neurotransmitters. Alteration of these chemical signals can greatly affect human behavior. Many hallucinogens either amplify or inhibit neurotransmitter function in the brain.

3. The effects of smart drugs remain open to criticism. A well-balanced diet supplies the brain, as well as other organs, with all of the basic raw materials needed to carry out it metabolism optimally

---

23. Volume Transmission in the Brain

1. The signaling along a neuron consists of one type. It is the action potential, a rapid means of communication. There are several types of communication at the synapse. Each is a slower means that along the neuron.

2. It is direct and specific along neuronal pathways. The endocrine mode of signaling is more diffuse, relying on hormones that circulate with the flow of blood.

3. It offers another mode of signaling when the normal means of synaptic signaling would not coordinate far-reaching sites in the brain.

24. Consciousness on the Scientific Agenda

1. They will involve the mechanisms of brain physiology and their translation into human behavior.

2. They connect the sensory (receptive) and motor (expressive) areas, integrating them for appropriate responses from the central nervous system.

3. They are abstract in nature. Scientists literally do not know what to look for aside from how to study them.

---

25. The Work That Muscles Can Do

1. The muscle shortens if its muscle fibers, or cells, shorten. The cell shortens if its myofibrils, specialized organelles, shorten. The organelles shorten if the myofilaments, protein molecules, react.

2. The calculation of work is a concepts of physics. Power, another concept of physics, is the rate at which work is performed.

3. It is difficult to apply the proper transducers and measurement equipment to humans in an active setting

---

26. Polarization Vision in Fish

1. This light is not visible but can be harmful. If exposed to it for long periods it can cause human skin cancer.

2. The eye offers the receptor cells that detect light. This stimulus must be converted to a signal in the optic nerve that is sent to a specific region of the brain for interpretation.

3. It is both. Particles of matter must oscillate in a direction that is perpendicular to the direction of transport of the wave they generate. Traveling with a frequency and amplitude, light also has energy properties.

---

27. Biomedicine Blasts to New Heights

1. It is really just another type of environmental stress worthy of study in order to expand the knowledge about body physiology.

2. They mass have been more readily available for energy. Also, glucorticoids produce sugars during stress conditions.

3. The number of cells in the muscle is not reduced. The diameter of the cells is, however, Therefore the muscle atrophies.

---

28. Sex and the Male Stick Insect

1. A sperm cell would not contribute to the development of the egg. The heredity of the offspring would be based on female chromosomes only.

2. The heredity would stem from the paternal parent only.

3. Chromosome analysis revealed the content of the fertilized egg. Electrophoresis revealed the enzymes involved metabolically.

29. Teasing Out Dietary Cholesterol's Impact

1. It has many natural roles, ranging from its contribution to the structure of cell membranes to a precursor molecule needed to make other steroids in the body.

2. It is not intrinsically toxic. The human body normally synthesizes it. The problem is when its level is too high in the blood, usually when surpassing 200 mg. per 100 ml of blood.

3. The extra cholesterol coats the inside of blood vessels causing a variety of cardiovascular problems ranging from blockages to the formation of blood clots.

---

30. Oral Contraceptives in Ancient and Medieval Times

1. They probably worked similarly to the modern birth control pill, preventing ovulation.

2. High levels of estrogen inhibit ovulation, thus removing the potential for fertilization of the female sex cell in the oviduct.

3. The statements are rooted in legend and incomplete historical records. Scientific proof is nonexistent.

---

31. Female Choice in Mating

1. The brilliant plumage could facilitate easier recognition of the male during mating.

2. The combative behavior, or the genes promoting this tendency, could promote survival of the species.

3. There was more interest toward studying natural selection or the properties of the male.

---

32. Germination of Fern Spores

1. Environmental changes in photoperiod, humidity, and temperature influence germination.

2. The changing properties of the wavelength of light, from blue to red, stimulate germination.

3. Light-absorbing molecules, messenger RNA, and complex proteins are involved in the process.

---

33. Twilight of the Pygmy Hippos

1. More examples of the coexistence of humans and the small mammals must be uncovered.

2. Changes in climate can alter necessary habitat requirements of either the mammals or other species that they depend upon. A significant change can make the habitat incompatible to support their future existence.

3. Humans could prey on the mammals directly or change their environment significantly enough to spell the demise of the small mammals.

---

34. How Many Species Inhabit the Earth

1. The undertaking is massive, facing attempts to tabulate an overwhelming number of different kinds of organisms. A great amount of coordination and communication is required by different scientists throughout the world.

2. The possible differences between two different kinds of organisms is not always clear. Are they reproductively isolated and distinct or just variations of the same species who did not have the opportunity to interbreed?

3. Biological diversity provides a variety of checks and balances to preserve a habitat. Species with different roles add stability to an ecosystem. Simple settings are less stable and more easily wiped out.

---

35. Evolution Comes to Live

1. The early primates did not live in an environment that favored fossil formation. The forest floor of these ancestors did not offer a source to cover up bones and preserve them.

2. Over some ancient time periods fossils are completely nonexistent. There is nothing to study or serve as a foundation for model building.

3. Model building is more in the realm of aesthetics, subject to the interpretation of an artist and not by a scientist. However, model building does make an important contribution to supplement science.

---

36. Erectus Unhinged

1. Members of different species cannot interbreed. Members of the same species can potentially interbreed. This is impossible to test or observe with ancient populations. Making inferences from fossils is difficult when the fossil record is often sparse and incomplete.

2. Even with modern populations it is often difficult to observe their ability to interbreed naturally. To test this potential is the laboratory creates an artificial situation that may never take place in a natural environment.

3. This approach states that new species evolve rapidly rather than through a series of gradual steps. Cladistic analysis may offer a more accurate explanation for human evolution than the long-held view of gradually changing species of the genus Homo.

---

37. Some Homonids Show Fidelity to the Tooth

1. One other line is the study of DNA samples. Similarities and differences in the genetic makeup of ancestors, compared to modern species, can lead to deductions about how close and remote ancestors are to modern species.

2. The article mentions that there was a scant sample of lower-body bones in the ancestral homonid. Fossil samples are often incomplete, within a species and among a succession of species.

3. A positive correlation between brain size and intelligence has not been proven. At most, it might prove a potential for development of intelligence.

204

38. Brushing the Dust Off Ancient DNA

1. A study of fossils reveals another level of organization and shows gross relationships that a study of molecules alone would not uncover.

2. More subtle relationships can be deduced at the molecular level that are not apparent at the organ-system level.

3. The majority of fossils are not preserved in conditions that preserve DNA for study.

---

39. Neandertals to Investigators: Can We Talk

1. The skull analysis shows that the motor organs were perhaps present. However, speech required more than just this. How well, for example, was the frontal lobe (modern speech center) developed in ancestors?

2. An analysis of the cerebrum and artifacts of language and symbols would add credence to the assertion.

3. It would undoubtedly be crude and simple by comparison, existing in a fledging stage.

---

40. Lab Insect Thwarts Potent Natural Toxins

1. Insects with a mutations thwarting the toxin are selected for survival. Only the microbes with a lethal toxin will have an advantage. This interaction could constantly change the makeup of each population.

2. Chemical agents can become concentrated in the consumers of a food chain, reaching lethal levels in the organism. Biological control is safer.

3. Genes code for the proteins in the plant, a more straightforward approach toward developing pest resistance.

---

41. Wilderness Corridors May Not Benefit All

1. Even if the survival rate is one per cent, it indicates that some wildlife is saved.

2. Populations would be fragmented into clusters, not being able to communicate over continuous tracts offered by corridors.

3. Wild animals do not behave as humans do. It is a mistake to impose human expectations on them.

---

42. Turtle Recovery Could Take Many Decades

1. It illustrates the principle of how organisms in an ecosystem are not independent but are related to each other through various types of interactions.

2. They will provide the future breeding component of the group.

3. It offers an escape hatch for the turtles while allowing the shrimp to remain trapped.

---

43. The Essence of Royalty: Honey Bee Queen Pheromone

1. Roles develop through chemical signaling. The structure of these signals, pheromones, must be understood at the molecular level.

2. Unlike the classical mechanism of a hormone traveling through the plant or animal body, a pheromone is discharged into the environment. It travels from organism to organism.

3. Other animal species communicate by chemicals discharged from one organism to another. Far less is known about this, however.

---

44. A Phantom of the Ocean

1. It destroys the activity of neurons, thus paralyzing numerous biological processes in the organism.

2. Its inactive, encysted stage on the ocean floor is its usual mode of existence. It is active only during the fish attacks.

3. The increase growth of algal populations takes oxygen from the water, necessary for the metabolism of most ocean species.

---

45. Diabetes Running Wild

1. There is a selective advantage toward survival and reproduction for the humans who can metabolize carbohydrates correctly. They are more likely to pass on their superior genes to future generations.

2. The term usually applies to contagious diseases. Diabetes mellitus is a genetic, noncontagious disease.

3. The frequency for the type II diabetes gene or genes should decrease over time.

---

46. The Ecology of a New England Salt Marsh

1. Temperature and the amount of sunlight are two possibilities.

2. Various types of competition along with predation and mutualism are likely.

3. The habitat specifically describes where a species lives whereas the niche describes its functional role such as feeding habits and patterns of mating.

---

47. Who's Cute, Cuddly and Charismatic

1. It is superficial and focuses on characteristics that often have minimal importance ecologically.

2. Invertebrates are necessary to start food chains. It is interesting to note that conservationists' efforts toward them are very weak.

3. It has led to their extinction, as they lack charisma and are not attractive to humans.

---

48. *Winning the Race in India*

1. The biological means is predation; humans are the predators and the rats are the prey.

2. A biological means is environmentally safe. Chemicals accumulate in the biomass of the upper members of the trophic structure of a food chain, often with undesirable consequences.

3. It would be less efficient, as only about 10% of the energy remains at each successive step of a food chain for use. More energy and biomass can be harvested from rice than from rats that each the rice as an additional link in the food chain.

---

49. Down Germany's Road to a Clean Tomorrow

1. At one time it led the world but currently is losing ground. Fears include that environmental protection also means weakening the economy. Efforts in Germany, however, disprove this claim.

2. The cost must be reduced. It is five times as expensive as using coal or gas.

3. There is no current infrastructure for handling this gas safely. An infrastructure of tanks and pipelines must be built.

---

50. From Tough Ruffe to Quagga

1. The possible predators seem to prefer other prey instead of eating the ruffe. Perhaps its behavior or its taste to potential predators is an adaptive defense mechanism.

2. They feed on Daphnia, a small crustacean at the beginning of many food chains. The biodiversity of this smaller water flea is decreasing by the grazing of the spiny water flea, threatening many food chains.

3. A new quagga mussel which may interfere with established food chains and predator-prey cycles.

---

51. Listen to the Mockingbird

1. Mockingbirds seek out mown lawns and ornamental shrubs indicative of northern suburbs.

2. There is little evidence for this. The arrival of the female mockingbird in the territory of the male is what evokes the male song pattern most often.

3. Hormonal changes coincide with the male vocal patterns. The nervous system must permit for memory banks of the patterns and the signaling system necessary for motor responses.

---

52. Taking Aim at a Deadly Chemical

1. If it does not affect them directly, they consider it irrelevant.

2. They are important predators that are necessary for the maintenance of balances and interactions of many ecosystems.

3. From a moral and ethical point of view, it is wrong to threaten the existence and health of any species.